THE POST-
CHORNOBYL
LIBRARY

Ukrainian Studies

Series Editor: Vitaly Chernetsky (University of Kansas)

Harvard Ukrainian
Research Institute

THE POST-CHORNOBYL LIBRARY

UKRAINIAN POSTMODERNISM OF THE 1990s

TAMARA HUNDOROVA

TRANSLATED BY SERGIY YAKOVENKO

BOSTON
2019

Library of Congress Cataloging-in-Publication Data

Names: Hundorova, T. I., author. | I͡Akovenko, Serhiĭ, translator.
Title: The post-Chornobyl library : Ukrainian postmodernism of the 1990s / Tamara Hundorova ; translated by Sergiy Yakovenko.
Other titles: Pisli͡achornobylʹsʹka biblioteka. English
Description: Boston : Academic Studies Press, 2019. | Series: Ukrainian studies | English translation of: Pisli͡achornobylʹsʹka biblioteka : ukraïnsʹkyĭ literaturnyĭ postmodern. | Includes bibliographical references.
Identifiers: LCCN 2019031932 (print) | LCCN 2019031933 (ebook) | ISBN 9781644692387 (paperback) | ISBN 9781644692394 (adobe pdf)
Subjects: LCSH: Ukrainian literature--20th century--History and criticism. | Postmodernism (Literature)--Ukraine. | Chernobyl Nuclear Accident, Chornobyl', Ukraine, 1986--In literature.
Classification: LCC PG3916.2 .H8613 2019 (print) | LCC PG3916.2 (ebook) | DDC 891.7/909004--dc23
LC record available at https://lccn.loc.gov/2019031932
LC ebook record available at https://lccn.loc.gov/2019031933

English translation copyright © 2019 Harvard Ukrainian Research Institute
All rights reserved.

ISBN 978-1-64469-238-7 (paperback)
ISBN 978-1-64469-239-4 (adobe pdf)

On the cover: Viacheslav Poliakov, from the "Lviv–God's Will" series, 2017.
via-poliakov.com
Reproduced by the author's permission.

Cover design by Ivan Grave.
Book design by PHi Business Solutions.

Published by Academic Studies Press.
1577 Beacon Street
Brookline, MA 02446, USA
press@academicstudiespress.com
www.academicstudiespress.com

Contents

Acknowledgements — vii
Translator's Acknowledgements — ix
A Note on Transliteration — xi
Preface — xiii

PART ONE
Chornobyl and Postmodernism — 1

1. Nuclear Discourse, or Literature after Chornobyl — 3
2. Nuclear Apocalypse and Postmodernism — 10
3. The Socialist Realist Chornobyl Discourse — 16
4. Nuclear (Non)-Representation — 22
5. Chornobyl and Virtuality — 28
6. Chornobyl and the Cultural Archive — 33
7. Chornobyl Postmodern Topography — 38
8. Chornobyl and the Crisis of Language — 44

PART TWO
Post-Totalitarian Trauma and Ukrainian Postmodernism — 49

9. Postmodernism: The Synchronization of History — 51
10. Ukrainian Postmodernism: The Historical Framework — 60
11. A Farewell to the Classic — 66
12. The "Ex-Centricity" of the Great Character — 83
13. Postmodernism and the "Cultural Organic" — 91
14. Postmodernism as Ironic Behavior — 96

PART THREE
The Postmodern Carnival   101

15. Bu-Ba-Bu: A New Literary Formation   103
16. The Carnivalesque Postmodern   109
17. Yuri Andrukhovych's Carnival: A History of Self-Destruction   119
18. After the Carnival: Bu-Ba-Bu Postmortem   136

PART FOUR
Faces and Topoi of Ukrainian Postmodernism   155

19. Narrative Apocalypse: Taras Prokhasko's Topographic Writing   157
20. The Virtual Apocalypse: The Post-Verbal Writing of Yurko Izdryk   164
21. The Grotesques of the Kyiv Underground: Dibrova—Zholdak—Podervianskyi   175
22. Feminist Postmodernism: Oksana Zabuzhko   186
23. Postmodern Europe: Revision, Nostalgia, and Revenge   198
24. The Chornobyl Apocalypse of Yevhen Pashkovsky   210
25. The Postmodern Homelessness of Serhiy Zhadan   218
26. Volodymyr Tsybulko's Pop-Postmodernism   237
27. The (De)KONstructed Postmodernism of Yuriy Tarnawsky   253

PART FIVE
Postscript   267

A Comment from the "End of Postmodernism"   269
A Commentary on the "End of Ukrainian Postmodernism"   290

Bibliography   303
Index   315

Acknowledgements

I would like to express my deepest gratitude to Academic Studies Press, in particular to its Slavic program for publishing this book. This publication has become possible thanks to the financial support of Harvard Ukrainian Research Institute (HURI), which is a leading institution of Ukrainian Studies. During my various tenures at HURI I had an opportunity to grow as a scholar and wrote the initial pages of this book. I am particularly grateful to George G. Grabowicz: his knowledge of literature, and the ability to analyze it having respect for the text, has consistently impressed me—I will always remain his student in this regard. I greatly appreciate Vitaly Chernetsky and Oleh Kotsyuba's friendly support and assistance throughout this project, from the inception of an English-language edition of the book and all the way to its publication. I am deeply grateful to the translator of this book, Sergiy Yakovenko, with whom I share not only my alma mater but also a long-standing love of modernism, Nietzsche, and literature of the beginning of the twentieth century. Finally, I am thankful to the Kyiv publishing house Krytyka which believed in me and published the original Ukrainian edition of my *Post-Chornobyl Library*.

Translator's Acknowledgements

I want to thank, first of all, Tamara Hundorova, my mentor and colleague, who entrusted the translation of her book to me. It has been a great pleasure for me, an intellectual challenge, and enormous responsibility. It is an equally pleasant task to thank Harvard Ukrainian Research Institute for this opportunity; in particular, I am grateful to Oleh Kotsyuba and Vitaly Chernetsky for their guidance and attention to my work. Finally, I thank the Academic Studies Press and especially Stuart Allen, the editor of my translation, for his patience, diligence, and inventiveness.

A Note on Transliteration

In the body of the text, we use the conventional transliteration for East Slavic names and locations. However, for the notes and citations we use a simplified version of the Library of Congress system of transliteration, omitting ligatures for those Cyrillic characters that are rendered by more than a single Latin character. Indication of the Ukrainian soft sign "ь" is omitted in the body of the text but is retained in the notes, where it is rendered with a prime. For example, Ольга Кобилянська is transliterated as "Olha Kobylianska" in the body of the text and as "Ol′ha Kobylians′ka" in citations.

Preface

There are many theories of postmodernism. From my perspective, it begins in 1946, when in the first issue of the almanac *Mystets'kyi Literaturnyi Rukh* (MUR, or Artistic Literary Movement), Jacques Hnizdovsky (Iakiv Hnizdovs'kyi) set in to talk about "Ukrainian grotesque." He maintained that postwar Europe was returning to the world of naive emotions, and that the new era was breaking out with an appreciation for Sancho Panza and a mistrust of the intellectual Don Quixote. The era of Don Quixotes is over, and under way is the epoch of the naive, practical, and emotional Sancho Panza, claimed Hnizdovsky—a graphic artist who in his art discovered an atomic plurality of worldview and in his paradoxical thinking echoed the artistic postmodern of the Dutchman Morris Cornelis Escher.

In that same issue of MUR, Viktor Petrov (Ber) discussed the new era of the split atom and envisioned that art would lose its modernist negative-experimental color and would become "positive, real, and natural, although its naturalness and reality will be the nature and reality of the technical world."[1]

The actual aesthetic thinking that I associate with the postmodernism of the end of the twentieth century is tied to the oeuvre of the American writer John Barth, who in 1996 labeled postmodernism as "endism": "endings, endings everywhere: apocalypses large and small,"[2] he wrote ironically and suggested putting "[a]n end to endings." One form of such an "end to endings" for Barth is a narrative, a story, which from the perspective of ending defers the end itself. Thus he talks about a woman who comes out of the house, approaches a man working in the garden in order to break to him some horrible news she has just heard over the phone. She knows the news, but he does not yet. This is

---

1   Viktor Ber, "Zasady poetyky (Vid 'Ars poetica' I. Malaniuka do 'Ars poetica' doby rozkladenoho atoma)," *MUR* 1 (1946): 21.
2   John Barth, "THE END: An Introduction," in *On with the Story: Stories* (New York: Little, Brown, and Company, 1996), 15.

where Barth finds the gist of the narrative—in the postponement of the trauma, in tricking both the addressee and the message, in suspending the communication by putting it sideways, behind, for a moment, not to freeze the emotions but to protect existence itself: "In narrated life," he says, "we could suspend and protract the remaining action indefinitely, without 'freeze-framing' … we need only slow it, delay it, atomize it, flash back in time as the woman strolls forward in space with her terrible news."[3]

To me, this kind of story is *postapocalyptic postmodern narrative*.

This suspension-in-play is, for me, embodied in the carnivalesque Ukrainian postmodernity, which begins simultaneously with the Chornobyl explosion in April 1986. Another version of postmodernism forms later, in the 1990s. It is characterized by a recombination of fragments and modes of writing, which shows similarities with the process of postapocalyptic existence. Overall, it is exactly the Chornobyl discourse that provokes the deployment of Ukrainian postmodernism because Chornobyl is not only associated with a socio-techno-ecological catastrophe that occurred in a certain time and place but also signifies a symbolic event that projects the postapocalyptic text about the postponement of the end of civilization, culture, and human into the post-atomic era.

My book about postmodernism focuses on the post-Chornobyl library. In my view, the post-Chornobyl library is a metaphorical image of culture that is threatened and salvaged at the same time, of culture that—like the ark, museum, temple, and list—is a bridge between real life and fiction, the past and the present, self and other, play and apocalypse, high and mass culture. I see Chornobyl as an event that legitimized the beginning of Ukrainian postmodernity, and the post-Chornobyl library as a number of texts, topoi, topograms, quotes, discourses, and canons that atomize Ukrainian culture and turn it into a process unfolding not only from the beginning to the end but also backwards, from the end to the beginning: a postapocalyptic narrative. The postmodern library is an archive that preserves culture and renders it relevant; it is also a field of the resurfacing of past complexes and old taboos. Polyphonic and polysemantic, the post-Chornobyl library is not reducible to a single overarching narrative: it is written on the margins of other texts and at the ends of others' stories, thereby filling in the gaps of the national culture.

This book results from observations that for the most part are detached from literary milieus and literary practices. I wanted to look at contemporary Ukrain-

---

3   Ibid., 28–29.

ian literature from a certain distance and to interpret what seems significant from the perspective of a witness without laying claims to fullness or definitiveness of my conclusions. I do not touch on the usual definitions of postmodernism, as I aim to build a specifically Ukrainian version of literary postmodernity.

There is no doubt that the Ukrainian literature that emerges in the 1990s is radically different from what preceded it. A separation from socialist realist thinking, along with an actualization of the aborted experience of the avant-garde and modernism of the 1920s, becomes the most important impulse for artistic experiment in the late Soviet era. At the same time, writers, first and foremost of the Eastern and Central European region, are introduced to the achievements and tendencies of Western literature and develop an interest in Western philosophy and cultural studies, previously unknown in the Soviet Union. The birth of Ukrainian postmodernity is also brought about by generational change. It depends on a new openness to the West; the informational-technological explosion which occurred due to the arrival of personal computers, video, and forms of mass communication; and the arrival of the "society of the spectacle" with its advertising industry, commodity, and the market of mass performances.

In this book, I want to sketch the main tendencies and directions taken by the postmodernist thinking in Ukrainian literature at the end of the twentieth century. I am also convinced that postmodernism does not appear in Ukraine suddenly, as an imported exotic fruit. Among the factors that contributed to its arrival are the experimental works of the Ukrainian avant-garde of the 1920s; MUR's intellectual prose fiction of the 1940s, which was saturated with Western existentialism; the Pop Art-inflected New York School poetry; Ukrainian whimsical prose—an offshoot of Latin American magic realism; and the adventurous, grotesque, and apocalyptic culture of the underground, which was incubated in the womb of socialist realism.

Finally, I took care to sidestep the traditional *victimizational* perception of Chornobyl. A third of a century after the event, it has acquired a new meaning, not only real-tragic but also symbolic and global. We live in a post-Chornobyl situation, in a postapocalyptic time, and this has turned out to be in Ukraine the gestation period of not only a new state but also of a new postmodern outlook and a new Ukrainian literature. This literature, polyphonic and polylingual, reformulates the whole history of the national cultural development; it teaches us to see otherness alongside identity; and it both offers itself to and receives from the vast intertext that is world culture. Thus, both life and literature exist in the nuclear age.

In some higher ironic sense, Chornobyl also bears witness to the fact that local tragedies and terrorist attacks are more frightening than a global catastrophe, and that, rejuvenated on the outskirts of civilization, the Chornobyl zone sometimes resembles the primeval Eden. It is postmodernity that has taught us to see such inversions.

Part One
CHORNOBYL AND POSTMODERNISM

CHAPTER 1

Nuclear Discourse, or Literature after Chornobyl

> Look under your feet—you won't be able
> to see the earth covered with our footprints
> take a careful look at the world
> don't tell me you didn't want it like that
> —Skriabin, "Chornobyl forevah"

In the second half of the twentieth century, the French philosopher Emmanuel Levinas proposed to look at the reflections on such apocalyptic situations as the Holocaust as a "representation in parentheses." This notion brought to mind the paucity of the modes of the description of reality, which slips out of control and does not yield to any definitive comprehension. Along with this, a point was made about the crisis of *literariness*, which was one of the main concepts of the modernist thinking associated with aesthetic progress in the twentieth century—more precisely, about the crisis of aesthetic auto-reflectivity and closedness of language on itself. Doubt arises that there is a unique—aesthetic—language, self-sufficient and different from the everyday, the one which allows for grasping and translating the content of fickle modern reality. The belief that artistic representation has a meaning tinged with an author's ethics also disappeared. Conjecture about an aesthetic experiment's ability—which the modernist work pretended to possess—to adequately represent reality was likewise devalued.

Ultimately, Levinas's phrase could be interpreted as an invitation to fictionality, intertextuality, and virtuality—discourse outside any representation of reality; and the philosopher perceived as an adept of postmodernist thinking, as his idea about "representation in parentheses" effectively undermines the aesthetic categories ascribed to literature in modernity. Levinas was seen as opposed to the idea that narrative was about a determinate event or situation—that it had a stable meaning; that it narrated a "truth" about an event or situation, and, further, had a broader ontological meaning that affirmed the human content of time-space, identified as *pleroma*[1] by Frank Kermode.

Nevertheless, a disciple of Edmund Husserl and Martin Heidegger, and the creator of an original existential philosophy that has at its heart an *ethics of responsibility*, Levinas can hardly be called an adept of postmodernism. On the contrary, he is an opponent of postmodernist deconstruction and postmodernist polysemy. His philosophy, based on an ethics of responsibility, revolves around the notion of "the other" and of the encounter face-to-face with this other—the encounter that is the foundation of humaneness. Such an encounter endows the self with a "bad conscience"—that is, a conscience "which comes to me from the face of the other" and "uproots me from the solid ground."[2] This consciousness positions humans before death and makes them ponder on the most important questions of being.

Levinas perceives the Holocaust as one of the greatest of humankind's tragedies, whose meaning defies comprehension. On the one hand, instead of creating definitive images, that is, "complete" history and "truthful" representation of the tragic event, what he calls for is to reveal this event as "an uncrossable gap," a "hole in history" which cannot be filled with any objective narratives.[3] Levinas shows a complete distrust of representation because it appeals to certain objectivity. On the other hand, a subjective, unconscious participation of I (or self) in history is combined with a responsibility for the other, thereby creating "the dia-chrony of a past that cannot be gathered into re-presentation."[4]

Nevertheless, however paradoxically, Levinas's ideas about the "hole in history" which cannot be filled with any narratives, and about "the dia-chrony of a past" that resists any representation, are echoed in postmodern theories about traumatic experience and its (un)representation in postmodern literature. Ultimately, one version of the origin of postmodernism is based on a supposition that the appropriation of traumatic events like the Holocaust or the

---

1 Frank Kermode, *The Sense of an Ending* (Oxford: Oxford University Press, 2000), 193.
2 Emmanuel Levinas, *Entre Nous: Thinking-of-the Other*, trans. Michael Smith and Barbara Harshav (New York: Columbia University Press, 1998), 148.
3 Robert Eaglestone, "From Behind the Bars of Quotation Marks: Emmanuel Levinas's (Non) Representation of the Holocaust," in *The Holocaust and the Text: Speaking the Unspeakable*, ed. A. Leak and G. Paizis (London: Palgrave Macmillan, 2000), 104.
4 Emmanuel Levinas, *Entre Nous*, 171.

atomic apocalypses of Hiroshima and Nagasaki produces aesthetics and poetics that are radically different from those that hitherto existed.

Indeed, traumatic events such as the Holocaust, the Great Purge, and the Holodomor acquire in the twentieth century a global symbolic meaning. Auschwitz, in particular, becomes a point of reference for Theodor Adorno, who sees in it a crisis of the "positivity of the apparent being" and the failure of the Enlightenment humanism. After all, he says, suffering "in the camps, without any consolation, burned every soothing feature out of the mind, and out of culture."[5] Later, Auschwitz will serve as a point of departure also for Jean Francois Lyotard: the distrust of knowledge and progress will prompt him to formulate the new postmodern situation. "Following Theodor Adorno, I have used the name 'Auschwitz' to signify just how impoverished recent Western history seems from the point of view of the 'modern' project of the emancipation of humanity," notes Lyotard and asks a rhetorical question: "What kind of thought is capable of 'relieving' Auschwitz—relieving [*relever*] in the sense of *aufheben*—capable of situating it in a general, empirical, or even speculative process directed toward universal emancipation?"[6]

This question lays a foundation for meditations about a "postmodernity" that no longer trusts thinking based on the idea that reality is "purposeful" and "complete." At the end of the twentieth century, Lyotard stated the general epistemological uncertainty of the time: "There is a sort of grief in the *Zeitgeist*. It can find expression in reactive, even reactionary, attitudes or in utopias—but not in a positive orientation that would open up a new perspective."[7] Therefore, proceeding from the word-symbol "Auschwitz," Lyotard theorizes a peculiar negative teleology of the postmodern. He perceives the postmodern world as founded on distrust of the progressive deployment of knowledge that works towards social and individual emancipation.

The treatment of "Holocaust" or "Auschwitz" as word-symbols does not mean an unethical reading of those traumatic events but, to the contrary, serves as a basis for discussion about the fate of morality and humanity in the modern world. I argue that the Chornobyl disaster is also an event of a traumatic significance—that "Chornobyl" becomes a word-symbol of the after-modern modernity which comes *after the catastrophe*.

---

5   Theodor Adorno, *Negative Dialectics*, trans. E. B. Ashton (London: Routledge, 1973), 365.
6   Jean-Francois Lyotard, "Note on the Meaning of 'Post,'" in *Postmodernism: A Reader*, ed. Thomas Docherty (London: Routledge, 1993), 49.
7   Ibid., 49–50.

It is no accident that a special issue of *Anthropology of East Europe Review* in spring 2012 dedicated to the commemoration and cultural representation of Chornobyl (and published after the accident at the Japanese nuclear power station Fukushima) is not so much preoccupied with the search for "scientific 'objectivity,'" as with exploring the "symbolic fallout" of Chornobyl—its place in collective memory and cultural artifacts, its integration into the everyday life and individual fates.[8] Such symbolic meanings of the Chornobyl catastrophe are exactly the subject of my analysis in this book.

On the map of the modern life, "Chornobyl" is a symbolic event that undermines the modernity that formed in the late Soviet era. It is characterized by disappointment in modernization under "socialism," which more than half a century ago was implemented by way of violence towards individual freedom and the exploitation of the intellectual and physical potential of humans. Chornobyl turned out to be a critical moment which precipitated perestroika and disenchanted the whole Soviet system. Mikhail Gorbachev, in an interview given twenty years after the Chornobyl tragedy, acknowledged that the catastrophe had turned from a technological into a social one because, in his words, "more than anything else, it made possible freedom of expression, since the system as we knew it could not exist like that any longer."[9]

It was exactly the Chornobyl factor that triggered the reevaluation of the totally rotten socialist system. Gorbachev, asserting post factum a connection between the Chornobyl accident and perestroika, and thus building a Chornobyl mythology, admitted that the breakdown "made absolutely clear how important it was to go on with the politics of glasnost, and I have to say that I started to think about time in before-Chornobyl and post-Chornobyl categories."[10]

During perestroika, various modernizing projects in the Soviet mode—social, national, technological, informational—were delegitimized. The accident focalized in itself all the disappointment with the Soviet system, the best testimony of which is *Chornobyl discourse*—the numerous official and unofficial proclamations, rumors, witness testimonies, and documental and fictional works about Chornobyl. Behind these narratives stand their immediate contexts: the lack of information; the death of firefighters; the mobilization of the military; the labor of miners, scholars, and nuclear scientists; the evacuation of people from the zone; and the fate of refugees.

---

8   Melanie Arndt, "Memories, Commemorations, and Representations of Chernobyl: Introduction," *Anthropology of East Europe Review* 30, no. 1 (2012): 2.
9   Mykhailo Horbachov, "Chornobyl's'kyi perelom," *Den'* 65 (2287), April 18, 2006, 1.
10 Ibid.

It is very important that the Chornobyl discourse is both post-catastrophic and post-traumatic. Not only does it serve as a means of discussion of modern times and real events but it also forms new modes of thinking. As Jean Baudrillard notes, "The catastrophe is the maximal raw event, here again more event-like than the event—but an event without consequences and which leaves the world hanging."[11] In fact, during the period of uncertainty after the catastrophe a new language emerges along with new forms of expression. It refers not only to events immediately connected with Chornobyl but also to various other spheres, and permeates various genres, styles, and types of speech—from the everyday to the artistic. All Ukrainian literature after Chornobyl in a way exists in an epistemologically new system: *after the trauma.*

As early as a few years after the accident Yuriy Shcherbak writes that the Chornobyl text must engender other artistic forms, different from, say, *War and Peace* and *And Quiet Flows the Don*. He points to one of such forms—the documental, as in testimonies collected by means of quotes and montage. A new Chornobyl style is taking shape, with graphically marked ellipses and pauses that pinpoint gaps and those places in witness testimonies that are incapable of capturing the impressions generated by experience. A new subculture is born— Chornobyl humor and Chornobyl jokes, similar to the Auschwitz cycle or witticisms about the Gulf War. In other words, "black humor," "gallows humor."[12]

How is one to represent the content of the post-Chornobyl reality? Speaking about postmodernism, Lyotard notes that very important for him is the moment of modernity's "working through" ("durcharbeiten") its own meaning and recalls the corresponding experience of the historical avant-garde of the 1920s and 1930s. In this way, Lyotard juxtaposes realism with the trans-avant-garde (which he associates with postmodernism) exactly from the perspective of modernity's "working through." Realism, in his view, presupposes that there is a certain organic unity—*reality*—which can be represented. However, as he points out, in the postmodern era "so-called 'realist' representations can no longer evoke reality except through nostalgia or derision."[13] Generally speaking, the trans-avant-garde aims not at demonstrating the organicity and wholeness

---

11 Jean Baudrillard, "Fatal Strategies," in *Selected Writings*, ed. Mark Poster (Stanford: Stanford University Press 2001), 195.
12 Larisa Fialkova, "Chornobyl's Folklore: Vernacular Commentary on Nuclear Disaster," *Journal of Folklore Research* 3 (2001): 181–204.
13 Jean-François Lyotard, "Answering the Question: What Is the Postmodern?" in *The Postmodern Explained to Children: Correspondence 1982–1985*, trans. Julian Pefanis and Morgan Thomas (Sydney: Power Publications, 1992), 3.

of reality but at recreating it through a juxtaposition of different projections, created by way of discourse, with the aid of language games, time shifts, fragmented identities, and masks. These artistic techniques are not employed to provide a so-called truthful image of a fact, such as the Chornobyl accident. Rather, they render the impression of shifts in reality, called forth by the event; and the most productive here is the moment of shift, rupture, and difference, which causes alternative states of consciousness.

Jacques Derrida's notion of "différance" is the most precise way to describe the postmodernist appropriation of post-Chornobyl reality. After all, when we talk about Chornobyl, we discuss what has been before and thereby differentiate it from what happens after. We talk about signs, premonitions, and patterns that caused the accident, and about the way the breakdown revealed those signs, made them symbolic. Different perspectives and views on what happened turn the narrative backwards and bifurcate consciousness. Reality is suspended, undermined, and contorted; it is made virtual because in the radioactive zone time, space, and human feelings in general become warped and acquire a different shape. The senses are no longer trustworthy, and reality itself becomes alien, diseased, and thus inimical to human existence. All this is reflected in the new post-Chornobyl linguistic consciousness, which bears witness to the shift from social-realistic certainty to postmodernist uncertainty and immanence. In the absence of essence—that is, a permanent ontological center—humans create themselves and their world through a language that is separated from the world of objects.[14]

When Chornobyl is regarded as the main event of the text that is generated by Chornobyl itself, it can be said that the post-Chornobyl discourse transforms all the real events and moods in Ukraine at the end of the twentieth century into a symbolic reality that has a higher meaning and aim and applies to a postapocalyptic survival. As a symbolic event, Chornobyl relates to diverse realities at various levels of abstraction: from the errors of the technological personnel to the human heroism of the firefighters putting out the nuclear fire right after the accident; from the social catastrophe to the techno-anthropogenic mutations; from the "spiritual Chornobyl" to the "new Middle Ages." In this way, a characteristic post-Chornobyl text emerges; and its nature is defined not so much by the singular object of description as by a circle of signs and meanings as well as

---

14 Hans Bertens, "The Postmodern Weltanschauung and Its Relation to Modernism. An introductory Survey," in *A Postmodern Reader*, ed. Joseph Natoli and Linda Hutcheon (Albany: State University of New York Press, 1993), 47.

symbols, associations, and motifs, connected with this object. Moreover, one of the main attitudes behind the Chornobyl discourse is distrust of the key ideals of the Enlightenment: universal progress and the development of science, reason, and technology, as well as the idea of socialism itself.

Of course, the Chornobyl text first emerges as works dedicated to the theme of Chornobyl. We are in possession of a rich Chornobyl library: works by known and unknown authors, poetry, prose fiction, essays and journalistic writing, memoirs, documents, and literary criticism.[15] It is worth noting that alongside so-called high culture, Chornobyl becomes one of the most beloved topics of popular culture, which is confirmed by a multiplicity of secondary school works, childhood recollections, patriotic-spiritual confessions, songs, and jokes dedicated to the accident. In its own fashion, Western postmodernism uses Chornobyl as a topos of "real reality" in detective stories, while in science fiction it is used as a virtual and fantastic place.

However, such a text encompasses also works that depict Chornobyl indirectly, irrespective of their lack of the Chornobyl topic proper. In these works, the traces of the accident are virtually absent, but the postapocalyptic situation is described. What is more important, they also use a new artistic optics, which emerges with the influence of Chornobyl imagery. These works consist mostly of nonlinear modes of narration depictions of explosive situations, explorations of parallel worlds, reiterations, transgressions, open endings, and their focus of action is on the surface of the text. The Chornobyl text is characterized by both multilingualism, in spite of the lack of communication per se, and the emergence of topological and allotropic thinking, signaling the invincible discreteness of the time-space. Another characteristic is intermediality, which serves as a means of expressing the inexpressible in the situation of "voicelessness." Chornobyl facilitates the switching of literature to virtual dimensions: by joining the real with the fantastic, it has ruined the ideal completeness of the world and undermined utopian faith in the expedience of rational semiosis.

---

15 See in particular Ivan Drach, "Chornobyl's'ka madonna. Poema," *Vitchyzna* 1 (1988): 46–62; Borys Oliinyk, "Sim," *Literaturna Ukraïna* 38 (17 veresnia 1988); Iurii Shcherbak, "Chornobyl'," *Vitchyzna* 4, 5, 9, 10 (1988); Volodymyr Iavorivs'kyi, "Mariia z polynom pry kintsi stolittia," *Vitchyzna* 7 (1987): 16–139; Volodymyr Morenets', "Poetychna epika Chornobylia," *Radians'ke literaturoznavstvo* 12 (1987); and Larysa Onyshkevych, "Echoes of Glasnost: Chornobyl in Soviet Ukrainian Literature," in *Echoes of Glasnost in Soviet Ukraine*, ed. Romana Bahry (York: Captus University Publishers, 1989), 151–70.

CHAPTER 2

Nuclear Apocalypse and Postmodernism

In the cultural memory of the end of the twentieth century, Chornobyl has become an apocalyptic event, and therefore it is no wonder that it is reflected first of all in the parameters of the apocalyptic discourse. Associated with a nuclear explosion, Chornobyl has become a symbol of global catastrophe—the destruction of the human world itself. "Chornobyl opened an abyss, something beyond Kolyma, Auschwitz, and Holocaust," said a village teacher, evacuated from the Chornobyl zone. "A person with an axe and a bow, or a person with a grenade launcher and gas chambers, can't kill everyone. But with an atom …"[1]

In our imagination, the scene of the Chornobyl catastrophe and the meaning to which it gives rise are most often associated with apocalypse as a metaphor of the destruction of the whole world. Another aspect of apocalypse—the birth of a new world (New Jerusalem) and the survival of humans (humankind, nation, planet)—is marginalized or even reduced in the Chornobyl discourse. However, as Meyer Howard Abrams points out, the biblical paradigm of the apocalypse has its author and presupposes that world history has a plot with an exposition, catastrophe (the Fall), crisis (the Incarnation and Resurrection of Christ), and conclusion (the Second Coming and the substitution of "the new heaven and the new earth" for the old world), which turns the tragedy of human history into a cosmic comedy.[2] Hence the Chornobyl discourse is associated with apocalypse as the mother of all catastrophes, with the only difference that it does not turn the "tragedy of human history" into a "cosmic comedy."

---

1   Svetlana Alexievich, *Voices from Chernobyl: The Oral History of a Nuclear Disaster*, trans. Keith Gessen (New York: Picador, 2006), 181.
2   M. H. Abrams, "Apocalypse: Theme and Variations," in *The Apocalypse in English Renaissance Thought and Literature: Patterns, Antecedents, and Repercussions*, ed. C. A. Patrides, J. Wittereich (Manchester: Manchester University Press, 1984), 342–68.

The apocalyptic perception of Chornobyl is superimposed upon the general millennial apocalyptic expectations at the turn of the twenty-first century. As James Berger notes, it even seems a little odd that at the end of the twentieth century apocalyptic feelings engulfed the whole world. Despite the tribulations of the time, there was no global crisis that would portend the end of the world: the ecological crisis seemed to be controlled by governments; gory local conflicts did not look like Armageddon. "Without the Cold War and the Soviet Union," says Berger, "there is no Evil Empire, no Antichrist, no immediate threat of annihilation."[3]

However, Berger names one important factor that worked against apocalyptic terror at the end of the millennium: "a kind of apocalyptic fatigue, or indeed, a widespread sense that the apocalypse has, in some sense, *already happened*. ... We know what the end of the world looks like. We know because we've seen it, and we've seen it because it's happened."[4] Indeed, at the end of the twentieth century, the apocalypse seemed to have happened, as Berger aptly points out, because "the images of Nazi death camps, of mushroom clouds ... of genocides in a dozen places, of urban wastelands and ecological devastation"[5] can be associated with nothing but apocalypse.

The whole modern era was permeated with the sense of an unending crisis—moreover, there was a wish for a final catastrophe. At the end of the twentieth century this crisis does not disappear; instead, a new feeling appears—that the final catastrophe has already happened and the crisis has already passed, and that the unending series of news about calamities, barely distinguishable one from the other, are nothing more than a new and complicated form of stability.[6]

Chornobyl was destined to unleash this chain of catastrophes at the end of the twentieth century. Chornobyl and the whole Chornobyl discourse invite us into the very heart of discussions about catastrophism. The meaning of Chornobyl is most clearly revealed in the context of the development of nuclear discourse, which is a product of the modern era; however, Chornobyl significantly changes this discourse by adding a post-traumatic tinge and by juxtaposing it with other catastrophic phenomena of the twentieth century such as the Holocaust, the Great Purge, and the world wars.

---

3   James Berger, "Twentieth-Century Apocalypse: Forecasts and Aftermaths," *Twentieth Century Literature* 46, no. 4 (2000): 388.
4   Ibid.
5   Ibid.
6   James Berger, *After the End: Representations of Postapocalypse* (Minneapolis: University of Minnesota Press, 1999), xiii.

It is well known that postmodernism is not only a type of consciousness and a cultural epoch but also a whole network of various discourses: cultural, political, artistic, and multimedia. Important philosophers of postmodernism—from Derrida to Baudrillard—explicitly associate postmodernism with the nuclear age. The very assumptions about the possibility or impossibility of a nuclear explosion gain a menacing character in the second half of the twentieth century—at the time of doubts about the nature and aim of human emancipation programmed by the Enlightenment. On the one hand, nuclear weapons testify to the greatest technological and "rational" power of humans over nature; on the other, they reveal the great barbarism underlying civilization—its potential to effect the total destruction of humankind. The nuclear bomb and nuclear consciousness cause misunderstandings and dissensions among people, nations, genders, and races, thereby disrupting communication processes. A new mode of communication emerges—that is, communication from the perspective of the end of the world, post-catastrophe, and postapocalypse.

Much has been written about the place of nuclear imagery in literature and culture in the nineteenth and twentieth centuries.[7] One thing is undeniable: nuclear apocalypse becomes an important cultural metaphor and a laboratory of images. Furthermore, nuclear apocalypse is linked with themes such as technological progress and the role of intellectuals in society, the nature of madness and the birth of a new individual, the perception of "the stranger" and the place of monstrosity in human history, historical catastrophes and futuristic images. In the course of the twentieth century, nuclear imagery realizes different narrative and psychological models in different literatures. For example, the Hiroshima witnesses link their impressions of the nuclear bombing with childhood images of the end of the world, separation, helplessness, and disappearance.[8] In German nuclear literature, nuclear apocalypse is associated with genocide, preceded by the Holocaust with its gas chambers and crematoria[9]. In Ukraine, atomic apocalypse is constantly associated with Chornobyl.

A political component has always been an important aspect of nuclear discourse, since nuclear weapons from the beginning have been *political weapons*. Although in the spring of 2010 US Vice-President Joe Biden stated that

---

7   See Spencer R. Weart's discussion in *Nuclear Fear: A History of Images* (Cambridge: Harvard University Press, 1988).
8   Ibid., 107.
9   See Wolfgang Lueckel, *Atomic Apocalypse: "Nuclear Fiction" in German Literature and Culture* (PhD diss., Graduate School of the University of Cincinnati, 2010).

"today, the danger of deliberate, global nuclear war has all but disappeared,"[10] nuclear discourse has been transformed rather than cancelled. A new danger has emerged: "the nuclear threats we face from terrorists and non-nuclear states seeking to acquire such weapons."[11]

It is a paradox of the end of the twentieth century, however, that, while a constant presence in international politics, atomic weapons did not cease to be perceived tragicomically on the level of mass consciousness. The fear of atomic war, compared with the "unbearable lightness of being" in the modern world, looks ridiculous in Vladimir Makanin's novella *Hole*, featuring—not without irony—a man raised on Soviet propaganda. He is afraid of atomic war and therefore builds his own bunker. "Afraid of atomic war, really?" the narrator remarks. However, the mass distribution—at the beginning of the third millennium—of postapocalyptic genres in literature, film, and the media proves that the postnuclear mindset has become firmly rooted in the modern human.

Starting from Hiroshima and Nagasaki, the danger of atomic holocaust becomes one of the most important topics of modernity: it is exploited by politicians and the media; it serves as a uniting factor during the Cold War; and it forms a planetary thinking in the age of globalization. Nuclear discourse has an important place in philosophy, where it facilitates the development of existentialist and futuristic mindsets. It is worth noting that atomic narrative is not only transferred to the future but is also superimposed on the past—for example, in writing about World War II. The Belarusian writer Ales Adamovich, known as an author of war fiction, after the Chornobyl accident wrote a postnuclear anti-utopia, *The Last Pastoral* (1987), a heart-wrenching story about the survivors of a devastating atomic war.[12]

By and large, nuclear narrative becomes a constituent of the cultural mindset of the Cold War period, when atomic catastrophe is perceived as a real threat (this is evidenced, in particular, by *The Emergency Plans Book*, designed for official use, which was declassified and turned over to archives only in 1998).[13]

---

10 Joseph R. Biden, Jr., "A Comprehensive Nuclear Arms Strategy," Voltairenet.org, 7 April 2010, http://www.voltairenet.org/article164856.html.
11 Ibid.
12 Adamovich has always been interested in the fate of a human after an atomic war. This interest is confirmed by his 2007 book of essays, reflections, and notes "... *Imia sei zvezde Chernobyl*'" (Minsk: Kovcheg, 2006) ["... And the Name of the Star Is Chornobyl"]. Larisa Mikhalchuk, *Bikfordov shnur*, http://www.br.minsk.by/index.php?article=30134.
13 Jonathon Keats, "Apocalypse Made Easy," Salon.com, https://www.salon.com/2002/02/07/doomsday, 8 Feb 2002.

Ultimately, from 1947 the apocalyptic clock tells a symbolic time that separates humanity from a global catastrophe, including a nuclear one.

After the Chornobyl catastrophe, the atomic apocalypse theme took a special tone, enriched by anticipation of the breakup of the Soviet Union and by expectations at the millennium's end. However, as early as in 1984 Frances Ferguson demonstrated that nuclear discourse, similarly to other great projects of modernity such as art, beauty, and progress, is based on sublimity. A nuclear explosion is something "unthinkable"—"the most recent version of the notion of the *sublime*."[14] Ferguson says that when in the seventeenth and the eighteenth century the aesthetics of the sublime reemerges as evidence of the uniqueness of individual consciousness, it represents a world in which the status of objects is gradually declining: it seems that objects are empowered not by the fact of their existence but due to their reflection in the human mind.[15] Such an unfathomable subjective dimension was ascribed to the fate of humankind, the existence of the planet, and the care about children not yet born. Thus the feeling of the sublime arises not simply due to an encounter with majestic natural phenomena, such as waterfalls, volcanoes, or hurricanes, but due to the fact that a self, striving for the unfathomable and the absolute, inexpressible yet always "present," crosses the limits of both itself and real objects, and thereby demonstrates its simultaneous strength and weakness.

Catastrophes are also endowed with the power of the sublime. As Douglas Kellner points out, "catastrophes confirm the power of the object over the subject, and delight people in their spectacular excess, just as humor which subverts the order of language and produces pleasure."[16] Susan Sontag, on the other hand, calls catastrophe "one of the oldest themes in art" that brings satisfaction and excitement. Moreover, as Sontag notes in 1965, "mass trauma exists over the use of nuclear weapons and the possibility of future nuclear wars. Most of the science fiction films bear witness to this trauma and, in a way, attempt to exorcise it."[17]

This nuclear sublimation is based on the fact that some destructive power—an atomic bomb—puts the existence of humans and of the whole world in danger; this power simultaneously causes wonder, fear, and a desire

---

14 Frances Ferguson, "The Nuclear Sublime," *Diacritics* 14, no. 2 (1984): 5.
15 Ibid., 4–10.
16 Douglas Kellner, *Jean Baudrillard: From Marxism to Postmodernism and Beyond* (Stanford: Stanford University Press, 1989), 161.
17 Susan Sontag, *Against Interpretation, And Other Essays* (New York: Octagon Books, 1982), 218.

to find a form that would make it possible to subjugate it and to *fathom the unfathomable*. If nuclear discourse is perceived as a form of sublimation after the trauma (catastrophe), it becomes understandable why talking about atomic war as a reality is simply impossible—because at stake here is the total destruction of the world, the individual, and language itself.

It is not a coincidence that Lyotard, reflecting on postmodernity, resorts to the notion of the sublime, which he equates with the "unpresentable": "We can conceive of the absolutely great, the absolutely powerful; but any presentation of an object—which would be intended to "display" that absolute greatness or absolute power—appears sadly lacking to us. These ideas, for which there is no possible presentation ... [o]ne could call them unpresentable."[18]

Nevertheless, artistic imagination aims at presenting the "unpresentable" and attempts to describe the object that resists description. A special role is reserved here for an aesthetic feeling of the beautiful combined with fear, which grows out of the realization of the impossibility to encompass (as a whole and simultaneously) an unending series of events, the mightiness of nature, an excess of creative imagination, and an atomic blast.

To present the "unpresentable" is one of the aims of the avant-garde, or trans-avant-garde, which Lyotard associates with postmodernism in terms of the repetition of the past. He notes:

> You can see that when it is understood in this way, the "post-" of "postmodern" does not signify a movement of *comeback, flashback,* or *feedback*—that is, not a movement of repetition but a procedure in "ana-": a procedure of analysis, anamnesis, anagogy, and anamorphosis that elaborates an "initial forgetting."[19]

The methods of depicting the "unpresentable," however, are quite specific: the avoidance of direct portrayal and representation; the representation of something, yet in a negative way; allowing sight while prohibiting looking; moving forward, reiterating; bringing satisfaction by causing pain. This is the principal difference between the trans-avant-garde (and the postmodern) and any realistic representation of the inexpressible.

---

18 Lyotard, "Answering the Question," 6.
19 Lyotard, "Note on the Meaning of 'Post,'" 50.

CHAPTER 3

The Socialist Realist Chornobyl Discourse

Among various works on the Chornobyl topic, Ivan Drach's "The Chornobyl Madonna" (1988) gained especial resonance in Soviet times. The "unpresentable" here has become one of the main rhetorical modes. The very intention to associate the Chornobyl tragedy with the symbolic image of the Madonna set rhetorical traps. On the one hand, it was possible to write about the Chornobyl Madonna as a "given," that is, as an already created image, worked upon by

> ... the gods of art
> Over millennia—from Rublev to Leonardo da Vinci,
> From the Vyshhorod Madonna to the Sistine
> From Maria Oranta to the Atomic Japanese Woman.[1]

The author's position in this case is boiled down to the role of a scribe: "to create her the way my pen is able to perceive her.[2]"

On the other hand, there was another possibility: the realization that no author is able to portray the majestic and the eternal: "You tried to write about Her, yet She writes using you, / As a pen, good for nothing, as miserable dust, as a pencil." Hence the poem turns right away to describe the sublimated and the extraordinary, in the face of which a creative individual is reduced. It causes a lack of faith in its own words ("I have no words. They are shot to the last word"); the direct speech, which is not mediated by the voices of others, perishes and is replaced by muteness. The lyrical subject of the poem admits that a poet can talk about the stately and the tragic only with gestures.

---

1   Ivan Drach, "Chornobyl´s´ka madonna. Poema," *Vitchyzna* 1 (1988): 43.
2   Ibid., 43.

In "The Chornobyl Madonna," Drach combines two rhetorical possibilities. First of all he appeals to fragments rather than to a whole, to the voices of "others" rather than to his own voice. Vasyl Kurylyk's painting and a postcard, the tale of a soldier in a construction battalion about the naked footprints of somebody else's mother, the remarks of an old woman with a cow in cellophane, who flees from the city to her house in the zone, the voices of a Chornobyl female tractor driver, and the Khreshchatyk Madonna fragment and multiply the image of the mythological Chornobyl icon, which becomes many-sided and multi-personal. The semantic amplitude constantly swings: from the abstract, sacral, and majestic to the human-like and real personality of the Madonna.

Drach's poem is eclectic and polyphonic to the highest degree. Not only does it render voices of certain characters (such as Kurylyk, the soldier, and the old woman) but also represents the voices of other authors who have written about Chornobyl: epigraphs from Svitlana Iovenko's "The Explosion"; Yavorivsky's *Maria with Wormwood at the End of the Century*; Anatolii Mykhailenko's "Roots and the Memory of the Zone"; Yuriy Shcherbak's "Chornobyl"; Borys Oliynyk's "The Chornobyl Trial"; and from Chornobyl folklore in general. These epigraphs make it possible to stick to the topic despite the fragmentariness of expression, introduce various Chornobyl plot peripeteias, and create the Chornobyl text as a "text of texts." Epigraphs from Shevchenko, Tychyna, and Symonenko broaden the circle of comparisons and analogies of the "text of the texts" while helping to maintain one dominant style—the elevated-mournful tone of the narrative.

A fragment about the Scythian Madonna adds a mysterious and apocalyptic overtone that relies on a juxtaposition of the past and the future, the sacral and the profane. In the flames of the descendants' fire, the womb of the Scythian mother yawns and a destructive Scythian comes out. At the same time, the Chornobyl apocalypse has absolutely real, profane dimensions: a grandson-apostate lifts up his hand against the old woman, and the stolen icon of the Chornobyl Madonna surfaces at a flea market in Paris.

An oscillation between opposing stylistic dimensions—realistic and symbolic, sacral and profane—defines the structure of the poem. Poetry is combined with prose, free verse with rhymed verse, and irony with sorrow and sarcasm. In particular, the "slavoslovy-lykhoslovy" ("adulators-detractors") get their share; from here and all the way to Cuba they simulate love for the Mother but abandon her right after "the black atom shook." This sarcasm and the accusation of others diminish the pathos of the poem; moreover, the tension of voices and the oscillation of styles gradually disappears, and Drach ends his poem rhetorically by appealing directly to the politician, the power engineer, and the scientist, and by accusing them of criminal acts. In this way, a rhetorical exculpation of the lyric subject takes place: by redirecting the fault to others, he hides himself in silence: "And I, I, an adulator, / … Have lost my depraved speech, / And now I am numb for ages."[3]

Thus, on the one hand, the poem states that it is impossible to represent Chornobyl directly. Instead, one can talk about it in fragments and indirectly—through symbols, signs, different styles, and alien voices. On the other hand, according to the socialist realist tradition, artistic expression must reveal the author and his ideological position; for this reason, in conclusion the poet strives to channel all the multi-directional voices into a single absolutely condemnatory word.

Svitlana Iovenko's novella *A Woman in the Zone* (1988–93) is another example of the Chornobyl apocalypse. In terms of genre, the novella combines a documentary tale with a quest and a psychological drama. It is about a journalist, Inna, who comes to the Chornobyl zone and seeks the truth about what happened at the moment of the catastrophe. While the lyric subject in Drach's poem discovers a disunity of truth that is impossible to grasp, the female character of Iovenko's novella, on the contrary, is not satisfied with partial, fragmentary truth, and strives to fathom the deep and *only truth*. As a journalist, Inna has gathered a lot of material and recorded many tales: "about many of them she could already write individual essays, but a strict sense of truth was telling her that the full truth lies not in this particular case, not in this particular human fate."[4] She already knows that Chornobyl is "a mosaic of faces, fates, and acts," yet she still seeks the "common great truth" because she is convinced that the small truths of individual partakers in the tragedy will reveal it.[5] This rhetorically reminds us of the Soviet narrative of the "great truth," so it is not a coincidence that the word "exploit" often appears in the novella.

The elevated and tragic—that is, romantic—"truth" finds its agent in Inna herself. Therefore, the novella becomes an embodiment of the sacrificial utopia of femaleness: "with her fateful involvement in the gigantic tragedy … she [Inna] has discovered herself new, truthful; she has found the lost meaning of life—to be a saving straw, to be an Ariadne's thread for those who are prone to

---

3   Ibid., 62.
4   Svitlana Iovenko, "Zhinka u zoni," in Svitlana Iovenko, *Liubov pid inshym misiatsem* (Kyiv: Ukraïns´kyi pys´mennyk, 1999), 150.
5   Ibid., 148.

succumb to the abysmal despondency and the criminal apathy of the cold-eyed bureaucrats."[6]

The originality of Iovenko's novella lies first of all in the fact that the Chornobyl apocalypse acquires a gender dimension. This approach is implemented through the gaze of a woman who constantly feels her remoteness from both the events and people, and, as the author maintains, it is probably due to her marginalization that the "woman in the zone" is capable of approaching the "great truth." Hence, having come to the zone as a member of a film crew, the journalist Inna becomes as though the center of a different vision—a new Chornobyl optics. This is defined by the vision of something that is concealed by cameras and that is not caught by cameras, about which numerous documentaries and records are silent. The female perspective becomes an optics of the unmediated Chornobyl reality. Thus Inna becomes part of the Chornobyl community, because she was "the only female voice among them, was a mother and a sister of every one of them." Moreover, she recognizes herself as a "Chornobyl mother, and perhaps a child of Chornobyl also."[7] The pathos increases, and soon the woman in the zone is associated with an angel, the law, and the "highest truth." Therefore, she is presented as a sacral center of the zone: "say, a woman in the Chornobyl reality is a special woman: too vigilant, watchful, as a wild animal that feels danger a mile away and is ready to ward off an offender."[8]

Consistently exalting Inna above other people, Iovenko symbolizes the Chornobyl reality and identifies it with a female perspective. Inna indeed listens yet does not hear the voices of others, their small stories, because she is preoccupied with her own story and her own suffering. After all, a tragedy underlies her soul—the death of her husband and son in an automobile accident. Her personal suffering lifts Inna up above everyone else, and she wishes to go through "this Chornobyl Golgotha alone." In order to perform her feat, Inna wants to recreate the last minutes before the explosion—that is, to gaze into the abyss and thereby to fathom the absolute, final "truth." She is captivated by, awestruck before the incomprehensible; and the chill of fatal danger gives her an opportunity to associate herself with the Mother of God: "O, if only she were the Mother of God, she would spread her protecting veil over the reactor, over towns and villages, over Ukraine and Belarus—she would save everyone."[9]

---

6   Ibid., 151.
7   Ibid., 155.
8   Ibid.
9   Ibid., 182.

Iovenko introduces a gendered existential theme into the topic of Chornobyl: her rebellious woman is fighting death itself. However, the pathetic-tragic presentation of the character crosses the boundaries of the authentic and becomes morally and artistically elusive. The consequence of it is the emasculation of tragedy: because Inna herself is identified with the whole "truth," she practically embodies the apocalyptic "truth" that is impossible to comprehend. Eventually, we witness how the expression turns into self-parody: Inna flees the zone, assuring herself that this is the zone of her pain because there she shed "a drop of her blood." Indeed, she got a sore foot, and from the "scratched, reddened ankle a drop of blood fell on Chystohalivka's ground."[10]

Both Drach's and Iovenko's Chornobyl texts attest to the fact that the (non)representation of the traumatic Chornobyl event is not only a theoretic-philosophical problem but also a moral and aesthetic trial for artists. This task is especially complicated in the framework of a socialist realist form of expression that seeks to comprehend and represent the objective and final "truth." This teleology leads either to a rhetoric of accusation against others or to a rhetoric of personal heroism.

Marko Pavlyshyn, reflecting on the new forms of representation generated by Chornobyl texts, suggests a new term—the "Chornobyl genre." The problem of genre, in his opinion, is conditioned by three moments of tension: "the first one is stylistic (between 'common language' and 'high style') and the second one is moral and synchronous (between criticism and apologetics); the third one is moral and diachronic (between today's and yesterday's assessment of the same phenomena)."[11]

Revealing the sense of these concepts, Pavlyshyn, in fact, points to the phenomenon of (non)presentation, which becomes topical in regard to the theme of Chornobyl. It is about a refusal to authorize a global event through personal experience, which leads authors to "the most objective modes of expression," such as authentic interviews or document collections. Other authors do not avoid literariness by escaping into documentary; instead, they openly resort to a "high style." One more danger lurks in the fictionalization of Chornobyl: narrative usually leads to the trivialization of the topic. "While documentary literature in its essence maintains respect toward the awe and grandeur of the

---

10   Ibid, 217.
11   Marko Pavlyshyn, "Chornobyl´s´ka tema i problema zhanru," in Marko Pavlyshyn, *Kanon ta ikonostas* (Kyiv: Chas, 1997), 177.

topic," writes Pavlyshyn, "fiction does not do that by definition and therefore is often disappointing due to the impression of inadequacy."[12]

Pavlyshyn concludes that the Chornobyl genre should be defined by a "constant reference to the specificity and the problematic nature of the author's position regarding the topic."[13] As analysis of Drach's poem "The Chornobyl Madonna" and Iovenko's novella "A Woman in the Zone" shows, in a Chornobyl text the "problems with the author's position regarding the topic" are supplemented by the problematic character of the very language aimed at the expression of the (in)expressible. Particularly with reference to this representational-linguistic complexity, the Chornobyl genre can gravitate not only toward pathos and tragedy but also toward ironic self-parody. This self-parody, which annihilates socialist realist forced optimism, is one of the sources of postmodernist aesthetics and poetics in the 1980s.

---

12 Ibid, 179.
13 Ibid.

CHAPTER 4

Nuclear (Non)-Representation

Nuclear explosion symbolizes in the minds of people, especially in the second half of the twentieth century, the idea of the end of the world—the apocalypse. "[A] nuclear catastrophe that would irreversibly destroy the entire archive and all symbolic capacity of culture," as Jacques Derrida stated (before Chornobyl) in 1984, can be regarded as "the only referent that is absolutely real."[1] The existence of culture in the atomic age is necessarily related to nuclear detonation and nuclear criticism, as Derrida calls reflections on the essence of the absolute referent, which signifies the finality of both culture and human in the world. However, the end of the world is only a part of the apocalyptic imagination. In reality, it is the world *after* the end of the world—the *post*-apocalypse—that is the real object of postapocalyptic imagery. Ultimately, the manifestation of signs, the transformation of that which is left, and the evaluation of that which is lost—all these demonstrate that the postapocalypse as such exists before the apocalypse.

The existence of a referent such as an atomic explosion stimulates the emergence of a special artistic discourse that unfolds not from the beginning to the end but from the end to the beginning, not *after*-the-end but *before*-the-end, and is the "mixing of voices, genres, and codes."[2] As a form of missive, such a discourse is able to change its destination, break an agreement, and dilute the meaning of the said and the heard.[3] The various repetitions, delays, exaggerations, as well as apostrophes and cryptograms, that are introduced into a text

---

1 Jacques Derrida, "No Apocalypse, Not Now (Full Speed Ahead, Seven Missiles, Seven Missives)," *Diacritics* 14, no. 2 (1984): 28.
2 Jacques Derrida, "Of an Apocalyptic Tone Recently Adopted in Philosophy," *Oxford Literary Review* 4, no. 2 (1984): 5.
3 Ibid.

change the missive itself (in particular, that lethal "missive-missile" carried by the atomic bomb); they swap the future and the past, lead to erroneous understanding, and separate the sender and the addressee instead of establishing contact between them.

A new discourse appears after the catastrophe as "the remainders of a recently destroyed correspondence. Destroyed by fire or by that which figuratively takes its place, more certain of leaving nothing out of the reach of what I like to call the tongue of fire, not even the cinders of cinders there are," says Derrida."[4] Such a discourse is born out of the erasing of the fragments of writing, out of missed (faded) letters, the absent name (signature), nontraditional modes of communication—for example, with the help of remaining after the catastrophe postcards-letters with destroyed words, phrases, and whole messages. "[W]hat is not said here (so many white signs) will never get there. ... Differing from a letter, a post card is a letter to the extent that nothing of it remains that is, or that holds. It destines the letter to its ruin,"[5] comments Derrida on the nature of the new writing in the postmodern age, in a specific way parodying epistolary literature.

In the new post-catastrophic form of writing, Derrida proposes abandoning the habitual communicative form of a message because it is unknown "Who is writing? To whom? And to send, to destine, to dispatch what? To what address?"[6] He suggests eliminating the communicative action itself, in which there is necessarily an author of the message and an addressee. Ultimately, all connections have been disrupted, languages destroyed, people wiped out; therefore, in post-catastrophic writing let it be "[t]hat the signers and the addressees are not always visibly and necessarily identical from one *envoi* to the other, that the signers are not inevitably to be confused with the senders, not the addressees with the receivers, that is with the readers (*you* for example)."[7]

When introducing the notion of "nuclear criticism," Derrida talks about the content of representation and the mode of writing in literature of the atomic age. He notes that such literature is directed to a fictional (asymbolic) referent: "The terrifying reality of the nuclear conflict can only be the signified referent, never the real referent (present or past)."[8] From this perspective, atomic war

---

4   Jacques Derrida, *The Post Card: From Socrates to Freud and Beyond*, trans. Alan Bass (Chicago and London: The University of Chicago Press, 1987), 3.
5   Ibid, 249.
6   Ibid, 5.
7   Ibid.
8   Derrida, "No Apocalypse, Not Now," 23.

(which fortunately has not happened, and may it never happen!) exists only in that which is told about it and where it is discussed; hence, it is first of all a type of *discourse*. This is the way atomic war is presented in public political discourse and in the discourse of experts/technocrats. "Some might call it a fable, then, a pure invention: in the sense in which it is said that a myth, an image, a fiction, a utopia, a rhetorical figure, a fantasy, a phantasm," says Derrida.[9] Confirmation that atomic war exists first of all as a discourse that influences politics can also be found today: nuclear rhetoric and rumors about atomic weapons actually provoked the war in Iraq.

According to Derrida, nuclear discourse appeals not to reality but to literature and is a textual figure. In general, all literature in the atomic age, in his opinion, necessarily reflects a threat for the human condition and the condition of culture—that is, for the symbolic order represented by culture. Derrida links the whole development of modern literature to the discourse about atomic war. He is convinced that the atomic situation means an *absolutely new and unique* state because at stake here is the existence of the archive of culture itself. For this reason, a prerequisite for literature's existence is a conception that it can be destroyed. This total annihilation of literature, hinted at by the atomic situation, is "the only referent" and "the only 'subject'" of all possible literature and all criticism.[10] Therefore, all literature is said to develop in an expectation of its annihilation, that is, the end of its ability to mean and represent something. Literature is also said to have to do with the atomic situation well before the invention of atomic weapons. Derrida states, "Literature has always belonged to the nuclear epoch, even if it does not talk 'seriously' about it. And in truth I believe that the nuclear epoch is dealt with more 'seriously' in texts by Mallarmé, of Kafka, or Joyce, for example, than in present-day novels that would offer direct and realistic descriptions of a 'real' nuclear catastrophe."[11]

By introducing nuclear criticism, Derrida implements several gestures toward modern literature: he splits up the nuclear situation, disseminates it in the form of a trans-historically perceived archive, and injects it into writing, removing the traumatic singularity of the situation itself.[12] In this way, Derrida maintains that modern literature is permeated with a pre- and post-traumatic nuclear symptom, which is first and foremost about a type of representation of

---

9   Ibid.
10 Ibid., 27.
11 Ibid., 27–28.
12 Paul K. Saint-Amour, "Bombing and the Symptom: Traumatic Earliness and the Nuclear Uncanny," *Diacritics* 30, no. 4 (2000): 69.

the nuclear (apocalyptic) message and about a danger for the literary archive and literature itself. This is where postmodern consciousness takes its origin; it is directed toward the expression of the ineffable, is permeated with the symbolism of the asymbolic, with the crisis of language in the circumstances of linguistic over-intensification, with an illusion of hyperreality and overproduction of signs. In addition, the syndrome of the anticipation of the end and the destruction of culture is a means of the development of particular postapocalyptic imagery, which unfolds between event and non-event, the real and the virtual, the written and that which will be written. It also is directed backwards: from the beginning to the end, from text to author, from Socrates to Plato, and vice versa. After all, at the time when you are trying to comprehend the hidden meaning of a text (a letter or a message), "it is the picture that turns you around like a letter, in advance it deciphers you, it preoccupies space, it procures your words and gestures, all the bodies that you believe you invent in order to determine its outline."[13]

Derrida, as we can see, *fictionalizes* nuclear discourse and talks about a (non)presentation and (non)linearity of artistic description, which is able to turn the end into fiction, an image. Thereby, he conceptualizes the nature of postmodern discourse, in effect, which shows itself as a postapocalyptic and post-catastrophic phenomenon. Since the "end" (a probable nuclear explosion) is occurring at every second in a sense, it crosses out everything that happens "after"; in consequence, logically ordered before-reality, which has now slipped from the subject's control, gains a chaotic character, while linguistic consciousness and the symbolic order of culture take a discrete and fragmentary shape.

This is especially characteristic of postmodern literature. The discursive space of postmodernism is hybrid, nonhierarchical, devoid of continuity, multidirectional, and hyperreal. Postmodern representation loses rational coherence and completeness because, firstly, the subject is neither capable of comprehending the whole of reality nor constructing it as a totality; and, secondly, it is not at all clear what reality is and where it can be found, since it is successfully imitated by hyperreal simulacra.

The otherness of nuclear postapocalyptic discourse, related to postmodernism, is especially evident in comparison with Frank Kermode's theory of apocalyptic discourse. It is well known that the apocalypse is one of the most powerful matrixes of artistic narrative in modern literature. Kermode's *The Sense of an Ending: Studies in the Theory of Fiction* (1967) is a key book for

---

13 From back cover of Derrida's *The Post Card*.

understanding modernist apocalyptic narrative. In this study, the British critic resorts to an analogy between apocalypse and the artistic form of a novel, in which he reserves a central role for the idea of "the end of narrative." In particular, Kermode emphasizes that apocalyptic form is the matrix that we superimpose upon time, events, and artistic narrative in order to humanize them and render them comprehensible. The meaning of the apocalypse lies in coordinating the past fixed by imagination and the past anticipated by imagination, and all of this is accomplished for those who are "in-between."

Kermode emphasizes that any story must be regarded in an existential manner, from the perspective of the beginning and the end. He considers the naive understanding of the end of the world, reflected in biblical apocalypse, essential to the self-awareness of the modern human. Moreover, in his opinion, the idea of the end of the world, like a shadow, is actually present in every work of fiction; it constitutes the plot and gives the reader an opportunity to be inside the story as a narrative. The inevitability of the end of a narrative resembles death. In the same way that death cannot be avoided, human life stretches between the beginning and the end; the story of a work unfolds between the beginning and the end in the same manner, and the reader's expectations are oriented toward the ending of the story, which makes meaningful all that is described.

Affirming the role of the apocalyptic model for the development of modern literature, Kermode underlines not only the orientation of the plot in the direction of the end but also the role of peripeteia, which performs an ironic function, undermining expectations and naive faith in the end. Peripeteia falsifies the reader's expectations and gives an opportunity to get to know the reality that is outside the work. Herewith, as Kermode states in the pre-Chornobyl era, the paradigm of the apocalypse allows us to see the world as making sense.[14]

Kermode's theory exerted a great influence on understanding literature of the "end." Further, he appealed first and foremost to modernist literature in which a person remains inside reality, reality makes sense, and the literary text is endowed with the capability for recreating and capturing it. Modernist apocalypticism differs from postmodernist postapocalyptic narrative in that in the latter the individual is situated outside reality, radically disrupted by the end or catastrophe, and reality itself has a meaning only at the moment of its repetition (pre-, post-, hyper-, or trans-reality). In short, the postmodernist literary text *reflects* reality, thereby erasing references to reality.

---

14 Frank Kermode, *The Sense of an Ending* (Oxford: Oxford University Press, 1968), 28.

Description from the perspective of after-the-end posits a different understanding of the world. For example, in Yuri Andrukhovych's novels, events unfold after the end (or its imitation)—after Stakh Perfetsky's suicide in *Perverzion*, for instance, or Otto V.'s suicide in *The Moscoviad*. In Yuri Izdryk's *Wozzeck*, the action also develops backwards, from the perspective of the disintegration of the world and the disruption of communication—a situation in which the protagonist finds himself after a love trauma. Likewise, the main event—a love fight-union between a writer and an artist—has already happened before the beginning of Oksana Zabuzhko's novel *Field Work in Ukrainian Sex*.

In John Barth's novel *The Floating Opera* (1955), the whole story, told by the protagonist, is tinged with hyperreality and irony precisely because the story is narrated as though after the catastrophe—after the suicide that the protagonist intends to commit. De facto putting existential and rhetorical pressure on readers, the author at the same time teases them, since it is not clear if they are being offered notes found after the protagonist's death or simply an imitation of them. Thus knowledge about the future relativizes the behavior of the protagonist, who demagogically admits, "*There's no final reason for living (or for suicide).*"[15] In addition, according to the logic of the black humor to which the author resorts, everything represented is but a fragment of reality. The reader is at the mercy of both reality itself and the author who conceals as much as he reveals. Likewise, the literature of the atomic age becomes elusive.

---

15 John Barth, *The Floating Opera* (Garden City: Doubleday, 1967), 245.

CHAPTER 5

Chornobyl and Virtuality

While Kermode sees artistic form as a lens for perceiving reality, Jean Baudrillard, one of the leading theoreticians of postmodernity, talks about the phenomenon of substitution—"*substituting the signs of the real for the real.*"[1] In his opinion, postmodernism means a disappearance of "the sovereign difference" between the real and the simulated; moreover, the imagined, the simulated, precedes the real, and the remnants of the real exist in the desert of the real. For this reason, virtuality, Baudrillard writes, substitutes for reality. This empire of signs, and the power of simulation he calls hyperreality, he associates with the postmodern.

For Baudrillard, nuclear discourse becomes analogous to simulation and endlessly generated hyperreality because simulation's "operation is nuclear and genetic, no longer at all specular or discursive."[2] It means that imitation does not reflect reality, as in a mirror, and does not abstract it in notions and symbols; instead, reality itself is generated as a result of nuclear disintegration—of "genetic miniaturization" and "the orbital recurrence of models."[3] In these processes, a special role is reserved for the "cooling" of reality and its characteristic emotions, feelings, and relations as a consequence of the devaluation of the sign, which "deadens" the reference.

Virtualization occurs at the cost of the prior "deadening" of reality with a subsequent sublimation (exaltation) with the help of images. Catastrophic virtuality is indicated here by proliferation. Accustomed to the staging of catastrophes like "the star wars," floods, and armageddons, we perceive real events through the categories of staging as well. As Baudrillard notes,

---

1 Jean Baudrillard, *Simulacra and Simulation*, trans. Faria Glaser (Michigan: Michigan University Press, 1994), 2.
2 Ibid.
3 Ibid.

the energy crisis and the ecological mise-en-scène are themselves a disaster movie, in the same style (and with the same value) as those that currently comprise the golden days of Hollywood. It is useless to laboriously interpret these films in terms of their relation to an "objective" social crisis or even to an "objective" phantasm of disaster. It is in another sense that it must be said that it is the social itself that, in contemporary discourse, is organized along the lines of a disaster-movie script.[4]

These reflections are worth remembering when we talk about Chornobyl. Undeniably, this event had multiple dimensions—political, social, cultural, psychological, and informational. It grew from the local to the global, turning from a historical event into an apocalyptic one, taking on a metaphysical hue, and gaining an existential or apocalyptic meaning. In contrast with Chornobyl, ordinary (pre-Chornobyl!) life acquired a new sense and turned into a symbolical revelation. As Yuriy Shcherbak noted two years after the catastrophe, "all those simple, usual images of life suddenly made a great impression on me, as though I received an epiphany, an understanding of some very important shift that has occurred in the minds in recent days."[5]

There was mention of Hiroshima and Nagasaki, but even more important were comparisons with the prophetic revelations of John the Theologian and Nostradamus. The Chornobyl accident also reminded of eschatological films-catastrophes. Shcherbak observed the effect of a peculiarly cinematographic "congelation" of reality: "the Dnipro, mountains, houses, and people—everything prosaic seemed to me then unusual, as though out of a science fiction movie. In those days I often especially recalled Stanley Kramer's film *On the Beach*—about how, after the atomic war, Australia is waiting for the appearance of the radioactive cloud."[6]

Later, in the process of the accumulation of tales and symbolic analogies, the Chornobyl event itself increasingly transmutes into a fantastic reality, in which factual details and circumstances become surrounded by inventions and details borrowed from horror movies.

No doubt, nuclear explosion was one of the most powerful images in the second half of the twentieth century—a demonstration of the power of the virtual-imaginary over the real and the everyday mindset of people. In the Soviet

---

4   Ibid., 43.
5   Iurii Shcherbak, "Chornobyl´. Dokumental´na povist´," *Vitchyzna* 5 (1988): 15.
6   Ibid.

period, pictures of atomic clouds and guidelines for civil defense hung in every building, and terms such as "shelter," "shock wave," and "radiation first aid" were repeated endlessly. As the emblems of the atomic age, they immersed people in the realm of nuclear imagery. The Chornobyl disaster, superimposed on these images, which Soviet people in the 1960s and 1970s were already accustomed to, reaffirmed the reality of the atomic era:

> This phrase ("the atomic era"), which used to be a ringing metaphor we repeated before the accident, then tuned into stern reality: the words "dosimetric control," "radiation," and "deactivation," all those "milliroentgens," "rems," "rads" and so on became a part of Kyivers vocabulary for good. ...[7]

However, Chornobyl had one more function: it led to the world of postmodern hyperreality. The Chornobyl tragedy sharpened perceptions of virtual dimensions. The "total," "objective," and "positive" reality, which the socialist realist imagery had propagated for so long, was destroyed. The Chornobyl accident destroyed the foundation of the "truth" upon which the socialist worldview was built. At the same time it revealed the possibility of a complete replacement of the real world by some other world. This was one of the effects of Chornobyl: it ruined the linearity of time-space, fragmented it into zones, and taught that the tactile senses could not be trusted. It also taught a distrust of nature. Trees in bloom, green grass, the air we breathe, vegetables, the water we consume—everything seemed false or inimical. Unseen, virtual rems and roentgens, and radiation that covered an enormous territory, remaining invisible and imperceptible to sight and smell, wrecked the physical presence of and trust in real things, imposed virtual images, and brought forth phantasms. Alongside the real Kyiv, there emerged the Chornobyl zone—a place of virtuality associated with monsters and mutants, with stalkers as depicted in the famous film by Andrei Tarkovsky, with an empty hole in which time went backwards, with the release of an abnormal energy, and with the real apocalypse. In time, the virtual Chornobyl became a favorite place for science fiction films and was used for numerous virtual games, most famously *S.T.A.L.K.E.R.* and its various iterations (*Call of Prypiat, Wind of Change, Shadow of Chornobyl,* etc.). The photorealistic "zone of exclusion" is here rendered according to its "real" prototype, and the virtual reality is superimposed on the "real" map of the city of Prypiat, the train station Ianiv, the Jupiter plant, the village of Kopachi, and so

---

7   Ibid.

on. Games such as *Counter-Strike Chornobyl* gained popularity; they depicted the zone, and imagined a whole series of wars in which the virtual world defeats the real one. In these works, the real world becomes invisible and is portrayed only through the form of virtuality: "one can see Chornobyl … virtually—wander a little, shoot someone," as one of the gamers confesses.

Irrespective of the fact that nuclear disaster raises a genuine question about the survival of humankind, culture, and the world, it also becomes a way in which humanity frightens itself. Although Hiroshima and Nagasaki were actual events, *images* of nuclear explosions, incessantly reproduced and aestheticized in the media, computer games, and movies, develop into fantasies of Armageddon.

One thing is undeniable: in modern culture, nuclear accidents such as Chornobyl or Fukushima turn from historical events into cultural (artistic) constructions and become not only real but also virtual phenomena. Chornobyl is a symbol which represents manmade catastrophe, the destruction of culture, a threat to human life, an exclusion zone, an ecological crisis, etc. It has entered into the vocabulary of modern culture as a synonym of "catastrophe." In Adriano Celentano's 2009 song "Sognando Chernobyl" ("Dreaming Chornobyl"), the refrain goes, "We will all blow up together." Here, Chornobyl is an analogue for various dangers: melting glaciers; nuclear weapons; and the fact that Venice is sinking. In the video, animals and mutants appear against a dismal background that includes photographic documentation of the site and a silhouette of the Chornobyl Nuclear Power Plant.

Atomic non-reality is a fantasy and rhetorical reality, represented in a stylistic and discursive manner. In futuristic and apocalyptic works, it draws on the "the imagination of disaster," to use Susan Sontag's term.[8] Sontag examines several models of science-fiction film scripts, which she wittily analyses. In addition, she notes that the "imagination of disaster" is not merely something found in the man-made anti-utopias common in horror and action films: it reflects "the deepest anxieties about contemporary existence … powerful anxieties about the condition of the individual psyche."[9] Post-catastrophic works are allegories of the processes of depersonalization in the modern world, as well as of the possible "collective cremation and disappearance, which can occur any moment without warning after the discovery of the atomic bomb."[10] However,

---

8   Sontag, *Against Interpretation*, 209.
9   Ibid., 220.
10 Ibid.

nuclear imagery cannot be boiled down to the depiction of a catastrophe and its consequences. It also has a discursive dimension—as a non-event, a hiatus, unlike events that are completely real. After all, to seriously "think the possibility of nuclear war," writes Christopher Norris, "a very real and present possibility, is to think beyond the limits of reason itself."[11]

---

11 Christopher Norris, "Versions of Apocalypse: Kant, Derrida, Foucault," in *Apocalypse Theory and the Ends of the World*, ed. Malcolm Bull (Oxford: Blackwell, 1995), 245.

CHAPTER 6

Chornobyl and the Cultural Archive

In Jean Baudrillard's opinion, nuclear reality is created through a situation of containment, the postponement of an event—most notably, the postponement of the reality of an atomic blast in the foreseeable future. In return, the politics of nuclear containment has established the most sophisticated system of the worldwide control that has ever existed.[1] After all, if the reality of an atomic explosion, which could be the last act of the human history, is suspended, then we can say that there is no absolute signifier of the end of history at all. For this reason, there is also no signified to which such a signifier refers—that is, a reality that should be protected from destruction. We find ourselves in a world of simulations, illusions, and signs, which do not refer to anything real; instead, they give only an impression of "presence," in the form "of kitsch, of retro and porno at the same time."[2]

In the context of philosophical and intellectual discussion about the nature of the postmodernist representation and postmodernism in the atomic age, Chornobyl becomes an indicative symbol—an event that turns history and culture backwards, rendering them threatened; an event that symbolizes the probability of an atomic blast and endows contemporary literature (in particular, Ukrainian literature) with a special capability for fragmentation, combination, erasing, and new development of meanings and cultural symbols. By and large, Ukrainian postmodernism is a post-Chornobyl text.

It does not mean that all postmodernist literature is directly linked to Chornobyl, however. Chornobyl is not only a real, but also a symbolic, referent for the whole of Ukrainian postmodernist discourse. After all, Chornobyl, as an

---

1   Jean Baudrillard, "Precession of Simulacra," in *Simulacra and Simulation*, 35,
2   Ibid., 41.

actual atomic explosion, established a discursive and fictional power over the imagination, foregrounded issues about the possibility of the existence of culture as such, united the past, the present, and the future in one moment of time, and redrew the geographical map of the world; it coincided with the disintegration of totalitarian Soviet consciousness, and became an (un)presentable apocalyptic event that undermined the socialist realist narrative of truth; and, finally, it generated artistic innovations at the threshold of the third millennium.

At the same time, Chornobyl undermined assumptions about the absolute power of hyperreality and the a-symbolism of atomic explosion. After all, the explosion *almost* happened. This almost-explosion demonstrated the limits of the technological power of humans over nature and showed our precarious existence in the time post-catastrophe, or postmodernity.

When I was in the Chornobyl zone, I had an opportunity to find out what life after a catastrophe looks like. Rather than ruins, preserved villages along the roads were overgrown with young groves that were difficult to penetrate. People who returned home, into the zone, are living as though in a postnuclear age, as though in a new Middle Ages, utilizing the remnants of civilization such as things abandoned by others. Clothes, food, and even language—everything has been reduced to the necessary minimum. However, this is not a postapocalyptic situation in which all traces of culture are destroyed, as Cormac McCarthy's *The Road* has it. Against all odds, the Chornobyl zone remains a part of the cultural world. Modern Chornobyl combines postindustrial reality, embodied in the majestic beauty of the power plant reflected in the water of the atomic lake, and a mediaeval, premodern reality, in which those who have chosen to stay live on scraps of culture in a primordial world remote from civilization. Chornobyl can be regarded as a metaphor of postmodernist difference, which distinguishes culture from the post-culture that is being born there.

Literature in the nuclear age, Derrida states, reflects a threat to the "archive," for culture as such. He emphasizes that in the literature of the atomic age the target is not only the fictitiousness of the referent (hence the importance of the performative, rhetorical aspect of literature). The target is also the function of literature as a cultural archive (a corpus of texts, names, author's rights, etc.). He states that without this archive the very institution of literature cannot function. "Literature" is the name of a certain body of texts, whose existence, possibility, and significance are threatened—maybe for the first and last time—by nuclear catastrophe.[3] "The death of the author," as described by Roland Barthes, in the

---

3   Derrida, "No Apocalypse, Not Now," 27.

Chornobyl world is perceived as very real; and the greatest irony of Chornobyl lies in the fact that culture becomes a thing that has no use in the zone. Libraries, along with other material valuables, are not in demand in the Chornobyl zone. In a school library, I saw scattered books that had been trampled on. The Chornobyl archive—the numerous cultural artifacts, photos, and letters, collected in the Zone and transferred to a museum—can hardly give a full account of the real life and culture of those who have lived on the Chornobyl land for centuries. We have to be satisfied with fragments and recreate their history from traces, remnants, and memories.

In addition, the cultural archive of the Chornobyl zone—from the legends and myths of the natives of Polissia to the socialist realist tales about the Soviet city of Prypiat—is turning into a topic that has now been much vulgarized in both high and mass culture. "High"—because authors who have been writing about Chornobyl want to give a more or less "truthful" account of something that cannot be expressed as "truth"; "Mass"—because the atomic apocalypse became a commodity long ago. Just wait a little and the sarcophagus will be taken to pieces for souvenirs, like the Berlin Wall. "Chornobyl kitsch" already exists as a cheap catharsis in the service of the press and literature. Theatrically scattered children's things beside gas masks were left behind by TV reporters and photographers. Such was the clichéd staging of innocence in an age of a man-made catastrophe that I saw in a childcare facility in Prypiat. In this way, Chornobyl has been turned into show, a postapocalyptic spectacle.

For Derrida, a nuclear explosion is, first and foremost, a *discourse*—that is, a type of narrative. In the Chornobyl discourse, a significant role is played not only by anti-nuclear symbolism but also by its anti-totalitarian and postcolonial varieties. Chornobyl initiates discussion about informational "openness," errors, and lacunas in the socialist system. It also acquires a postcolonial significance and instigates discussions about the metropolis and province, power and subjugation. It also has a geo-cultural aspect. In particular, Chornobyl as a text tells the tale of a radioactive cloud that, crossing the border of the two great military blocs, destroys the wall between two systems—the Soviet and the Western. Thus Chornobyl draws a map of a new Europe. "After Chernobyl, the Berlin Wall no longer exists," insisted Jean Baudrillard.[4]

Baudrillard views Chornobyl as a geo-cultural marker that reveals the difference of cultural identities, specifically the "Western" and "Eastern European."

---

4   Jean Baudrillard, *The Illusion of the End*, trans. Chris Turner (Stanford: Stanford University Press, 1994), 45.

As Baudrillard states, it appears that previously accepted ideas about the hermeticism of the West is in reality a myth. The West is transparent. In fact, the newly discovered cultures (most significantly, the countries of the Eastern bloc), which rushed to the West, threaten the very existence of the West, which still believes in compatibility of all cultures. In this way, Chornobyl opens a painful wound in Western European identity—the rejection of the "other."

At the same time, Chornobyl also reveals the limits of the Western cultural archive. As a true advocate of the West, Baudrillard sees a connection between the onslaught of postmodernism and the threat of the destruction of European culture, history, and memory. The crash of the socialist system has only reassured Western intellectuals that the democratic and humanistic West has to "erase the past" from the memory of the nations that have had a totalitarian experience. However, Baudrillard, who in the 1970s was a partisan of revolutionary theory and politics, notes that a backward process is also occurring: the totalitarian past pulls the present into its void. What we are witnessing resembles how "the dead and the survivors of the camps have sucked our last vague desires for culture, law and morality."[5] Therefore, in Baudrillard's opinion, the victory of the West over communism has led to the devastation of its own history and established the postmodernist turn toward history, understood not as a succession of facts but as a *recombination* of various discourses and archives.

Beside this recombination of history (discourses), at the meeting of the postmodern West with the post-totalitarian cultures of Eastern Europe, there also occurs a simultaneous mutual exchange and undermining of values. Western consensus, in Baudrillard's opinion, is sustained by "an epidemic of democratic values—in other words, this is a viral effect, a triumphant effect of fashion."[6] In this way, the heavy burden of history is growing. However, at the same time the "democratic rewriting" of history in the West, as Baudrillard ironizes, is akin to stain removal or cosmetic surgery—"cosmetic surgery elevated to the level of the political, and to Olympic performance levels."[7]

Reflecting on the end of the last century, Baudrillard announces the sell-off of the cultural archive. The revival of past or passing forms of history, this attempt to escape the apocalypse of virtuality, is a utopian wish. The more we want to locate the real and the referential, the more we are immersed in simulation and the counterfeit, he says. Resurrected values are evanescent, unstable, and—like

---

5   Ibid., 49.
6   Ibid., 44.
7   Ibid., 43.

fashion and the stock exchange—prone to fluctuation, argues Baudrillard. Any rehabilitation of old boundaries, structures, and former elites will never have such an authentic meaning, such an "aura," as in the past. If one day the aristocracy, say, regains its position, it will, nevertheless, be "post-modern."[8]

Therefore, the postmodern is like playing with bits and pieces, a recombination of established forms of history and culture. The postmodern rehabilitation of the cultural archive, according to Baudrillard, amounts to cryonics—that is, a deep freezing of the past. In order to become immortal, things must be frozen in advance. In the same way, history must also be cryonized, as in Disneyland, where a prefiguration of real things and of all preceding historical periods takes place. Thus, as Baudrillard sees it, Western postmodernism and the wish to salvage the Western literary archive in the atomic age propose a cryonization of culture.

Cryonics, congelation, bitter winter, informational explosion—all these things are the effects of nuclear detonation. It is not a coincidence, then, that Baudrillard identifies the informational explosion at the end of the twentieth century with an atomic blast. Their essence is identical, he says: the disappearance of human feelings and emotions, and their repetition in a "congealed," mass-media form. In this way, one of the possible versions of postmodernist writing is formed—the one that refers to a media "reality" of the imitated. Therefore, Western postmodern theory, by appealing to nuclear discourse, not only probes the possibilities of postmodernist writing but also warns about its boundaries.

---

8   Ibid., 117.

CHAPTER 7
Chornobyl Postmodern Topography

Postmodernism in Ukrainian literature unfolds topographically, as an interlacing of ideal places, geographical maps, texts, and discourses on the surface of the national culture's body, with its centers, peripheries, hybrids, and nomads. For me, Chornobyl is the symbolic event and place that virtually unites this whole topography. We live in a post-Chornobyl age, because it is Chornobyl that has changed our worldview, having freed it from romantic pathos and Enlightenment expediency; having turned upside down the existent dimensions of time and space; and having questioned the possibility of expressing the ineffable. Frederique Lemarchand aptly points out that

> [w]e seem to be witnessing the emergence of a *new world*. The Renaissance, as a whole epoch, was marked by the advent of a new world under the badge of economic and geographical expansion, the discovery of a new world due to the use of maps and compass. The world that emerges with Chornobyl initiates a different era, the era of a reduced inhabited world, the era of a world that regresses, a world formed by technical 'progress,' first nuclear and—a little later—genetic, including manipulations of living beings.[1]

The new Chornobyl reality cannot be measured by maps, although after the accident there was marked interest in cartography, with the map of the radioactive pollution becoming a popular text. More important, though, is the easiness with which in the minds of people who experienced Chornobyl the passage from a geographical map to a symbolic one takes place. Remote travel destinations

---

1 Frederique Lemarchand, "Topos Chornobylia," *Dukh i Litera* 7–8 (2001): 374.

lose their appeal over journeys to the recesses of nearby, and undiscovered, mental expanses. Impressions from the concrete content of the accident and from the suffering that it caused quickly become overgrown with symbolic realities borrowed from cultural texts of various ages and nations. Eventually, a virtual version of Chornobyl develops. In this way, faith in an observable, authentic reality is undermined. All-pervasive, indiscernible, unlimited, radiation reaches the most unexpected and remotest places, destroys geographical markers, integrates with consciousness, and generates a nonorganic, artificially designed body.

With Chornobyl, the very understanding of corporeality changes. The body is no longer closed: it becomes deformed, hybrid, or even entirely counterfeited. By spreading arbitrarily and uncontrollably, Chornobyl's radiation does not project itself as a map with established and clear-cut boundaries; instead, it creates "zones" of pollution, manifests as "stains," and therefore does not correspond with either modern cartography or postmodern risography. Space becomes like a punch card with separate holes—dead zones. Lemarchand rightly notes that

> [i]t is even worse that the border with the polluted zone is most often perceived as the boundaries of a *shelter* inside which people feel protected from *danger*—not only because the catastrophe has already occurred but also because the goalposts that referred to things, creatures, and nature have not been ruined entirely but rather have mixed up."[2]

Out of this new mixture in the Chornobyl zone, there arises new life: life after the catastrophe. About five kilometers from the fourth nuclear reactor live Sava Hryhorovych Obrazhai (from Semykhody) and Olena Dorofiivna (from Khoromne). Their villages are nonexistent, their history is erased; so, having occupied a few homes abandoned by their proprietors, the man and woman, like Adam and Eve after the expulsion, got together and began to make a home in the zone. They dress in others' clothes, use others' belongings, and cultivate somebody else's land. A horse, a dugout, scavenged clothes and utensils that have been collected from all around—they occupy half of the yard. Everything has begun not from nothing but secondhand, from the residue of others' lives, from someone else's used stuff. This is how a new Middle Ages has emerged—a life among the fragments of human civilization. This is how a new culture has emerged—a post-cultural one.

---

2 Ibid., 373.

It ruins time, distorting both the near and the distant; it disrupts the connection between the inside and the outside, because there is no direct correspondence between the two anymore; it reshapes nature, generating the most unexpected hybrids in a state of the "dream of mind." It becomes "impossible to enjoy the country landscape that fulfils ... the criteria of Western romantic aesthetics with its woods, rivers, flowers, play of light and shadow, abundance of domestic and wild animals, and so on, without being frightened by the thought that, contrary to the illusion, it is all about a new world created by a nuclear catastrophe—'the world after the apocalypse.'"[3] Svetlana Alexievich depicted this world with all its horrors.

Of course, Chornobyl is first of all associated with the real event that occurred on 26 April 1986. However, from the perspective of time, the Chornobyl catastrophe gains a universal symbolism. It introduces Ukraine into the context of the global postmodern world of the end of the twentieth century, when "local" complexes and conflicts all of a sudden turn out to be in tune with the logic of late capitalism, which, according to Fredric Jameson, characterizes the advent of the postmodern era. The point of the matter is that ideas of the modernization and Westernization of Ukrainian culture, announced from time to time in the course of the twentieth century, are now synchronized with the crisis of Western modernity as such.

It would be a simplification, however, to think that postmodernism was exported to Ukraine from the West and that it cancelled the actuality of modernist ideals in Ukrainian national self-consciousness. Ukrainian postmodernism, like its other local and national variations anywhere in the world, was born out of a unique complex of ideas, circumstances, and realities, as well as out of a combination of global and local preconditions. As Halina Janaszek-Ivaničkova notes, the most important aspect of the postmodernist turn in Eastern Europe is a change in the social and philosophical outlook of the former socialist countries. Postmodernism has blazed a trail for intellectual pluralism and blasted open a whole worldview, thereby revealing "reality" as an unending performance in constant need of legitimization; it has decisively overturned determinism, has introduced the notion of truth as a creative human activity that unfolds through interhuman contact, has facilitated the comprehension of the world through a variety of ideas, opinions, forms of linguistic expression, and has restored a taste for trans-avant-garde art.[4]

---

3 Ibid., 376.
4 Halina Janaszek-Ivaničkova and Douwe Fokkema, eds., *Postmodernism in Literature and Culture of Central and Eastern Europe* (Katowice: Śląsk, 1996), 10–11.

In its turn, Western postmodernism is no longer steady and changeless. Its earlier versions differ from its later ones; it undergoes a constant modification, absorbing the forms of an opposing culture of the postcolonial type, the culture of marginal groups, and newly created and creolized languages. In other words, postmodernism itself reacts to the heteroglossia and deterritorialization that are characteristic of the modern world. A powerful source of postmodernism's self-development is a today's widespread cultural hybridization and the growing role of regionalism, along with the interplay of high and popular cultures. All these factors in a similar way also influence the development of postmodernism in Ukraine.

Opponents of postmodernism constantly accuse it of a total relativism, something alien to the politically engaged countries of the postcolonial "Third" and the post-socialist "Second" worlds. However, postmodernism is not at all apolitical—something confirmed by the logic of its development in the modern literature of Asia, the Balkans, and the post-socialist countries. After all, as in the 1980s Jameson defined the peculiarities of the development of Third World culture through the notion of "national allegory"—in essence a political reading of literature. Jameson maintained that in the West literature belongs to the sphere of private rather than public life and is a realm of individual taste and solitary meditation rather than public debate and means of liberation. He noted that in Third World literatures the history of private life and individual fate is always an allegory of political struggle and the public culture and society.[5]

It would be appropriate to note here that the Chornobyl tragedy has also become a public national allegory, specifically, in Ukraine and Belarus, and Ukrainian post-Chornobyl postmodernism has turned out to be an allegory of the crisis of the totalitarian mindset. It is a fact that Chornobyl accelerated the process of reconstruction in the Soviet Union as well as actualized anti-totalitarian semantic paradigms such as the incompetence of the communist party, the suffering of the people, and the destruction of the environment.[6] However, it is important to remember the broader, global context of Chornobyl. It is more significant than its anti-totalitarian content and addresses the question of postapocalyptic survival in the modern world at large. In this sense, Chornobyl is a global event.

---

5 See Imre Szeman, "Who's Afraid of National Allegory? Jameson, Literary Criticism, Globalism," *The South Atlantic Quarterly* 100, no. 3 (2001): 803–27.

6 Alexander J. Motyl, *Dilemmas of Independence: Ukraine after Totalitarianism* (New York: Council on Foreign Relations, 1991), 44.

As an event in the cultural text of the end of the twentieth century, Chornobyl joins together ontological, ethical, ideological, and ecological problems of the late totalitarian era. In addition, it releases the pent-up force of catastrophism, which causes a reevaluation of values and triggers an apocalyptic consciousness. Absorbing Chornobyl catastrophism, anti-totalitarian thinking in the late Soviet era not only fuels the revision of socialist ideology but also prompts a rethinking of history the emergence of a new geo-cultural map and fills in the "blank spots" of culture. All of this occurs not from the perspective of progress but by finding the fullness of the present in the past. This is how cravings for the future are devalued, and how grand narratives based on a desire for heroism, adventure, and idealism are replaced by the feeling that every move is a transgression which not only alters the object but also sets limits on it.

In Ukrainian history at the end of the twentieth century, Chornobyl, in effect, became the point of departure for the legalization of the postmodern situation. The event acquired ideological significance. In a socio-cultural sense, it marked the crisis of totalitarianism and overturned the Soviet worldview. Ideas already prevalent in underground attitudes towards the official mindset came into the open, became public, and thereby facilitated the devaluation and depreciation of totalitarian consciousness at both "high" ideological and "low" everyday levels. The transgression of cultural and ideological codes acquired not only an anti-totalitarian tone but also an apocalyptic one. The reading of the prophetic symbols of Chornobyl, under the sign of the "star wormwood" mentioned in John the Theologian's *Book of Revelation*, was in tune with a trust in gnostic philosophy that was characteristic of the end of the century and promoted the deployment of a new futuristic mythology.

Politically, the "star wormwood," which had risen above the expanses of the Soviet Union, prophesied its imminent demise. The map of Europe underwent changes; the values of modernization and the assumptions behind Euro-logo-phallocentrism—that is, of the absolute of the "European white man"—were reevaluated.

"After Chernobyl, the Berlin Wall no longer exists,"[7] noted Jean Baudrillard on the threshold of the third millennium. As a matter of fact, Chornobyl inaugurated the era of a united Europe. It opened Ukraine to the West, and the West became visible to Ukraine, since it was there that talk about the real sense of the Chornobyl accident, and about the dangers that it presented to the world, began. The opposition of self and "other," so important in Cold War

---

7 Baudrillard, *The Illusion of the End*, 45.

rhetoric, was suspended: on the one hand, the Western discourse of centrism was being dismantled; on the other, an era of new nationalisms was coming. The world found itself in a strange situation. That which philosophers were discussing in the West—the crisis of modern ideals, virtual wars, the future apocalypse, for example—suddenly materialized, in a condensed form, in one event: Chornobyl. Along with Chornobyl, postmodernism erupted on the outskirts of Europe.

CHAPTER 8

Chornobyl and the Crisis of Language

The post-Chornobyl, postmodernism, undermines the ideological space of totalitarian culture and in addition accomplishes a decentralization of literature and a de-heroization of its heroic narrative. Postmodernist Ukrainian authors find their element in puns, stylization, and linguistic irony, all of which give them the opportunity to escape from the authority of official culture and to liberate themselves from the idols and masks of the totalitarian past. At the same time, literature becomes dialogical and even polylogical. It contains a variety of discourses, language forms, and jargons; spoken language here plays the role of testimony, and speech acquires the form of existential corporeality.

In addition, the Chornobyl postmodernist text not only testifies to the destruction of boundaries and demarcations between the real and the virtual but also ruins the symbolic order and language itself. It becomes impossible to use old words to describe the totally different reality that is unfolding after the atomic apocalypse. Slipping away from any comprehension, such a reality "can only be "*symbolized* or *signified*, that is, represented as an invisible presence."[1] It seems that this invisible presence can be signified only through nonverbal (post-verbal?) existence.

As a matter of fact, the crisis of language—perhaps, the most significant effect of Chornobyl—appears first of all semantically. Roger and Bella Belbeoch draw our attention to the way the meaning of the word *sarcophagus* is distorted in Chornobyl discourse. "Sarcophagus" is the nickname for the reinforced concrete construction that contains the remains of the reactor. However, for the ancient Greeks *sarcophagus* (that is, *engraved coffin*) meant *flesh-eating* because sarcophagi were largely made from limestone—a material that dissolves, eats,

---

1   Lemarchand, "Topos Chornobylia," 372.

human remains. The Chornobyl sarcophagus, on the contrary, preserves the corpse. Having constructed a case around the accident, people hope to protect themselves outside its limits.

Folklore was quick to capture the grotesque contradiction between words and reality in Chornobyl discourse, and the following joke was born:

> —This sarcophagus in Chornobyl: is it similar to the ones in Egypt?
> —Of course not! In Egypt, there's a sarcophagus around the mummy. In our case, it's vice versa: the sarcophagus is surrounded by millions of mummies!"[2]

Chornobyl embodied anti-totalitarian attitudes, giving birth to an unheard-of outburst of a mass joke creation and brought out into the open the underground culture. As Myron Petrovsky writes, "the creators and bearers of jokes maybe for the first time in the whole history of Soviet rule stopped hiding in the city underground and started—with the impudence of desperation—to do their thing publicly, out loud."[3]

Joke became a mini-narrative, targeting the public Soviet masquerade and official lies as well as ideology: "the people's little apocryphal story turned against the big and deceitful state myth." The collection of "mushrooms from around Chornobyl," jokes from the time of the nuclear catastrophe, recorded by Bohdan Zholdak, bears witness to the carnivalization of the apocalyptic madness. It shows how Japan's Hiroshima becomes Ukraine's "kher-iz-namy" ("we are fucked up"); how a two-headed *khokhol* becomes the coat of arms of Chornobyl; how the epitaph "I died of radiation" evolves into "And I, from information"; and how AES (Atomic Electric Station) rhymes grotesquely with "long live CPSS" (Communist Party of the Soviet Union). In short, the collection says, "vsia planeta kryie matom nash radians′kyi myrnyi atom"—"the whole planet curses our Soviet peaceful atom."[4]

After all, Chornobyl is not only a crisis of language. It is also a crisis of representation as such. This condition was felt by both the authors and the critics of Chornobyl literature. Marko Pavlyshyn has formulated this dilemma most clearly: "Is it pertinent to show such a global event through the lens of only one's own

---

2   "Hrybochky z-pid Chornobylia. Anekdoty chasiv iadernoï katastrofy. Zapysano z narodnykh vust Bohdanom Zholdakom," *Dukh i Litera.—Iegupets′. Khudozhnio-publitsystychnyi al′manakh Instytutu iudaïky* 8 (2001): 159.
3   Ibid., 144.
4   Ibid., 145, 146, 148, 149.

limited (and by whom authorized?) consciousness?"⁵ Pavlyshyn mentions, in particular, the traps hiding in various modes of the description of the Chornobyl genre: in the rhetorical virtue of "'humilitas' (modesty). "As one would say," he writes, "I don't build my fictional voice from the Chornobyl material—I only allow the facts to speak in their own voice";⁶ and in making the Chornobyl topic trivial or epic-like. Likewise, Svetlana Alexievich writes that Chornobyl is a phenomenon that cannot be depicted mimetically and thus triggers a search for the representation of the unrepresentable. Oksana Zabuzhko, who translated Alexievich's *Chernobyl Prayer: A Chronicle of the Future*,⁷ aptly points out that the author

> at large tends to abstain from direct speech, from the classic situation of the narrator as a mediator between characters and readers, with a right to evaluative commentaries; she sincerely and most genuinely withdraws any aspiration for her 'own truth,' which would edit the partial 'truths' of the characters, and she interviews them not at all to get a confirmation for her own ideas, as the 'socialist' method teaches.⁸

The Chornobyl event, which emerges in her book out of the narratives of a number of witnesses, exists only as *personal* truth rather than a fictionalized event—and there is no language in which it can be told. For example, Mykola Khomych Kalugin, a father who lost his daughter, testifies that Chornobyl exists for him as pain, and narrative itself is treason. "When I talk about this," he says, "I have this feeling, my heart tells me: you're betraying them. Because I need to describe it like I'm a stranger … to suffer like this."; "I want to bear witness: my daughter died from Chernobyl. And they want us to forget about it."⁹ Such talk goes beyond the limits of conventionality and fictionality. However, even this is almost impossible since the silence or broken speech of the "Chornobyl person" is itself dependent on rhetoric. While it is very hard to cross the boundaries of literature, it is possible to define these boundaries.

The post-Chornobyl crisis of language and representation is especially apparent against the background of classic literature—for example, in Vasyl

---

5   Marko Pavlyshyn, "Chornobyls'ka tema i problema zhanru," 177.
6   Ibid., 177.
7   The English translation is *Voices from Chernobyl: The Oral History of Nuclear Disaster*.
8   Oksana Zabuzhko, "Postskryptum: monoloh perekladacha pro podzvin pokynutykh khramiv," in Svitlana Aleksiievych, *Chornobyl': khronika maibutnioho*, pereklad i pisliamova Oksany Zabuzhko (Kyiv: Fakt, 1998), 188.
9   Alexievich, *Voices from Chernobyl*, 33.

Stefanyk's short story "A Good News" or Kotsiubynsky's "Apple Blossom." In both cases, fathers talk about the death of a daughter: one by aesthetically recreating her death, the other by avoiding its direct description. However, literariness—that is, conventionality, narrative, and the right to represent—remains untouched. In Alexievich, it is the other way around: the man refuses the right to narrate, talks nonlinearly and nonobjectively (in other words, he becomes almost speechless), and he is aware that his testimony is odd, almost impossible to articulate. This is why he just talks about himself as a "Chornobyl person" and does not describe what he actually experienced:

> "We lived in the town of Pripyat. In that town.
> "*I'm not a writer. I won't be able to describe it.* My mind is not enough to understand it. And neither is my university degree. There you are: a normal person. A little person. You're just like everyone else—you go to work, you return from work. You get an average salary. Once a year you go on vacation. You're a normal person! And then one day you're turned into a *Chernobyl person*."[10] (Emphasis added)

Indeed, "Chornobyl man" has become not only a tragic subject but also a sensation, a *wonder*. Chornobyl, a catastrophe that was perhaps for the first time on a worldwide scale, evidences how catastrophism becomes an object of reproduction, how it is replicated in horror movies, and how it evolves into a mass-culture fantasy. Chornobyl has become a media image just like 9/11 and the Christmas tsunami of 2004. In the postmodern era of simulacra, catastrophes are easily appropriated and retold in an entertaining fashion: they become cartoons or horror movies. Protesting against this, a Chornobyl person, Mykola Kalugin, a witness who refuses to speak and becomes symbolically "dumb," resists the fictitious (the sham) Chornobyl and challenges it with his own "little narrative"—his own truth. His testimony is directed against Chornobyl kitsch—the sale of memories or souvenirs of the tragedy and replications of the apocalypse found in popular culture and the media. "They turned Chernobyl into a house of horrors, although actually they just turned it into a cartoon. I'm only going to tell about what's really mine. My own truth," says a witness of Chornobyl.[11] Apocalyptic personalization and the birth of a new language are, therefore, among the effects of the Chornobyl narrative.

---

10   Ibid., 31.
11   Ibid., 32.

Part Two

POST-TOTALITARIAN TRAUMA AND UKRAINIAN POSTMODERNISM

CHAPTER 9

Postmodernism: The Synchronization of History

Postmodernist discourse employs primarily three principal ideas: Jean-François Lyotard's critique of metanarratives and his statement that the new postindustrial society is able to adequately express itself only through the language of "linguistic tricks"; Jean Baudrillard's theory of simulacra; and Jacques Derrida's critique of Western metaphysics. A good few interpretations and conjectures about postmodernism have formed within this triangle of ideas. Perry Anderson in *The Origins of Postmodernity* offers, perhaps, the most profound analysis of the provenance of postmodernism and states that the origin of postmodernism is not only the triad of Lyotard-Baudrillard-Derrida but, first and foremost, *boundary 2: a journal of postmodern literature*, which, starting from 1970, has consistently developed the concept of postmodernism. Precisely due to this journal, discussions about the new style, which was emerging in the literature of the 1970s and was different from modernism, entered public discourse and grew into a fully philosophical theory of postmodernism.

The notion of catastrophe also appears in various theoretical discussions about postmodernism. For instance, Fredric Jameson begins his book on postmodernism as follows:

> The last few years have been marked by an inverted millenarianism in which premonitions of the future, catastrophic or redemptive, have been replaced by senses of the end of this or that (the end of ideology, art, or social class; the "crisis" of Leninism, social democracy, or the welfare state, etc., etc.); taken together, all of these perhaps constitute what is increasingly called postmodernism."[1]

---

1 Fredric Jameson, *Postmodernism, or, The Cultural Logic of Late Capitalism* (Durham: Duke University Press, 1991), 1.

This sense of an ending is also founded on the Chornobyl atomic accident, which revealed disappointment in the various ideological and cultural conceptions built on the idea of progress.

However, distrust of the opposition of "I" and the "other," which is initiated by postmodernism, coexists in the Ukrainian context with a glorification of identity ("samosoboiunapovniuvannia" ["filling oneself with oneself"], as Vasyl Stus puts it). The existential "self-filling of self with itself" characterizes the pathos of modernization of the organic myth of national culture defended by the "60ers." For this reason, the postmodern situation in Ukraine, and discussions around it, naturally intensifies reflections on the 68ers' legacy, and postmodernism itself appears to be, among other things, a generational debate—between the "80ers, "who differ among themselves in their sympathies and antipathies toward the "60ers," and the "90ers," who are in opposition to both.

It must be emphasized that although the opposition between self and other weakens in postmodernism, this destabilization is not originary and does not cancel distinctions. Rather than sharp antagonism, postmodernism reveals the dynamic and multidirectional play of differences between identity and difference. The undermining and destabilizing of the relationship between self and other is achieved by means of "repetitions of repetitions"—by replaying seemingly unequivocal situations with a certain alteration of roles and situations of the subject. Repetition is an important factor of postmodernist consciousness because postmodernism does not conceal that it is constructed from quotations. It is worth noting that in Ukrainian postmodernism repetitions and the processes of "othering" (the isolation of the other) are not devoid of resentment, abuse, negation, and narcissism, while self-expression coexists with the technique of repetition, quotation, hybrid superimposition, and discursive collage. Postmodernists generally gravitate toward a romantic conception of characters and, most strikingly, the multiplication of their identities, masks, and so forth. After all, E. T. A. Hoffmann's character already

> identifies himself with another person, so that his self becomes confounded, or the foreign self is substituted for his own—in other words, by doubling, dividing and interchanging the self. And finally there is the constant recurrence of similar situations, a same face, or character-trait, or twist of fortune, or a same crime, or even a same name recurring throughout several consecutive generations."[2]

---

2   Sigmund Freud, "The Uncanny," http://web.mit.edu/allanmc/www/freud1.pdf, 9.

The situation is aggravated by the fact that in postmodernist texts the repetition of different sides of the self, and of her repressed desires, not only acquires a demonic character and threatens to return to the real world but also becomes a desired disappearance in the maze of quotations and repetitions.

On the one hand, after Chornobyl, which coincided with the millennium, aesthetic thinking altered in a fundamental way and, perhaps, cannot be considered a mere repetition of the romantic era. The very world looks disparate and incomplete, with contaminated areas and dead ends, where it is impossible to escape to an exotic country. The only possible escape is into virtuality. Reality, on the other hand, becomes a "zone" into which another life is breaking from the virtual past and future. At the same time, history is perceived not as a linear concatenation of events leading to a bright future and not as a mythological cycle of various eras. History, it appears, can be reversed or distorted; there can be parallel and alternative histories. The West's postmodern experience of historical oversaturation and historical fatigue is transformed in Ukraine in a unique way. It manifests as a disappointment with the great imperial and totalitarian narratives of history along with a romantic mythologizing of the national history. Ultimately, all these attitudes are reflected in the perception of Chornobyl, which embodies the apocalyptic irony of the postmodern age—that is, being in a state of *post*-modernity.

As a postmodernist narrative, Chornobyl also united, in the cultural outlook, a real event—a suicidal nuclear explosion in the very heart of Ukraine (and Europe)—with a profound uncertainty and distrust of the kind of rational and autonomous self that can coherently act upon the world. Between the real and the mental there was something terrible edging into view—something unconscious, sensuous, and primordially frightening, which undermined faith in aesthetic sublimation and the harmonious coordination of the real and the imaginary. All of this coexisted with the sense of the falsity of social, cultural, and technological institutions. In this way, Chornobyl initiated an amnesia with respect to the past (what good was it to remember what happened before modernity?) and plagiarism (what good is there in citing authors if the only thing that exists is a warehouse, a museum of unnamed things/ideas?). Chornobyl also engaged in a play with local, virtual, and fragmentary realities, which, it seemed, were able to fill in the gap that had suddenly opened along with the Chornobyl "timelessness." At the same time, its logic resisted comprehension, targeted the limits and zones of misunderstanding, and introduced in the realm of that which Lyotard calls a "paralogy" an endless and uninterrupted production of illusions with no pragmatic or rational sense.

Altogether, these conditions are related to the new postmodern ideas which formed in Western thought under the influence of existentialism, post-structuralism, and deconstruction, and which permeated Western culture of the postwar period. Especially important were those ideas developed in the 1970s and 1980s, when practices and objects that had been decisively rejected by high modernism (which was growing in self-awareness through their negation) were gaining legitimacy. Among those practices is, first of all, dependence on the market, mass culture, and media—that is, on the other. In short, the recognition of the other undermines the autonomy and self-isolation of high modernist art.

It is important, however, to note that the same chain of cultural-philosophical discussions that in the West facilitated postmodernism, existed in a significantly reduced form in the former Soviet Union, if at all. One can recall the Tartu School (especially Yuri Lotman), Mikhail Bakhtin's theory of carnival, the popular forms of the Soviet underground, the phenomenon of jokes, and the dissident culture of resistance. Yet the general intellectual-philosophical context in which the post-Soviet postmodernism was being born was totally different from the Western one.

However, according to Boris Groys's highly original theory, in the post-Soviet Russian sphere postmodernism itself is associated with "post-Sovietism," while socialist realism is understood as a "half-style" or a "proto-postmodernist technique of appropriation" in the service of the essentially modernist ideal of historical exceptionality, internal purity, and autonomy from everything external.[3] On the whole, postmodernism in the West, in Groys's opinion, is an attempt to preserve high culture, but this time not through a rejection of the other but by flirting with it and taking into account the whole communicational situation of contemporary culture. The difference between Western postmodernism and Russian "post-Sovietism" lies, for Groys, in the fact that "the only technique of postmodern culture" in the West is appropriation—that is, the transfer of elements of low, mass, commercial culture into the realm of high culture;[4] whereas postmodernism in its post-Soviet version means relegation of high culture (that is, socialist realist culture), by low culture. From this perspective, the tension between the Soviet and non-Soviet disappears: it vanishes the sphere of the other, which must be defeated and appropriated. As a

---

3 Boris Grois, "Polutornyi stil´: mezhdu modernizmom i postmodernizmom," *Novoe literaturnoe obozrenie* 15 (1995): 44–53.
4 Ibid.

consequence, post-Soviet postmodernism gravitates toward self-appropriation, which is precisely what Russian conceptualism attempts.

Groys boils postmodernism down to a purely stylistic phenomenon and to the resolution of the oppositions between high and low culture, the West and the non-West, the modern and the non-modern. He does not account for philosophical-ontological features of postmodernism like its plurality of perspectives. By and large, though, in the late Brezhnev era, Western postmodernism breaks into the Soviet Union not so much in the form of high modernist culture as, instead, in the form of "low," commercial culture—in short, as "consumer society." Video, supermarkets full of gaudy stuff, McDonald's, computer games, and so forth, symbolize, first and foremost, the Western world. As Andrzej Stasiuk aptly points out, the image of the West was created by imagination, building it like "Lego" out of "trash, garbage, waste."[5] There, various things could be found in combination, often secondhand: "Postage stamps, comics, automobile commercials, Louis de Funes movies, chewing gum with picture stories, record covers, albums, cigarette packets, *Wrangler, Lee, Rifle* flickered before our eyes like visions from the postindustrial paupers' bible."[6]

Indeed, in the late Soviet era Western postmodern culture is first and foremost associated with objects. *Things* become characteristic of Western postmodern society, which seemed hermetically closed in on itself. Commodity fetishism ("thing-ism") as a cultural phenomenon provoked by postindustrial capitalism seems attractive and desirable for the Soviet everyman brought up on socialist realist *thinglessness*. Yet even embryonic perestroika, which makes the boundaries between the two worlds more penetrable, as well as the apocalypse of Chornobyl, which, however paradoxically, unites these two worlds, principally transforms Soviet consciousness. Instead of socialist optimism, perestroika plants in the Soviet mind ontological and historical uncertainty just like the Western iteration, and thereby prepares ground for the deployment of post-totalitarian versions of postmodernism.

One more peculiarity of the development of postmodernism in Ukraine is the Western *museification* of culture, which gave birth to the postmodern technique of quotation. At the national level, this becomes the idea of rediscovery of a new ideal of plenitude—a complete and unedited national culture which would include all the expropriated and banned authors, works, and cultural

---

5   Andzhei (Andrzej) Stasiuk and Iurii Andrukhovych, *Moia Europa. Dva eseï pro naidyvnishu chastynu svitu* (Lviv: Klasyka, 2001), 61.
6   Ibid.

eras and styles (such as the baroque, the modernist period, and the era of the avant-garde). In this way, Ukrainian post-totalitarian postmodernism liberates the national culture from the complex of incompleteness. For all its faults, the museification of culture is intertwined with cultural restoration. Indeed, at the same time, cultural artifacts begin to exist not in museum collections, as it is the case in the West, but in virtual collections and in new canons and sub-canons, which can be altered repeatedly.

This new conception of the mode of existence of national culture is based on the binary of completeness/incompleteness, which plays a cardinal role in the processes of cultural development in Ukraine in the 1980s and 1990s. Among all the newly discovered texts and conceptions in the cultural field in this period, Dmytro Chyzhevsky's *History of Ukrainian Literature* has the greatest resonance—most importantly, his thesis about "incompleteness" of Ukrainian literature. As is well known, Chyzhevsky links this incompleteness to the emergence of Ukrainian modernity in the romantic era. The early nineteenth century sees the birth of new genres and styles due to the newly established Ukrainian nobility seeking to satisfy its intellectual needs through Russian language and culture. Therefore, as Chyzhevsky notes, ideas of the folk and of the folksy starts to dominate in Ukrainian literature, in which the genres and styles of high literature were otherwise insufficiently developed. Chyzhevsky published his *History* in the US as early as 1956, but the waves of dissatisfaction with the idea of "incompleteness" reached Ukrainian shores only half a century later. Almost all literary scholars at the beginning of the 1990s touch on this issue.[7]

In fact, the active process during perestroika of bringing back to the national culture earlier forgotten or banned authors, and the growing interest in specific styles and eras (for instance, the baroque and futurism), bore witness to a desire to demonstrate the "completeness" of the national culture and its styles and functions. This process occurred, however, not according to the rules of modernism: the "re-filling" did not comply with the modernist notion of reduction and hierarchy. The point of departure of the modernist canon, formulated at the beginning of the twentieth century by T. S. Eliot, was the idea of the *organic* unity of aesthetic culture—a unity which has an ethical character and can reorganize itself every time a great work is produced. However, at the end of the twentieth century in Ukraine it was difficult to talk about an organic

---

7   See Dmytro Chyzhevs′kyi, *Istoriia ukraïns′koï literatury. Vid pochatkiv do doby realizmu* (Ternopil: Femina, 1994).

hierarchy as a basis for building a high canon for three reasons. Firstly, a canon had not been created yet in Ukrainian culture. Demotic and socialist realist versions of the canon could not satisfy the intellectual reader any longer. Secondly, it was hard to establish a unified canon because the processes of disintegration and decentralization—generational, stylistic, and ideological—were very powerful in the culture. It was more appropriate to talk about multiple canons rather than a single one. Thirdly, the very process of renewal was developing into the collecting of artistic phenomena of different styles and artistic value, in which the intellectual novellas of Valerian Pidmohylny were close neighbors with Dmytro Bilous's patriotic poems. Postmodernism taught that "organic unity" could only exist in the form of plurality.

It is worth noting that postmodernism was not in the least a phenomenon brought in from the outside: it was emerging out of the needs of the development of Ukrainian literature itself. In Ukrainian society at the beginning of the 1990s, multivalent layers of culture, local and international, are discovered. On the one hand, the modernist and avant-garde works that had been concealed in Soviet times; on the other, unknown, not-yet-translated Western "high" literature and philosophy, as well as mass literature. At the same time, one notices a special appreciation of such styles as the Ukrainian baroque of the seventeenth century, with its emblematics, cultural syntheses, and love for "charms," as well as of the neoclassicism and futurism of the 1920s, which were rebarbative in their contemporary historical context but which became approachable at the end of the twentieth century due the creative experimentalism of the time, in particular the cultivation of cultural analogies and their thoroughgoing destruction.

The distinctiveness of the reception of these historical styles is worth emphasizing here. Rather than signify the renewal of cultural continuity and the integrity of culture itself, they served as a way to synchronize different cultural levels and stylistic layers. These styles were regarded as themselves renewable; there was a seduction in being *neo*-baroque or *neo*-avant-garde. The modern and premodern, the baroque and postmodern coexist as equals. A paradoxical situation emerges: Ukrainian literature and culture, without experiencing the full process of modernization and without developing high modernist forms, all of a sudden encounters the kind of postmodern inquiry that resurrects the premodern forms of culture that were especially powerful in earlier Ukrainian culture—Ivan Vyshensky, Hryhory Skovoroda, school drama, mystery plays ("vertep"), and sideshows. Thus, at the end of the twentieth century Ukrainian literature actively cultivates forms of baroque and neo-baroque thinking, while the modernist and avant-garde impulses toward form creation synchronize with the postmodernist plurality of languages, fictional worlds, identities, and styles. In this way, Ukrainian postmodernism, hybrid on the one hand, synthetic on the other, takes shape.

At the same time, it is important that the postmodernist canon reflects the existence of Ukrainian culture at the meeting of two geo-cultural paradigms—the one that came from the West and the one that had developed on the national ground; different ideologies—post-totalitarian (regional) and globalized; different artistic practices—modernist and socialist realist; and different cultural strata—high and popular culture. This transitional aspect of postmodernism gave it a special meaning and turned it into cultural critique. Most significantly, postmodernism helped disclose prohibited places and "dead" zones in the national literature, and lifted taboos surrounding sex, uncensored language, and patois. It also rehabilitated various classes of "deviant", such as the mentally ill, drug addicts, and alcoholics. All of these liberated zones were perceived as forms of a rebellious carnivalesque behavior and as challenges to the artificial socialist realist discourse. Culture itself also looked like a hybrid body with gaps that needed to be filled. In the euphoria of perestroika it seemed that "blank spots" in culture could be dealt with by publishing materials that had been previously banned, renewing in this way the wholeness (completeness) of the national cultural canon.

By the end of the 1990s, it became clear that "blank spots" consisted not only of unknown authors and unknown texts, but also of unknown contexts and modes of interpretation. Thus it was necessary to deconstruct those socio-cultural ideologies that had supported previous processes of canon creation—in particular, such socio-cultural phenomena as socialist realism, populism, and high modernism. The post-socialist academic scholarship proved to be unprepared for such a deconstruction, which prompted scholars in the humanities actively to learn about Western theoretical schools and concepts. It also triggered the development of popular modes of criticism, which were oriented not toward critical thinking but toward the dissemination of formulas and clichés.

This rethinking of the socialist realist canon encountered a problem. As Roksana Kharchuk aptly points out, "at the beginning of the failure of socialist realism, the worst possible scenario was realized: works of high artistic quality, previously banned, are placed among mediocrities taken from the previous canon. Therefore, interesting poets like Geo Shkurupii and Yevhen Pluzhnyk are placed next to mediocre prose fiction authors such as Ivan Le and Petro

Panch, while Mykhailo Stelmakh and Oles Honchar retain their position as classic that resist any critical reevaluation."[8]

One feature of Ukrainian postmodernism at the end of the twentieth century is a noticeable amnesia with regard to socialist realist literature: its prevailing concepts and styles are not deconstructed and are therefore transferred into a new post-totalitarian discourse. A serious analysis of socialist realist literature starts in Ukraine only at the beginning of the twenty-first century, with the publication of Tetiana Sverbilova and Liudmyla Skoryna's *Ukrainian Drama of the 1930s as a Model of Mass Culture and a History of Drama in Figures* (2007), Iryna Zakharchuk's *War and Word: The Military Paradigm of Socialist Realism* (2008), Tamara Hundorova's "Socialist Realism: An Ideological Kitsch," in *Kitsch and Literature: Travesties* (2008), and Valentyna Kharkhun's *The Socialist Realist Canon in Ukrainian Literature: Genesis, Development, and Modifications* (2009). This is not unusual. A similar thing happens in Russian literature. Conceptualism and social art, along with the work of Boris Groys, Igor Golomshtok, Evgeny Dobrenko, only initiate serious analysis of socialist realism in the early 1990s. However, to be fair, Ukrainian authors/postmodernists extract from the socialist realist paradigm its demotic-patriotic component soon enough, and it quickly becomes an object of postmodernist irony.

---

8   R. B. Kharchuk, *Suchasna ukraïns´ka proza. Postmodernyi period. Navchal´nyi posibnyk* (Kyiv: Vydavnychyi tsentr "Akademiia," 2008), 31.

CHAPTER 10

Ukrainian Postmodernism: The Historical Framework

Discussions about postmodernism in Ukrainian literature started at the beginning of the 1990s, although the phenomenon itself appears earlier, in the 1980s.[1] Such a belated—compared with literary practice—employment of the term "postmodernism" is not exceptional. In the West, "the postmodern" is also first received as an uncertain term. Only in the 1980s and 1990s does it take roots and gain a broader foundation, mainly due to the work of Jean-François Lyotard, Fredric Jameson, and Ihab Hassan. In Russian literature, the notion of "conceptualism" is used to denote the new literary trend, and only later is it associated with postmodernism. In Ukraine, the new trend is initially associated with the "neo-avant-garde."

Discussion was about the emergence of a new literature—characterized by irony and spectacle—which parodied ideological clichés, revived baroque playfulness, and foregrounded new kinds of characters and roles. For some critics, postmodernism held the promise of a Ukrainian bestseller; for others, it was an alien and inorganic phenomenon, imported from the West.[2] However, this brief discussion about the fate of postmodernism in Ukraine also laid bare a quite different question: was Ukrainian literature modern?[3] Following on from this

---

1 It is remarkable that both literary critics and academic literary scholars simultaneously began to talk about postmodernism in Ukraine (see, in particular, Natalka Bilotserkivets's "Bu-Ba-Bu ta inshi. Ukraïns′kyi neoavangard: portret odnoho roku," *Slovo i chas* 1 [1991] and Halyna Syvachenko's "Zrushennia kordoniv: postsotsializm chy postmodernizm?" *Slovo i chas* 12 [1991]).
2 Bilotserkivets′, "Bu-Ba-Bu ta inshi. Ukraïns′kyi neoavangard: portret odnoho roku": 42–52; Marko Pavlyshyn, "Ukraïns′ka kul′tura z pohliadu postmodernizmu," *Suchasnist′* 9 (1993): 79–83; Marko Pavlyshyn, "Shcho pere-tvoriuiet′sia v *Rekreatsiiakh* Iuriia Andrukhovycha?" *Suchasnist′* 12 (1993): 115–27.
3 Oleh Il′nyts′kyi, "Transplantatsiia postmodernizmu: Sumnivy odnoho chytacha," *Suchasnist′* 19 (1995); Marko Pavlyshyn, "Zasterezhennia iak zhanr," *Suchasnist′* 10 (1995): 111–19.

question was another one: did Ukrainian literature have a full-fledged modernism—that which in the West represented an alternative to postmodernism? It seemed that answers to these questions would facilitate talk about the legitimacy or illegitimacy of postmodernism in Ukrainian literature. Hence, in its early reception, postmodernism in Ukraine is regarded, on the one hand, as an alternative to modernism and, on the other, is perceived as a universal, or what Jameson calls the "cultural logic of late capitalism." This last circumstance generates discussions about the very possibility of talking about postmodernism in socialist societies, which are too remote (at least ideologically) from "late capitalism."

However, paradoxically, in the postmodern situation exactly the actualization of modernism appears to be important for Ukrainian cultural consciousness. This is because in Ukrainian literature in the twentieth century the forms and concepts of high modernism, oriented toward the autonomy of art, did not have an opportunity to develop in full because they were stunted by socialist realism. Avant-garde tendencies, marked by attempts at the aesthetization of life, were undeveloped too, in spite of existing within underground literary culture. At the same time, Western postmodernism in the 1990s is theoretically considered as a reaction to high modernism, while in North American literature it is seen as a continuation of Pop Art and the avant-garde.

Even when it comes to the traditions of modernism and the avant-garde in Ukrainian literature, they are not enough to be considered a source of the national version of postmodernism. During the nineteenth and twentieth centuries, Ukrainian literature was developed in unusual circumstances. Ukrainian literature's existence within the structure of the so-called "Great Russian" culture,[4] beginning especially from the romantic period, reduced it to the level of a colonial literature. Ukrainian literature, on the one hand, rejects the forms of high culture because the "small Russian" elite, especially active in the eighteenth century, identifies itself with Russian culture (with respect to both language

---

4   The notion of "common Russian" literature is used in the sense proposed by Mykhailo Drahomanov in "Literatura rosiis´ka, velykorosiis´ka, ukraïns´ka i halyts´ka" (Lviv: T-vo im. Shevchenka, 1874). In particular, Drahomanov regards "common Russian" literature as a universal cultural formation, common for nations that comprise the Russian empire, and juxtaposes it with national versions of Ukrainian and Russian literatures and the geographical-cultural version of the so-called Galician literature. Such a hierarchy looks like a relation between a colonial literature and a metropolitan literature. Later Piotr Struve talks about the existence of "parallel cultures" based on the development of, on the one hand, "common people's" (i.e., Russian) culture and, on the other hand, local, ethnographical cultural versions (P. Struve, "Obshcherusskaia kul´tura i ukrainskii partikuliarizm," *Russkaia mysl´* 1 [1912]).

cultural expectations). On the other hand, the desire for self-preservation and national self-determination, starting with Kotliarevsky's *Eneida*, results in the domination in Ukrainian literature of the first half of the nineteenth century of "low" parodic forms, which coexist with a romantic conception of populism and a distinctly national and Enlightenment tendency.

Throughout the nineteenth century, Ukrainian literary culture was oriented toward the preservation of a unified demotic canon,[5] a literature of the "common people," which was actually a realization of the Enlightenment model of popular culture. Even the modest forms of modernism and high culture, which appear in Ukrainian literature at the end of the nineteenth and beginning of the twentieth centuries in combat with the demotic canon, are eliminated in the era of socialist realism, which essentially borrows the demotic conception of popular literature and at the same time ideologizes it and renders it functional.[6] All of this is annihilated by the modernist and avant-garde developments in Ukrainian literature in the first decades of the twentieth century. As a consequence, the fate of modernism in the national literature is at the heart of literary-critical discussions at the end of the twentieth century.[7]

However, the particularity of postmodernism lies in the fact that, as Douwe Fokkema notes, "postmodernism can be called the first universally accepted literary code, and can be interpreted in terms understood in all cultures, regardless if they went through the previous period of modernism."[8] In Ukraine, postmodernism becomes simultaneously a reflection upon, the completion of, and a criticism of Ukrainian modernism. At the same time, it also continues and develops the avant-garde tendencies of the twenties and thirties (it is not a coincidence that the most popular authors during the 1990s are the futurist Mykhail Semenko, the formalist Maik Yohansen, and the modernist Viktor Petrov [V. Domontovych]). However, the Ukrainian version of postmodernism

---

5   I discuss this in more detail in my book *Proiavlennia slova. Dyskursiia rannioho ukraïns′koho modernizmu. Vydannia druhe, pereroblene ta dopovnene* (Kyiv: Krytyka, 2009), 73–150.
6   See Tamara Hundorova, "Sotsrealizm iak masova kul′tura," *Suchasnist′* 6 (2004): 52–66.
7   See Solomiia Pavlychko, *Dyskurs modernizmu v ukraïns′kii literaturi* (Kyiv: Lybid′, 1997); Tamara Hundorova, *Proiavlennia slova. Dyskursiia rannioho ukraïns′koho modernizmu. Postmoderna interpretatsiia*; Iaroslav Polishchuk *Mitolohichnyi horyzont ukraïns′koho modernizmu. Literaturoznavchi studiï* (Ivano-Frankivsk: Lileia-NV, 1998); Vira Aheieva, *Poetesa zlamu stolit′. Tvorchist′ Lesi Ukraïnky v postmodernii interpretatsiï* (Kyiv: Lybid′, 1999).
8   Douwe Fokkema, "Metamorfoza postmodernizmu. Europejska recepcja amerykańskiego pojęcia." *Postmodernizm w literaturze i kulturze krajów Europy Środkowo-Wschodniej*, ed. H. Janaszek-Ivaničková and Douwe Fokkema (Katowice: Śląsk, 1995), 22.

is a phenomenon of a postmodern type. It actively uses popular culture, street forms, and commercials, and becomes performative and multi-stylistic.

In the symbolic, for the 1990s, *Pleroma 3'98: Mala ukraïns´ka entsyklopediia aktual´noï literatury. Proiekt "Povernennia demiiurhiv"*, the emergence of postmodern discourse in Ukraine is interpreted as a need and

> synthesis of postmodern concepts and the non-postmodern situation in Ukrainian art at the end of the eighties and the beginning of the nineties, against the background of the supremacy and degradation of T[estamentary]-R[ustic]-discourse and the cultural shock caused by the rapid growth of the informational openness of society in Ukraine.[9]

It is also worth noting that the neo-avant-garde and postmodernist intentions in Ukrainian literature of the 1990s coincide. This is, perhaps, one of the differences between Ukrainian postmodernism and, for instance, Polish postmodernism, in which the neo-avant-garde work of Witold Gombrowicz, Tadeusz Różewicz, and Czesław Miłosz during the 1960s and 1970s precedes postmodernism and prepares the ground for its arrival.[10]

Ukrainian literature can be regarded, with certain reservations, as a distant analogue of the pre-postmodernist phenomena of the "whimsical" prose of the 1970s (Vasyl Zemliak and Volodymyr Drozd), as well as that of the underground of the 1980s.[11] At the same time, artistic postmodernism in post-Soviet literatures is not in direct and absolute opposition to socialist realism,[12] since some prototypes of early postmodernism appear within the limits of socialist realist totalitarian culture, most notably in its unofficial manifestations, as well as in the underground. These alternative cultural practices are dominated by irony and fictitiousness, which not only help to undermine ideological clichés

---

9   Volodymyr Ieshkiliev, "MP-dyskurs u suchasnii ukraïns´kii literaturi," in *Pleroma. Mala ukraïns´ka entsyklopediia aktual´noï literatury. Proiekt "Povernennia demiiurhiv"*, vol. 3, eds. Volodymyr Ieshkiliev and Iurii Andrukhovych (Ivano-Frankivsk: Lileia-NV, 1998), 91.
10 See Krzysztof Uniłowski, *Polska proza innowacyjna w perspektywie postmodernizmu: Od Gombrowicza po utwory najnowsze* (Katowice: Wydawnictwo Uniwersytetu Śląskiego, 1999), 13–49.
11 Ieshkiliev, "MP-dyskurs u suchasnii ukraïns´kii literaturi," 91.
12 In particular, Roksana Kharchuk states that "in socialist realist literatures postmodernism was first and foremost an alternative to socialist realism. This is the first most essential feature that makes postmodernism of the former socialist realist literatures different from its Western version, in which it was, however mild, an alternative of modernism" (Kharchuk, *Suchasna ukraïns´ka proza. Postmodernyi period*, 6).

but also provide an opportunity to jarringly disturb the meanings of the real world, to multiply said world and combine it with others. The main difference between postmodernism and socialist realism, however, is in the former's orientation not toward *totality* but toward plurality with regard to values, identities, and styles.

Therefore, the emergence of postmodernism in Ukraine, as in general within the territory of the former Soviet Union, has its unique points. Firstly, aesthetic devices that can be related to the forms of early postmodernism, such as irony, quotation, the creation of alternative worlds, and so forth, were already in use within the system of socialist realism. These forms are created not only by the underground (Moscow conceptualists in Russia or Kyiv, ironic prose in Ukraine) but also by so-called official literature, by authors belonging to the Writers' Union (Vasyl Zemliak, Volodymyr Drozd, Viktor Miniailo, Valery Shevchuk, and Pavlo Zahrebelny in Ukraine; Yuri Trifonov, Vasily Shukshin, and Andrei Bitov in Russia). Secondly, the very term "postmodernism" is born and established much later than the artistic phenomenon it signifies, and only with the spread of this term in the West do authors accept or reject it. (The position of John Barth, for instance, is remarkable in this regard: he associated himself with the traditions of "black humor" and later regarded his identification as a postmodernist quite suspiciously.) In the case of Russian conceptualism we witness a different tendency. As Mikhail Berg points out,

> The very recent transition of the leading authors of conceptualism from the system of the underground to the Establishment is at the same time marked by their transition to the level of world art. It was then that they understood that the term "conceptualism" had become obsolete and that they needed to call themselves what they had not wanted to call themselves ten year back, so they began to call their texts postmodernist.[13]

In Ukraine, we witness something very similar: while born in the middle of the 1980s, the Bu-Ba-Bu group becomes associated with postmodernism only later. It was initially regarded mostly as an avant-garde movement oriented toward neo-baroque models and the decapsulation of art. Likewise, critics do not yet consider as postmodernists the authors of the "Kyiv Ironic School" of

---

13 Mikhail Berg quoted in *Postmodernisty o postkul´ture. Interv´iu s sovremennymi pisateliami i kritikami*, ed. Serafima Roll (Moscow: Elinin, 1998), 43.

the 1970s (Volodymyr Dibrova, Bohdan Zholdak, and Les Podervianskyi) or the 1990s texts of Podervianskyi and Zholdak's, despite their use of patois, irony, collage, and social critique.

Following the example of the editors of *Pleroma 3'98. Mala ukraïns'ka entsyklopediia aktual'noï literatury*, critics have started to differentiate in Ukrainian literature two types of postmodern discourse: the postmodernism of the 1980s and the postmodernism of the 1990s, or the "post-Hessean" and "post-Borgesian." Nevertheless, the generational divide, tactically important in the nineties, does not reveal the full idiosyncrasy of postmodernism's development in Ukrainian literature, and the names of Hesse and Borges were selected purely arbitrarily. Instead, it is appropriate to talk about the quite ramified structure of Ukrainian postmodernism that combines a carnivalesque wing (mainly the Bu-Ba-Bu group) and an apocalyptic wing (itself divided into two types—the "rhetorical" apocalypse of Izdryk and Taras Prokhasko and the "metaphysical" apocalypse of Yevhen Pashkovsky). We can also isolate feminist postmodernism (Oksana Zabuzhko) and popular postmodernism (Yuri Vynnychuk). The Kyiv Ironic School, established in the depths of the underground of the 1960s and 1970s, and the neo-avant-garde of the 1990s (Serhiy Zhadan, Andriy Bondar, and Volodymyr Tsybulko) also gravitate toward postmodernism.

CHAPTER 11

A Farewell to the Classic

The postmodernist period is marked by the desacralization of the classic. It is not about an absolute reevaluation of the literary classic and its functions. From now on, the classic is not an absolute, and innovation is not predicated upon a rejection of the classic tradition, as it was the case in modernism and the avant-garde. In the postmodern era, the classic becomes relative; it is replaced by particular canons and vogues. Even the very notion of "great" literature becomes irrelevant. Readers' interests increasingly gravitate toward "sub-literature" and popular culture; after all, the time has come when the classic itself is consumed in the wrapper of mass culture. Postmodernism tries to reorganize the cultural mindset, but in a way that allows for the preservation of a sense of the primacy of the classic. "Ukrainian culture today needs IV fluids, enemas, pills, and reanimation. And this, I suppose, is one of the attempts to represent the corpse of Ukrainian culture as a living, complete being," states Yurko Pozayak.[1]

While revising all existing canons, postmodernism utilizes the principle of repetition, substituting the logic of hierarchization with that of simultaneity. What does this mean? For postmodernists, literature is neither a closed canon nor something historically remote in time: it exists simultaneously, synchronically, as a trace—that is, in those texts, plots, and names that construct the archive, library, museum, anthology, and list.

For example, Andriy Bondar, in his "poem that will never be translated into other languages" ("virsh iakyi nikoly ne perekladut´ inshymy movamy"), carries out a ritual of postmodernist revision—a rewriting of the "dictionary" of literature that simultaneously inscribes it in the underground and avant-garde. In short, at the same time he both erases and writes a new canon. He revises the vocabulary of the "Ukrainian literary encyclopedia," deriding official literary scholars—old-fashioned, time-serving, conformist—"in patched gray suits

---

1 See www.aup.iatp.org.ua/news_arch.php.

in the dusty rooms of the institute of literature of the academy of sciences of Ukraine, who fill with their senile bodies in the sick Ukrainian time-space." By inscribing his "healthy" list into the "sick" time-space, Bondar playfully relinquishes and erases his name ("but who am I to judge them for all that"), leaving, however, the names of a *new* literary canon that he highly esteems:

> there is no boris vian but there are vianu tudor viviani raphael
> there is no danilo kiš but there is egon erwin kisch
> there is no maurice blanchot but there is blasco ibanez
> there is no sacher masoch but there is iurii zakhovai (died in Lviv)
> there are no georg heym gottfried benn jean jenet
> and thank god,

But

> thank god there is something about abelard and ariosto
> about tom wolfe and virginia woolf
> beside an article about roman mykhailovych vul
> the author of the literary-critical sketch
> "M. Gorky in the Crimea"
> but there is nothing yet about tom wolfe the second
> (what are two tom wolfes for?)[2]

Reflecting on the possibilities of postmodernism in Ukrainian literature, Viktor Neborak underlines that the condition of the emergence of a postmodernist work is play with a "previous literary tradition," the shaking up of "the immobilized canon of modernism," and the presence of "certain concentric circles, such as a canon of national literature, a European literary canon, and so on."[3] In his opinion,

> if this approach were to be applied to Ukrainian authors (without reference to a specific name), then they, if they consider themselves postmodernists,

---

2   Andrii Bondar, "virsh iakyi nikoly ne perekladut' inshymy movamy," in *MASKULT* (Kyiv: Krytyka, 2003), 30–31.

3   "Bubabists′kyi khronopys Viktora Neboraka: shche odna intryha z pryvodu suchasnoï ukraïns′koï literatury," *Literatura plius* 9–10 (2001): 34–35, www.uap.iatp.org.ua/litplus/lit34-35.php.

would play on a Ukrainian canon that fits the European tradition. But there is a problem with a Ukrainian canon! The postcolonial canon does not exist! There are certain more or less important authors. In other words, everyone now creates his or her own canon. Let us try to imagine, what kind of a canon, approximately, from the point of view of Ukrainian literary tradition, could satisfy Andrukhovych as a prose fiction author. There, to be sure, Valery Shevchuk should find his place. Valery Shevchuk himself mentions among his predecessors in the twentieth century some representatives of the "executed renaissance"—Valerian Pidmohylny, etc. Before them, perhaps, Ivan Levytsky, but with his non-textbook works—the novel *Clouds*, etc. Still earlier—Skovoroda. Before Skovoroda—baroque, Cossack chronicles, Ivan Vyshensky. … In other words, there is a need to build such a canon up into *Primary Chronicle* and to write a novel that plays with all this. And, to be sure—in the worldwide literary context. And only then will there appear a text that is postmodern from the point of view of Europe.[4]

Thus Viktor Neborak states the absence of a "postcolonial canon" in Ukraine which, he is convinced, prevents Ukrainian postmodernism from a full-fledged existence. However, in reality, as early as in the 1990s in Ukraine there are various canons, the postcolonial included. Solomiia Pavlychko in her work on modernism formulated a fairly strict canon regarding Ukrainian literature. George Grabowicz purposively builds a postcolonial canon in *Toward a History of Ukrainian Literature* (1997). There are new editions of the known works of Serhii Iefremov, Mykola Zerov, and Dmytro Chyzhevsky, each of them proposing different versions of the canon. *Pleroma. Mala ukraïns'ka entsyklopediia aktual'noï literatury* also puts forward a canon of the new literature of the end of the twentieth century. Oksana Zabuzhko, Vira Aheieva, and Nila Zborovska build a feminist canon of Ukrainian literature. The mission of postmodernism, according to all of these commentators, consisted in the task of legitimizing a plurality of canons and introducing into them an element of play. In other words, legitimizing heterogeneity, irreducibility, and autonomy.

Intertextuality is one of the mechanisms that helps make it real. In general, postmodernist play is characterized by the manipulation of narrative perspectives, the erasing of differences between fiction and reality, the creation of multiple realities, heteroglossia, quotation, multiple voices and languages, hybridization, creolization, and so forth; play with literary canons, a conscious game involving

---

4   Ibid.

the symbols and signs of cultural memory, the breaking of taboos—all these practices are important sources of imagery also for Ukrainian postmodernist authors. The national classic experiences a "winnowing." It forfeits its sacral meaning and becomes a playground. It is profaned as in, for instance, Andriy Bondar's ironic, negative canon and in Andrukhovych's perverse canon. In the latter, the author renames Ukrainian classical authors according to their "deviations." Such a canon, in fact, becomes a list and is explicitly mystified. Andrukhovych frankly admits that he is not interested in the literary history of Ukraine or Ukrainian classic authors, and that the image of literary Ukraine he creates (delivering talks or writing novels) is "quite mystified."[5] "I was preoccupied with a certain type of fiction," he writes."[6]

In the post-Bu-Ba-Bu period, Andrukhovych still enjoys deriding advocates of the national classics, and thus shocks his compatriots. "Literature could have been different," he reiterates indefatigably. He also suggests his own radical recipe for a "modernization of the classics," a recipe that can be looked at as preparatory material for a postmodernist novel-pastiche. In this work, a canon of classic Ukrainian literature would be both created and played with. The protagonist of his novel, *Twelve Circles* (2001–3) represents, then, *Ukrainian Classics: Reread and Augmented*[7]—a series of modernized textbook plots superimposed on the reality of the brutal 1990s:

> *Kaidash's Family* in his version became a settling of scores within a mafia clan, the struggling seminarians in *The Clouds* to the point of nausea and hallucination smoked the hash they procured from southern sugar refineries, and "The Horses Are Not to Blame" ended with a gang rape of Arkady Petrovych Malyna, a liberal landowner, by the squadron of Cossacks he had summoned to his estate.[8]

In time, Andrukhovych associates the national literary tradition not with the "great family" but rather with a pop-band with a conventional name, "Bad Company": it unites authors not according to their canonized images but according to the principle of cliché. Thus the literary canon becomes a list of personified perversions:

---

5 Ibid.
6 Taras Prokhas´ko, *Inshyi format, Iurii Andrukhovych* (Ivano-Frankivsk: Lileia-NV, 2003), 37.
7 Yuri Andrukhovych, *Twelve Circles*, trans. Vitaly Chernetsky (New York City: Spuyten Duyvil, 2005), 94.
8 Ibid., 95.

> Hrytsko, as all tenors, is a pederast.
> Ivan is a reveler, freemason, and fraudster.
> Taras is a drunkard and a loafer, especially at service.
> Panko is a compulsive scribbler, while Marko is a hermaphrodite.
> Panas is an asshole, and Borys is a pedant.
> Iakovych is a hopeless atheist, a seer.
> Leska and Olka are lesbians.

Mystifying Ukrainian literature, Andrukhovych takes up the task of altering the image of the Ukrainian poet, and all his characters—bohemian poets—ultimately want to identify themselves according to the stereotype of the demonic artist. This creation of the "demonic image of the cursed poet, prone to various temptations,"[9] is fully displayed in *Twelve Circles* through the character of Bohdan-Ihor Antonych. Andrukhovych himself underlines that he aimed to "ruin the stereotype of Antonych as a respectable child from a respectable family."[10] After all, all protagonists from Andrukhovych's previous novels also belong to the archetype of a "cursed poet," which Andrukhovych tries to make Antonych turn into.

Bubabists recognized Ivan Kotliarevsky as a prototype of such a modern poet, and carnivalesque Ukrainian postmodernism regards him as its predecessor. "Back to Kotliarevsky!" proclaims Andrukhovych. "Two hundred years ago this overly gay wandering teacher for the children of the nobility, a connoisseur of alcohol and not only postal cards, quite unexpectedly to himself became the author of the first Ukrainian book."[11] Perhaps modern postmodernists are attracted to the creative liberty with which the author of *Eneida* treats Virgil's mythic source, yet at the same time they are to drawn to Kotliarevsky's dignity when he defends his authorial rights in polemics with the publisher Maksym Parpura. They are also attracted, of course, to the archetypal character of Aeneas as a wanderer and eternal traveler. Such a character corresponded with the very idea of a poet—"to be constantly tempted and not to evade the temptations."

In the course of the last two centuries, *Eneida* has been a test of self-determination for both "modernists" and "populists" (from Panteleimon Kulish

---

9 Prokhas´ko, *Inshyi format, Iurii Andrukhovych*, 41.
10 Andrukhovych, *Twelve Circles*, 95.
11 Iurii Andrukhovych, *Dezoriientatsiia na mistsevosti. Sproby* (Ivano-Frankivsk: Lileia-NV, 1999), 99.

up to Mykola Zerov). In the post-Bu-Ba-Bu period, Viktor Neborak reads *Eneida*, emblematic for the Bu-Ba-Bu era, not through the archetype of the bohemian but in terms of the overcoming of a marginal status of existence and the need to put down roots.[12] Having disrupted his game, Neborak found for himself a way, different from the postmodern one, to escape the vortex of transformations and masquerade mutations. His book *Perechytana "Eneïda". Sproba sensovoho prochytannia "Eneïdy" Ivana Kotliarevs´koho na tli zistavlennia ïi z "Eneïdoiu" Vergiliia* (*Eneida* Reread: An Attempt at a Sensible Reading of Ivan Kotliarevsky's *Eneida* in Comparison with *The Aeneid* of Virgil [2001]) is an epilogue to Bubabism as a socio-cultural phenomenon and at the same time its critique. Neborak opposes uprooting and putting down roots, games and rituals, marginalization, and resistance to demonic evil, clearly positioning himself on the side of ethics and foundations. It seems improbable but the poet-Bubabist characterizes Aeneas—the archetypal for the Bu-Ba-Bu hero, who was "khlopets´ khoch kudy kozak" ("quite a Cossack for a lad")[13]—as an offender and a bearer of evil, while reading the whole of Kotliarevsky's *Eneida* through a lens that rejects bohemian marginalization and eternal wandering.

Indeed, during the period of their greatest activity, the mythology of the Bubabists is based not on the topoi of putting down roots, not on Skovoroda's focus on the self, and not on the Shevchenko myth, but on the glorification of bohemian poet-wanderers like Aeneas. The point is to change the archetype of the whole of Ukrainian culture at the end of the twentieth century. Newly discovered in the 1990s, philosopher Mykola Shlemkevych talked in the middle of the twentieth century about the "lost Ukrainian person" and identified several types of mentality, among which he especially emphasized the Shevchenko version of an active personality that recreates life according to the laws of righteousness. It can be said that at the end of the twentieth century the Shevchenko cultural-anthropological model of righteousness, which was important for the Ukrainian sixties, for instance, loses traction and is supplanted by the playful model of burlesque, which surfaced in Kotliarevsky at the end of the eighteenth century. In opposition to academia, the postmodernists create their own image of Kotliarevsky.

"A daredevil-merrymaker with an Apollonian halo around his head, a conqueror of the bodies of the sexy Didos and of the hearts of barely kissed

---

12 Viktor Neborak, *Perechytana "Eneïda". Sproba sensovoho prochytannia "Eneïdy" Ivana Kotliarevs´koho na tli zistavlennia ïi z "Eneïdoiu" Vergiliia* (Lviv: Astron, 2001), 10.
13 Translated by Wolodymyr Semenyna.

Lavinias, a joker-friend and a capable hellion with a skillfully "accrued yet hidden for a while from the jealous eye" intellect—this is the type that the wise Ivan Petrovych molded in his Poltava alchemist lab," states Neborak."[14] All the while, instead of discoursing on the demise of the Kotliarevsky tradition ("kotliarevshchyna"), Andrukhovych talks about "an endless journey to immortality" and of the "clownish laughingstock and buffoon," designed for burlesque.[15]

George Grabowicz's essay about the semantics of the "Kotliarevsky tradition"[16] in the middle of the 1990s in a paradoxical way diagnosed the actuality of the Kotliarevsky theme for contemporary Ukrainian culture. Grabowicz regards the Kotliarevsky tradition as one of the metahistorical models of Ukrainian literature's self-awareness—a model based "on covering-up the real 'self' behind the figure of the narrator, on escape into a pretended simple-mindedness or naivety, on self-defense by means of intimacy, and on the voice of the group,"[17] and considered it first from the postcolonial perspective. It implies both the restrictions such a style imposed upon Ukrainian literature and the infecting of the Great Russian literature of the imperial center with an anti-canon in the guise of the low burlesque style of Gogol. Grabowicz also suggests that the deep poetics of socialist realism feeds on "the tradition and archetypes of the Kotliarevsky tradition."[18] However, one can argue that the Kotliarevsky tradition becomes an anti-canon of the socialist realist canon: it infects post-Soviet culture with premodern jocular culture, burlesque, adventure, and masquerade, in order to escape socialist realism's modality. But the Kotliarevsky tradition does not only subvert: it also renews the national literary tradition.

There were quite a few causes of this process. Bubabists rejected the existing didactic models that had hitherto characterized national-cultural self-awareness in modern Ukraine—both the demotic and the modernist. In addition, in its essence Bubabist carnivalization is a distinctly post- and anti-co-

---

14 Viktor Neborak, "Iurko i Sashko, Sashko i Iurko," in *"Bu-Ba-Bu" (Iurii Andrukhovych, Oleksandr Irvanets', Viktor Neborak): Vybrani tvory: Poeziia, proza, eseïstyka* (Lviv: Piramida, 2007), 41.

15 Iurii Andrukhovych, "Zahybel' kotliarevshchyny, abo zh Bezkonechna podorozh u bezsmertia," in *"Bu-Ba-Bu" (Iurii Andrukhovych, Oleksandr Irvanets', Viktor Neborak): Vybrani tvory*, 101.

16 See Hryhorii Hrabovych, "Semantyka kotliarevshchyny," in Hryhorii Hrabovych, *Do istoriï ukraïns'koï literatury* (Kyiv: Krytyka, 1997), 316–32.

17 Ibid., 332.

18 Ibid.

lonial phenomenon[19]. Here the figure of Shevchenko plays a key role, alongside Kotliarevsky. The demotic model was built around the myth of Shevchenko as the father and prophet of folk writing (Mykola Kostomarov; Panteleimon Kulish); the modernist model, on the contrary, rested on the aesthetic exaltation of the figure of the artist as a religious and creative individual (Mykola Yevshan). The avant-garde model was also deployed around the myth of Shevchenko, but it developed through the act of a radical severance—the symbolic burning of *Kobzar* and, through that, the rejection of the popular and philistine image of the poet. The Bubabists were, evidently, closer to the avant-garde version of cultural criticism. In addition, like Kotliarevsky with regard to Virgil, they introduce the elements of playfulness, burlesque, and buffoonery into the processes of self-awareness, and openly shape the classics with the help of masks and fiction.[20]

Postmodernists do not avoid Shevchenko; on the contrary, his poetical world finds itself at the center of the postmodernists' clandestine and ludic engagement with the national classic. In fact, Shevchenko is precisely the "cursed poet" with whom postmodernists are trying to identify. The fact is that the all-pervading parabola of, let us say, Andrukhovych's most popular novels is openly postcolonial and unambiguously refers to Shevchenko. In particular, the archetype of the bohemian poet—the main protagonist in all of Andrukhovych's novels—is linked not only to Aeneas-the-playboy but also to the Cossack prototype of the wandering poet-kobza-player, who serves as a virtual lyrical subject for Shevchenko.

Shevchenko actually reveals and molds the colonial symbolism of Ukraine, employing it as a cosmological image of space, which is divided into the metropolis and provinces, and developing it on the basis of a historiosophical myth of the fate of Ukraine. This symbolism is identical with the poet's own fate and, as well as through the natural-philosophical emblematics of poetic landscapes, is intertwined with the ideal topology of Ukraine. The metaphorical image of colonial Ukraine includes the opposition of the oracular language (identical to "Lord the Word") and language as conversation (a means of direct contact among people in Shevchenko's works).

In *The Moscoviad*, Andrukhovych uses the central topos of Shevchenko's mystery "Dream"—a trip to the center of the Empire—and creates a grotesque

---

19 See about this in more detail in Tamara Hundorova, "Postmodernists′ka fiktsiia Andrukhovycha z postkolonialnym znakom pytannia," *Suchasnist′* 9 (1993): 78–83.
20 Significant here are not only the fictionalization of real figures, such as Bohdan-Ihor Antonych in Andrukhovych's *Twelve Circles*, but also the concoction of fictitious writers, as, for example, in Yuri Vynnychuk's novel *Malva Landa*.

portrait of the capital that very much resembles the puppet-like image of the "muzzle-punching" in Shevchenko. Yet most important is that implicitly patriotic imagery in Andrukhovych is related to Shevchenko's cry for Ukraine. After all, in the postmodern era, Shevchenko also gains a new image. Grabowicz's interpretation of the symbolic autobiography of Shevchenko, including his homo- and auto-erotic subtext, is quite consonant with the rewriting of the classics in Ukraine during the 1990s.

Implicit and explicit quotations from Shevchenko can be found in Viktor Neborak, Serhiy Zhadan, and Volodymyr Tsybulko; the Kapranov brothers even publish their *Kobzar 2000*. Zhadan is harsh and sarcastic when referring to Shevchenko:

> Having hurt his feet on the stubble,
> old Perebendia by the hedge
> complains that turning dissolute
> the wicked children in these hard times
> have lost all shame and sold out Ukraine,
> have abandoned spirituality and fight,
> and overall do fuck knows what.[21]

Neborak, on the other hand, appropriates Shevchenko's irony, quoting phrases/clichés from Shevchenko and creating out of them a panorama of the "movement of masses" at the end of the twentieth century:

> I was thirty,
> I herded asses with my eyes,
> in the time of independent movements of masses
> I first tipped up and then dipped down
> on various stages; Soviet state
> learned the language, Cossacks
> scurried across the Sich; the overseas
> brother clutched his brother
> in tight embrace; thirty
> I was; I prayed to God—
> about what? as though someone
> cast a spell on me. ...[22]

---

21 Serhii Zhadan, *Balady pro viinu i vidbudovu. Nova knyha virshiv* (Lviv: Kal'variia, 2001), 72.
22 Viktor Neborak, *Litostroton. Knyha zibranoho* (Lviv: LP-vydavnytstvo, 2001), 298.

However, ironic takes on Shevchenko, and pastiches of his work, are not so much a means to destroy the classics as they are a way to renew their intimate and personal contents, earlier diluted by uncritical, almost ritualistic worship. Shevchenko becomes probably the most popular author for remakes, and Neborak composes pastiches based on the most famous textbook works of Shevchenko: "I was thirteen" and "A Cherry Orchard by the House." He repeats a number of Shevchenko's phrases, preserves his rhythmic and intonational line breaks, and reproduces the rhetoric of the lyrical yearnings the poet voices. He does not *struggle* with the poet; he talks about his accord with him:

> What next? Each receives his own.
> Some look upward, some—downward,
> some break up a stone, some drink beer,
> some enjoy striptease—
> I don't care, I
> don't care, days are passing,
> and nights, I hit the bottom …
> —But I do care!!![23]

Thus Shevchenko becomes one of the main figures in the dialogue between Ukrainian postmodernists and the national literature. Not only does Kotliarevsky return, but Shevchenko's thinking is renewed—a thinking which developed in the force field of romanticism and contained a multi-vector polyphony of literary models and, indeed, occurred within the framework of postmodernism. This event encompassed romantic-grotesque imagery, feminine mythopoetic symbolism, postcolonial tricksterism, bohemian carnivalism, and even expressionist imagery. Unfortunately, classic Ukrainian literature in its development did not fully unfold all these potential versions of imagery offered by Shevchenko's work, and criticism actualized only one aspect of Shevchenko—the demotic.

"In the sweet Darusia I recognize modern Ukrainian literature," notes a reader reflecting on a new novel by Maria Matios which takes up the symbolic mythology of Shevchenko:

> This is she, ukr-modern-lit, for the majority of Ukrainian readers either
> "mute" or "whisper[ing]" to herself something unintelligible." This is she,

---

23 Ibid., 298.

ukr-modern-lit, [who] wanders at night in the cemetery of the Ukrainian classic and calls to the graves, "Dad! Mom!" The Ukrainian man of letters nowadays is repressed by his individuality because he respects our customs, and customs require concealing individuality. He behaves as though he is not able to be god on the pages of his own book, as though fate has destined him to be a pupil or hired hand of the omniscient Third Person—a sponsor, an editor, a "cultural elite," a vogue. ... Every year the store of popular genres is being filled with hackneyed items, the scrapheap of counterculture is growing, and new figures are lying down in the morgue of "serious" literature, which doesn't have readers. All the while the real talent wanders through the village called "Ukr-modern-lit," and he has a headache from the candy given to him for his great patience.[24]

In its adjusted to mass culture version of the Kapranov brothers, Shevchenko's *Kobzar* is turned into erotically tinged anecdotes for men and women. These tales have the mostly familiar (or slightly modified) titles of Shevchenko's works and are divided by gender: the soft kind for women ("The Witch," "The Mermaid," "Little Kateryna," "Petrus"); the hard kind for men ("Dream," "Haidamaka," "The Night of Little Taras," "Varnak," "Prychynna," "The Desecrated Grave"). It is an anthology of wonders, adventures, and jokes for mass consumption. Neither intertextuality nor the pastiche of the classics works here; what we have is a provocation or, more precisely, an imitation of the book of archetypes which *Kobzar* represents in Ukrainian consciousness.

Volodymyr Danylenko in *Tinkle-Jingle* (*Dzen´ky-Bren´ky*) creates a museum containing not just one author—Kotliarevsky or Shevchenko, for example—but a museum containing every Ukrainian literary classic. He fills this gallery of Ukrainian classic authors with parodies of their characters, starting with Nestor-the-Chronicler and ending with "Stelmakh, Honchar, Pervomaisky, and Rostyslav Sambuk."[25] After all, an end to the taboo on names and on the personification of the Soviet literature in the images of the real Ukrainian authors can be found in Andrukhovych's *Recreations*, in which Bilynkevych, an instructor at the city committee of the Young Communist League, and by extension a KGB agent, is trying to find out "what the poet Mykola Nahnybida

---

24 Sheliazhenko, "Solodka Darusia, heroi ukrsuchlitu," http://gazeta.univ.kiev.ua/actions.php?act=print&id=532.
25 Yuri Andrukhovych, *Recreations*, trans. and with an introduction by Marko Pavlyshyn (Edmonton-Toronto: CIUS, 1998), 42.

is working on right now."²⁶ The poet-carnivalist Nemyrych replies that, "'You know, old man, I believe he died some thirty years ago. ... It is, however, altogether possible that he's working on something, though up till now there has been no news of him.'"²⁷

Danylenko does not talk about the burial of Soviet literature; he places classic authors in a kind of a vault that recalls Café Aeneas in the building of the Ukrainian Writers' Union. There is Shevchenko again, but this time in his portrait version of a "peasant in a cap":²⁸ "The doors opened with a rumble, and Shevchenko barged into the hall. 'Here I am!' he exclaimed merrily. Taras Hryhorovych was tipsy, in a sheepskin cap and coat, and in boots lavishly smeared with tar."²⁹

In his parody, Danylenko reminds us about certain stereotypes the figures of the Ukrainian classics are associated with in mass consumption. He profanes literature by identifying it with the national cuisine: it both cases, the targets are "our pride" and our "tastes"; so, as the narrator says, "we know what we have. And we have a great history of blood pudding and potato pancakes. However, naturally, borsch with pampushkas is not the same as headcheese with horseradish, and crunchy bacon and young potatoes with sour cream are even better."³⁰ Danylenko's quasi-literary burlesque is based on the transformation of major and minor classics as well as modern authors—Serhiy Zhadan, Yuri Andrukhovych, Oles Ulianenko, Viktor Neborak, and Yevhen Pashkovsky—into dolls; he grotesquely plays up figures and roles, ironizing the literary republic and introducing recognizable realities. Quite in accordance with Freud, he goes as far as the ritual consumption of the father Shevchenko:

> "Gentlemen," Pluzhnyk started to fidget intelligently, "let us divide the poet among us. There will be enough for everyone." Taras Hryhorovych suspiciously stared around at those present when Andrukhovych and Svydzyns′kyi smeared him with cream, sprinkled him with nuts, poppy, and raisins, until the prophet turned into a big cake with a pink flower on the bald spot. Then the cake was cut into pieces and shared among all. "Now everybody has his own Shevchenko," Korniichuk started to champ, taking a bite of a piece of the sweet sheepskin.³¹

---

26 Ibid.
27 Ibid.
28 Volodymyr Danylenko, *Misto Tirovyvan* (Lviv: Kalvariia, 2001), 162.
29 Ibid., 162.
30 Ibid., 205
31 Ibid., 200–1

This farce, played up by a modern writer in the literary field, not only desacralized the figure of the Ukrainian man of letters but also gathered together and synchronized the whole literary field, reminding readers of famous literary disputes and about past and present Ukrainian authors and critics.

The 1990s are remarkable not only for their ironic rewriting of literary and cultural clichés but also for their serious rereading of the *classics*. For instance, Oksana Zabuzhko frankly admits her responsibility to the "literary lineage" and the fact that for her "to write about Shevchenko means a colossal energetic injection."[32] What her study about the national myth in Shevchenko and her novel *Field Work in Ukrainian Sex* have in common is the mutual postcolonial tonality. The interpretation of the national fate from the perspective of colonial consciousness of a Ukrainian woman, the character of Zabuzhko's novel, foregrounds Shevchenko's signifiers of femininity, which the poet personalized in the figure of the unfortunate single mother. In Zabuzhko's work, there is also an echo of demonic masculinity—the basis of the Cossack myth in Shevchenko.

The "dictionaries" of culture, rewritten in Ukrainian postmodernism, are based on ideologemes and mythologemes that have a didactic and popularizing character, and that proved to be uniquely tenacious in the Ukrainian outlook of the nineteenth and twentieth centuries. In postmodernism, they are critiqued based on linguistic (that is, rhetorical) games, which feature inversions and transgressions (the breaking of boundaries and the transfer of meanings). The postmodernist inversion of the official canon and high culture usually rests on a special kind of parody and internal quotation and is realized for the most part in the direction of mass literature.

However, it seems that this rewriting can be viewed only through the so-called aggression of popular and mass culture, which essentially capture the classics and travesty them. Mass culture in the postmodern technological era is produced from above, through remakes, pastiche, and a conscious exploitation of the achievements of so-called high culture. In addition, postmodernist games with the classics are fairly complex and multi-layered. Not only the boundaries between high and low cultures are destroyed in this process but also—through selection, the lifting of taboos, and cleansing of stereotypes—the classics are renewed and reshaped. This process can be associated not with the avant-garde revenge upon, and destruction of, the classics but instead with a sophisticated postmodern renewal of the classics as a cultural field. This field becomes a place for an unending game of various dispositions and agents.

---

32 Taras Prokhas′ko, *Inshyi Format, Oksana Zabuzhko* (Ivano-Frankivsk: Lileia-NV, 2003), 27.

The postmodernism of the Bubabists and their symbiosis of high and mass cultures also have a sophisticated lineage. This is not so much an avant-gardist inversion of the national culture similar to Semenko's "burning" of *Kobzar* as it is an ironic "filling-in" of the gaps and recesses of the national classic, which exists first and foremost as a form of virtual Ukrainian literature. Such a virtual literature turns into a private library; as Viktor Neborak notes in the dedication to his "Chrysler Imperial," "the dandy Kafka transformed me into a library of my own crazy stuff."[33]

Especially significant in the relationship between postmodernism and the classics is a severance from the so-called "rustic syndrome" and the "testament-rustic discourse" (Volodymyr Yeshkilev)—in other words, from the demotic tradition of Ukrainian literature. Postmodernists execute this retaliation, carnivalizing literature, which they perceive as either "empty" ("porozhnynna") or "almost dead."

After all, postmodernists are not alone in renewing and rewriting the classics. Viacheslav Medvid, a modern writer who can barely be considered a postmodernist, proposes an alternative form of the reception of Ukrainian literature. In particular, he recommends against rejecting the "rustic aesthetic," instead proposing that it be combines with the ethical and psychological directives of high modernism.[34] Such literature must, in his opinion, consciously repeat the demotic classic tradition, rather than depart from it. Medvid suggests renewing the publication of the journal *Osnova* (*Foundation*), founded by Panteleimon Kulish, and again imitating, for instance, Teslenko. Thus his cultural project rests not on overcoming of the demotic syndrome but on giving it a modern form.

Panteleimon Kulish in his time offered the most comprehensive formulation of the concept of demotic Ukrainian literature. It bore all the signs of invented, artificially constructed cultural identity. The basis of this conception lies in the romantic identification of "self" with "the people," the idealization of moral values innate to "our country folks," and the right of the Ukrainian author to speak "not only on his behalf" (Kulish's "Foreword to Community [A View on Ukrainian Literature]"). The cultural catalyst herewith is the identification of educated Ukrainian artists with "the great collective demotic person."[35]

---

33 Viktor Neborak, *Litostroton. Knyha zibranoho*, 291.
34 Ieshkiliev and Andrukhovych, eds., *Pleroma. Mala ukraïns'ka entsyklopediia aktual'noï literatury*, 72.
35 Panteleimon Kulish, "Prostonarodnost'" v ukrainskoi slovesnosti," in Panteleimon Kulish, *Tvory v dvokh tomakh*, vol. 2 (Kyiv: Dnipro, 1989), 523.

Although the later Kulish is oriented toward the creation of high literature, in the second half of the nineteenth century he asserted the demotic (popular) over the high character of Ukrainian literature. Ukrainian literature, Kulish preached, is

> not the creation of a narrow circle, which arranged for itself an exquisite life among the masses whose only choice is to submit to its taste, its concepts, and its very life; rather, it is the voice of the mass, but one that does not imitate educated people, whose interests do not include the necessities of life, but rather draws into its healthy environment the most developed educators, with all their erudition, with all the glitter of their artistic intelligence. All these people, with all the personal advantages over the demotic element, feel and realize with their minds that the power of the folk's spirit lies not in their educated and artistic circles but in the mass of which one individual person is insignificant but in its collectivity, with those of his own kind, preserves the laws of literary taste and the folk meaning for the most developed and original representatives of our people.[36]

In fact, exactly such form of enlightenment and demotic speaking "not on its own behalf" is the target of Ukrainian postmodernist authors' attack, as they state their right to be free in their creative work and in their literary games with the classics.

After all, Viacheslav Medvid at the end of the twentieth century is not so much interested in the ideal essence of the populists—"populus," "people,"—as he is interested in the linguistic-ontological nature of the people's being, immersed as it is in the real soil of the everyday, primarily small-town or rustic, yet openly articulated as rational, intelligent, and modern. His characters give voice to a stream of consciousness while preparing swill for piglets; they speak up the most exquisite passages, combining the most perverse desires when growing out of their childhood.

The traditional demotic prose of the nineteenth century was openly metonymical—parts of life, of characters, reflected in the pages of a novella or a short story, were supposed to be parts of life (ethnographically, socio-graphically expressive). New modern prose, which is Medvid's goal, must, by contrast, be metaphorical. It also is quotational in essence, analytical rather than playful; it is existential and evocative of the famous classical plots and situations of

---

36   Kulish, "Kharakter i zadachi ukrainskoi kritiki," 517–18.

Ukrainian literature. Medvid's short story "Village as a Metaphor" is typical of this type of prose. Constructed as a narrative about a boy whose father sends to his aunt for milk, and for a drink, the story focuses on the spiritual awakening of the protagonist, with whom the author himself is ready to step "beyond the edge of something unknown again."[37]

The clan theme is one of the most significant for Medvid. Like Faulkner, Medvid creates his imagined Yoknapatawpha, which he portrays in detail and with many voices; it becomes a place where the world, characters, and language are newly born. The thingness and reality of everyday being do not conceal but, on the contrary, lay bare the eternity of time. Time-space is structured by metaphorical situations, which often take root in childhood. One such situation is initiation. At the beginning of "Village as a Metaphor," the boy is entirely subjected to the power of the female realm. Doing chores, helping his mother, he is under his mother's wing, in the female side of the world. By entrusting him with a "mission," his father shows him the symbolic link that ties together the entirety of human culture, its ups and downs, its corporeal and spiritual nature. Either avoiding women, of whom he is both afraid and attracted to, falling under the power of his aunt, or discovering what sin is, in the end the boy gets to know the other world and returns home changed, different from what he was.

In this way, the village unfolds as a metaphorical cultural space organized around the male-female axis. This journey through the village landscape (which represents the space of culture) at the same time not only means the emergence of a teenager but also becomes the time-space of the birth of language—his own language. From the start, language flows through the protagonist as an unconscious and inadequate stream of consciousness—the stream of the other, social-cultural unconscious, according to Jacques Lacan. The boy is not yet able to decipher the symbolism of this unconscious. He sees only the material world, which is familiar and known to him: the bucket, the calf, the fence, the bottle, that which he feels with his hand, and so forth. Hence his internal language is a foreign stream of consciousness, which would more naturally be at home in Joyce or Faulkner. In the end he begins to talk in his own voice, and about himself, in language that is not merely borrowed from the dictionary of culture. Art hunts down and overcomes childish fear, says Medvid in his essay "Empire Ludens," and differently rewrites the Ukrainian demotic classics with its Pylypkos, Fed′kos-scapegraces, and socialist realist "Taras's ways."[38]

---

37 Viacheslav Medvid′, "Selo iak metafora," in Viacheslav Medvid′, *Desiat′ ukraïns′kykh prozaïkiv* (Kyiv: Rok kard, 1995), 82.
38 Viacheslav Medvid′, "Imperiia ludens," *Kurier Kryvbasu* 119–21 (1999): 16.

Although theoretically and declaratively the neo-modernist tradition in Ukrainian literature at the end of the twentieth century (realized in the pose of Valery Shevchuk, Viacheslav Medvid, Yevhen Pashkovsky, and in the poetry of Vasyl Herasymyuk, Ihor Rymaruk, and Oksana Zabuzhko) is opposed to postmodernism, and postmodernists, in their turn, associate the neo-modernists with a traditionalist rustic syndrome, the true situation is much more complicated. Neo-modernism exists not apart from the postmodernist mainstream but in relation to it,[39] because the powerful force field of postmodernism, pulling in authors of various trends and styles (regardless of whether they recognize themselves as postmodernists or not), influences their poetics and feeds their imagery with quotations, irony, and intermezzos.

---

39 Volodymyr Yeshkilev also points to this in the glossary to Ieshkiliev and Andrukhovych, eds., *Pleroma. Mala ukraïns′ka entsyklopediia aktual′noï literatury*, 82.

CHAPTER 12

The "Ex-Centricity" of the Great Character

Postmodernism rejects the logic identity and replaces it with paralogy and plurality. By appealing to the other (that is, to the cultural ethno-national multiplicity of the world, to class social, and sexual difference, to alternative modes of cognition, as well as to various styles) postmodernism insinuates that the world is chaotic and illogical. And this is not about transforming it into a universally organized unity and diluting the plurality in some general and homogeneous totality. It is about preserving differences and combining them in a nonaggressive manner.

It means that the postmodernist character does not want to defeat, appropriate, or colonize the other with its actions, gaze, or existence. Rather, it wishes to demonstrate, and play up, the difference between itself and the other. And there, various levels of the postmodernist mindset are possible: from sophisticated, ironic games to manipulations of the other's consciousness, although the inadmissibility of the mutual reduction and autonomy of each individual consciousness is obvious.

A peculiarity of the postmodernist situation in Ukraine is the fact that in the post-totalitarian situation the encounter with the other turns out to be a very traumatic event. This is the reason for the presence of the radical avant-garde shade of the reevaluation of values and resentment. In the general cultural aspect, Ukrainian postmodernist reflection turns into revision and becomes an attempt to inscribe its own history into a given text (narrative) of Western culture. In relation to their own national classic and contemporary authors, Ukrainian postmodernists are not devoid of offensiveness and a certain revanchism.

The postmodern subject in the works of Ukrainian authors is not lacking in trauma either, originating in modern times in war and reconstruction and in the post-carnival and postapocalyptic madness of the world. For example, Zhadan's homeless poet-trickster from the collection *Big Mac* and in the novel *Depeche Mode* in his encounter with the other world is inflicted with the wound of fatherlessness. Andrukhovych's and Zabuzhko's characters survive the loss of love and even the death of a beloved person. Izdryk's characters suffer from schizophrenic depersonalization. Taras Prokhasko's characters live in a time without plots and without cause and effect, and reside in a hermeneutics of niches. All of them are barely surviving after the stormy 1990s; they see the weakness of their parents, and feeling an existential rootlessness they create their own method of self-defense—by finding people like themselves. Parents, especially, catch it. "What's happening to all of them?" speculates Baz in *Depeche Mode*:

> They also obviously began as decent cheerful inhabitants of our towns and villages, they obviously liked this life, they couldn't at first have been the depressing jerks that they have become now at the age of fifty to sixty. If so, where are the roots of their personal great depression, what are its causes? Obviously, the cause is sex, or Soviet rule—personally I can't think of another explanation.[1]

The broken family, the stepfather instead of father, who, in addition, dies, the lost connection with the brother, the mother and the troubled relationship with her—all of this precludes the post-totalitarian subject from a desire to mature and provokes contempt toward the world of adults.

Zhadan's protagonist is a half-orphan like this. In addition, he appears to be also homeless: he lacks both parents and a family home, which ought to provide a shelter for him. As a result, he becomes a migrant, a person with no permanent dwelling place, and he enjoys this. The character is an eternal tramp, who easily associates himself with any vagabond, revolutionary, immigrant Turk, that he meets on the road. At a loose end, a stranger at the feast of European wellbeing, he wanders through the stations, trains, and hostels of Germany; he meets people similar to himself—outsiders and artists and old and toothless former hippies. He despises the bourgeoisie with their big mobile phones, and the global market; but at the same time he calls for a symbolic murdering of John Lennon and, along with him, faith in the tempestuous sixties, which are

---

1 Serhiy Zhadan, *Depeche Mode*, trans. Myroslav Shkandrij (London: Glagoslav Publications, 2013), 59.

imagined as a time when, at least virtually or emotionally, one could hide from depressing modernity.

This life is like a river flowing against its own current, and just like the protagonist of *Depeche Mode* who is "moving forward through tight hard air," "without any goal, without any desire, without any doubt, without any faith in success."[2] Hence such a postmodern "homeless" can tell a story about himself only through what he denies rather than through what he loves:

> I don't believe in memory, I don't believe in the future, I don't believe in providence, I don't believe in heaven, I don't believe in angels, I don't believe in love, I don't even believe in sex—sex makes you lonely and vulnerable, I don't believe in friends, I don't believe in politics, I don't believe in civilization—but it's better to take things less globally: I don't believe in church, I don't believe in social justice, I don't believe in revolution, I don't believe in marriage, I don't believe in homosexuality, I don't believe in the constitution, I don't believe in the holiness of the pope in Rome; even if someone were to prove to me the holiness of the pope in Rome I wouldn't believe in it, on principle I wouldn't.[3]

Communication with the other for this kind of postmodern individual appears to involve parting from family bonds rather than learning an alternative. Beside Zhadan's asocial protagonist—an "orphan," who very rarely breaks through to the other because he lives in a world deprived of communication (conversation, understanding)—there is also Yuri Andrukhovych's protagonist, who is a bohemian poet and a narcissist who sees the world in his own mirrored reflection through which he wanders as in a maze. The whole world, subjected to constant transformations, is the product of the fancy of these characters. Thus in the novel *Perverzion* other characters become the projection of his phantasms and another's words (quotations): Layla Sheila is "a snake-bodied, winged lamia, a sukkumb called Lilith"; Alborak Gabriel is "a jin with no name, who had stolen from the prophet his horse and appropriated his name"; Paul Oshchyrko is "the spirit of the African forest Dada"; Gaston Dejavu is "Marquis Frog, a thin-legged rattler"; Riesenbock is a faun coming out of goblin forests; Mavropule is both drunk Saturn and the spirit of Bahaf. Indeed, Stakh Perfetsky, Andrukhovych's many-faced and many-named postmodern protagonist, theoretically justifies his

---

2 Ibid., 191–92.
3 Ibid., 190–91.

appearance in the world, where reality does not exist: "the boundless quantity of our versions about it, each one of which is erroneous, but all of them, taken together, are mutually contradictory."[4]

In this sense, selfhood is not a constant identity but merely an empty place where many masks meet and part. The postmodern subject becomes a rhizoid, perfunctory human. It is worth noting that Andrukhovych's character, wandering through space, is not looking for a father, as Zhadan's lyrical subject is, but rather resides in a presymbolic, maternal, syncretic and shapeless world, which easily succumbs to various modifications and stylizations. Such a protagonist is looking for ideal love in a time called "post-loving." In fact, both bohemian and the teenager are immersed in a fluctuating fluidity of being subjected to metamorphoses. The mutable and decentered self is feminine in nature, and these qualities shape postmodern consciousness. Andrukhovych states, playing up the femininity and narcissism of his characters, "This is nothing like 'feminine'—this is nothing but 'narcissisticism.' Therefore I … write about the 'femininity of poets,' precisely them."[5] It is paradoxical, however, that Andrukhovych's bohemian protagonist, who is a particle of the presymbolic, maternal matter that gives him his protean nature, at the same time presents himself as an overman and on his way quite brutally destroys all manifestations of femininity.

The increasing amorphousness and chaos of life itself is reflected in the fluidity of one's experience, its openness to others' experiences, and its general decentering. The protagonist of Izdryk's *Wozzeck* is a quintessence of such a decentering: the split self exists as I, You, and He. I is the thinking subject; You is the corporal image formed by the surface of the body that divides the living from the inanimate; and He is the alienated I-as-other Spirit (God?) that the self perceives from the outside and is the author of the work you yourself are writing. Izdryk's postmodern world is a-grammatical: language does not serve understanding. Instead, it records the separation of people and their suspension in reiterations and regressions. Grammatical forms make possible or, conversely, prevent "the meeting of the first and the second person" in one's self. Language, which ought to be a link between being and time, is falling apart and going to pieces—both graphically and semantically, as in the prayers at the end of "Night" (chapter one) and "Day" (chapter two). Thus, communication with the other is rendered highly complex, because there is a fear of seeing a foreign

---

4   Yuri Andrukhovych, *Perverzion*, trans. and with an introduction by Michael M. Naydan (Evanston: Northwestern University Press 2005), 223.

5   Iurii Andrukhovych, "Vona robyt′ mynule zhyvym i nezavershenym," *Komentar* 2 (2003): 5.

face ("it might prove to be familiar") and danger in showing the self's true face ("it might be noticed"). The world becomes similar to a book because "those who enter this world, this book," says He, "will never find a way out and will wander amidst nonexistent things, among myriad copies of their own 'I' that burn in the fire of incessant annihilation."[6]

The post-Chornobyl world is postapocalyptic. It appeals to the end—not only the end of communication but also the end of the world, simultaneously postponing all ends and luring the postmodernist character into a maze of quotations, discursive practices, and texts/palimpsests. Yet this wandering in the world of vacancies, transgressions, collage-quotations, and traces also points to, for the postmodernist, the thread of a rope—potential salvation—that stretches above the abyss. The Nietzschean buffoon and tightrope walker called state sliding on the surface of things.

By and large, the Ukrainian postmodernist text of culture as a post-Chornobyl text was born out of a traumatic worldview. It would be worth recalling here Mikhail Epstein's theory about the traumatic nature of postmodernism, which, in his opinion, is a sensuous reflection on the enormous pressure of ideology (in the case of socialism) or information (in the case of Western society). To explain this condition, Epstein turns to the perception of atomic explosion. He asks, "What is the most adequate reaction to the explosion of an atomic bomb? Is it detailed observation of it—or the loss of the very ability to see? Is not blindness the most exact testimony of those events which outweigh the capacity of perception and thus leave their trace in the form of traumas and contusions, scars and scratches, that is, signs drawn on the body and therefore read as a testimony?"[7]

Traumatic perception, linked to the impossibility of interpreting reality, generates postmodern sensitivity, in which the real is mixed up with the invented, and the world of signs looks more real than reality itself. Epstein is right when he says that

> postmodernism is a culture of light and swift touches, contrary to modernism, which was characterized by the figure of drilling, deep penetration, and undermining the surface. For this reason, the category of reality,

---

6   Izdryk, *Wozzeck*, trans. and with an introduction by Marko Pavlyshyn (Edmonton: CIUS, 2006), 106.
7   Mikhail Epshtein, "Informatsionnyi vzryv i travma postmoderna," http://old.russ.ru/journal/travmp/98-10-08/epsht.htm.

> similar to any deeper dimension, appears to be rejected—because it presupposes a reality that is different from an image, from a system of signs. Postmodern culture is satisfied with the world of simulacra, traces, and signifiers, and perceives them as they are, not attempting to get to the signified. Everything is perceived as quotation, as a conventionality behind which it is impossible to find any origin, beginning, or descent.[8]

Epstein's point is that for postmodernists reality simply disappears; instead, they are fascinated with the play of signifiers, with the maze of meanings, and with interpretations of quotations and traces. Roots and depths are replaced by a rhizomatic interlacing of traces and significations. Epstein juxtaposes this perception with an informational explosion and an experience of trauma: when there is no adequate reaction to a traumatic phenomenon, reality disappears and the world of fantasy, defending the organism, comes out into the open. Moreover, the imagined, symbolic world molds the real world.

Due to the fact that the informational explosion is augmented by the trauma of Chornobyl, in Ukrainian postmodernism imagery also explodes. Such imagery not only compensates for the gap between imagination and reality and has a surplus, compensatory meaning, it also produces a new picture of the world. In the work of Ukrainian postmodernists there is a large palette of grotesque, monstrous, and sexual images that draw a new field of meanings, new and unusual models of behavior, and new characters. Hence, imagery, language, and signs are primary and alter perception in the post-Chornobyl era. The reality of post-Soviet daily life in postmodernists' texts is recorded through numerous neologisms, in linguistic heterogeneity, and by employing foreign words, idioms, and slang. Since it is impossible to fully interpret the original trauma from its mere trace, there occurs a delay in representation, a gap in communication, and a growth of associations, gaps, and fragments; there is a sliding on the surface and joy in guessing, which leads not to the discovery of the really real but to a labyrinth of past meanings and images.

This representation based on "traces" was achieved by means of stylization and irony—the two most popular modes of postmodern writing in Ukrainian literature. They renewed the value of indirect and unusual language; they brought back a taste for it. It is postmodernist pastiche that facilitates the recognition of the other (as difference in texts, the law, behavioral models, and the author's self), at the same time erasing it. Postmodernist stylization unites two

---

8   Ibid.

texts, two intentions, without trying, as in parody, to reject another's speech, which the author employs, as Julia Kristeva notes, "without running counter to its thought—for his own purposes; he follows its direction while relativizing it."[9]

Irony, in its turn, is also a form of play; it also is a breakthrough to the other, namely, "a stepping out of 'real' life into a temporary sphere of activity with a disposition all of its own,"[10] as Johan Huizinga stated in his classic work on play *Homo Ludens* (1938). In accordance with the modernist era, in which he lived, Huizinga noted that in play activity, there exists a "spatial separation from ordinary life";[11] and that play, exactly at the heart of said closed space, governed by its own laws, occurs. One interesting paradox in the deployment of postmodernist theory is worth our attention. The discovery of ideologically postmodernist works, especially the scholarship of Huizinga, becomes the basis for the legalization of the postmodernist mindset at the end of the twentieth century. However, postmodernism itself, on the contrary, underlines the simultaneity of existence of fictional and real worlds. Postmodernism is oriented not toward the unity of reality but toward the constant production of a plurality of worlds. This is aided by the ludic interaction of levels and metalevels which generates situations that may be comic (jocular) or pre-comic (ironic),[12] but which are never immutable and absolute.

Thus, the discovery of the other has become one of the most important signs of the postmodernist situation in Ukraine: other languages, identities, sexes, nations, and behavioral models, all irreducible to existing ones, and, in consequence, a desire for a plurality that precipitates the ruin of the socialist realist narration that advocates the monologic.

It is significant that the destruction of such a monologism is linked to the demise of the concept of the great hero. In modern Ukrainian literature, the postmodernist ironic playfulness and stylization are often a shelter for characters that primarily have a marginal social status and do not keep within the framework of the great hero. Among these outsiders are bohemian superhero, a poet-alcoholic, a schizophrenic lover, a self-taught philosopher, a sportsman, and a narcissistic intellectual woman. Their epistemological position lies in

---

9   Julia Kristeva, "Word, Dialogue, and Novel," in *The Kristeva Reader*, ed. Toril Moi (New York: Columbia University Press, 1986), 44.
10 Johan Huizinga, *Homo Ludens: A Study of the Play-Element in Culture* (London: Routledge, 2002), 12.
11 Ibid.,18.
12 Patrick O'Neil, *The Comedy of Entropy: Humour, Narrative, Reading* (Toronto: University of Toronto Press, 1990), 83.

epistemological inquiry about the sense of the world, and their unbearable lightness of being is felt in the way they break the limits of the real world, often escaping into a verbal play which is the space of freedom. Such inquiry for such characters lies in changing narrative masks and linguistic invention—two ways in which they construct their own world. This invention consists of simultaneously exposing and concealing the self—that is, offering a series of manifestations that the characters themselves can believe in or that can show the reader that I is an other. Hence, in spite of the growing subjectivity of the postmodern character, he or she constantly tries to mystify, escape, and conceal identity. Language here helps these characters: they believe that they can constantly recreate language, but in reality it is language that recreates them.

The very appeal to the other in search of one's own identity brings postmodernist authors closer to each other. In the post-Chornobyl text, the openness with respect to the other reveals itself in de-heroicizing and de-romanticizing artistic expression, through stylizing the "already-said," and by the "naive" experiment with possible and probable realities. From now on, the postmodern individual puts on the mask of the superhero and embodies the nonhierarchical and fragmented consciousness of the outsider character.

CHAPTER 13

Postmodernism and the "Cultural Organic"

From the start, the first discussions about the fate of Ukrainian postmodernism revolved around a central theme—a theme which has actually been the same for the past two centuries, under various guises, and which has aimed at determining the content of Ukraine's project of modernity. This theme is the preservation of the national-cultural tradition as well as the feasibility and, more importantly, boundaries of its transformation. At the end of the twentieth century this opposition takes the form of the polarity of "organic" and "inorganic" culture, in which the first is associated with the national tradition while the second is associated with the postmodernist elements of the West. Decanonization, which gains in strength in the post-totalitarian period, splits into ideological-didactic and performative actions. The first action is implemented primarily within the framework of academic scholarship and refers to filling in so-called blank spots"; it also presupposes a certain rotation within the canon—supplanting one author with another in the secondary school and university curricula. The second action, most clearly expressed in Volodymyr Yeshkilev and Yuri Andrukhovych's *Pleroma 3'98. Mala ukraïns′ka entsyklopediia aktual′noï literatury*, is oriented toward a canon-simulacrum, which is created first of all from the position of the relevance of literature (belonging to modern movements, friendships, and milieus).

Quite a few critics even in the post-Soviet era see literature as a means of education and ideology. In this context, postmodernism is perceived as mass culture, which threatens national unity and undermines the didactic function of literature. At the same time, the perception of the nature and function of mass culture firmly rests on the old Marxist principle of the "cultural industry." This principle, as a matter of fact, was formed as early as in the first part of the twentieth century; and opponents of postmodernism presented the ideas of the

Frankfurt School of philosophy (the leader of which was Theodor Adorno) in a vulgarized form. The paradox here lies in the fact that in the course of the twentieth century seemingly antagonist aesthetic and cultural trends—socialist realism and high modernism—both reject mass culture and perceive it as nonart, or kitsch. For critics of a modernist orientation everything that does not belong to the sphere of high culture has no value and therefore is disqualified as mass culture. For advocates of socialist realism, mass culture symbolizes bourgeois art. For critics of a nationalist orientation, mass culture puts the existence of unified national culture in danger, although Benedict Anderson clearly demonstrates that it is mass ritual, such as newspaper reading ("the almost precisely simultaneous consumption ['imagining'] of the newspaper-as-fiction," replicated by thousands and millions of others) that creates "that remarkable confidence of community in anonymity which is the hallmark of modern nations."[1]

However, quite a few modern Ukrainian critics are essentialist and think that mass culture is the embodiment of amoral and low artistic values which threaten the preservation and development of the organic model purportedly lying at the foundation of Ukrainian national culture. Olena Logvynenko, a leading *Literaturna Ukraina* (*Literary Ukraine*) commentator, the official organ of the Union of Writers of Ukraine, recognizes the mission of Ukrainian culture to train the "intellectual-cognitive capacities" of the reader and states that the main shortcoming of mass literature is that it is "devoid of the high mission of the authorial responsibility to the people and its fate."[2] Logvynenko, evidently, would not agree with Harold Bloom, who writes that the canon can be anything but program of social salvation because the reading of Homer, Dante, Shakespeare, and Tolstoy does not make us better citizens.[3]

Oxana Pachlovska, in her turn, states that it was the sixties who created the "genetic code of the new Ukraine," having sublimated the "authentic features of the egalitarian and Europocentric nature of modern Ukrainian culture."[4] For her, Ukrainian culture is the "ethical, critical, and aesthetic consciousness of society," and only the "imbalance of the cultural dynamic in Ukraine during the

---

1 Benedict Anderson, *Imagined Communities: Reflections of the Origin and Spread of Nationalism* (London: Verso, 1983), 39–40.
2 Olena Lohvynenko, "Shcho za fasadom 'uspishnoho pys'mennyka': spozhyvats'kyi ehoïzm chy osobysta vidpovidal'nist'," *Literaturna Ukraïna* 8, no. 4953 (February 28, 2002): 3.
3 Harold Bloom, *The Western Canon: The Books and School of the Ages* (New York: Riverhead Books, 1994), 15–16.
4 Oksana Pakhliovs'ka, "Ukraïns'ki shistdesiatnyky: filosofiia buntu," *Suchasnist'* 4 (2002): 65.

twentieth century" is a block to "'experiencing' literature as an organic given and as unity."[5] In this respect, her position echoes Ivan Dziuba's warning about the danger in the current situation of losing the so-called national and historical specificity of Ukrainian culture, ruining its "naturalness of culture."[6]

Ihor Kravchenko even talks about the necessity of controlling literature by filtering and regulating it in order to prevent the "erasure of the difference between literature and its imitations" and, more importantly, to escape the superimposition of foreign canons upon the natural culture.[7] He perceives mass culture as a "subversive means" with regard to the Ukrainian mentality, while postmodernism means for him literature as "something entirely self-sufficient, which is limited to itself, does not need any reader, and, it seems, even an author—both are replaced by the notions of discourse and text."[8]

Something akin to religious criticism emerges. Such criticism is directed at the evil within contemporary literature—postmodernism, that is, which is unambiguously associated with mass culture. The majority of Ukrainian critics emphasize the danger and foreignness of the phenomenon of postmodernism. For Serhiy Kvit, for instance, "these -isms and posts-" are features of the current "time of turmoil" and its art of "guilty conscience," which bring an "unnecessary fantasy, contaminating language and, along with it, national thinking."[9] All of this "ruins Ukrainian culture."[10] Probably the most trenchant accuser of postmodernism, Oleksandr Iarovy, equates postmodernism with Stalinism and fascism. For him, it is a "double lie," "the logic of the full stomach," anti-Word and anti-literature.[11] "The destruction of crosses in 1917, the attack on Shevchenko, and the imposition of postmodernism"—"all of these are related phenomena for me," says Iarovy; and he sees the future of art in the renewal of the sacred "Word, which will not teeter on the line between imitation of the archaic and kneeling before the decaying 'art' of the declining surrounding civilization."[12]

---

5 Ibid., 83.
6 Ivan Dziuba, "Metod—tse nasampered rozuminnia," *Literaturna Ukraïna* 3 (January 25, 2001): 3.
7 Ihor Kravchenko, "Priorytet krytyky," *Literaturna Ukraïna* 2 (4899) (January 18, 2001): 3.
8 Ihor Kravchenko, "Chas diiaty (Literatura iak haluz′ ukraïns′koï kul′tury)," *Dnipro* 7–8 (2001): 92.
9 Serhii Kvit, "U mezhakh, poza mezhamy i na mezhi," *Slovo i chas* 3 (1999): 63–64.
10 Ibid.
11 Oleksandr Iarovyi, "Skazhu, iak ie," *Literaturna Ukraïna* 8 (4905) (March 1, 2001): 7.
12 Oleksandr Iarovyi, "Lyst samomu sobi. Kviten′ 19, 2001," *Literaturna Ukraïna* 15, no. 1412 (April 20, 2001): 3.

While critics summarize mass literature as a the "light, detective story's excitation of readers' nerves and erotic stimulation,"[13] the representatives of mass literature themselves, such as the Kapranov brothers, are unhappy that modern Ukrainian literature resembles a fan club of authors writing for one another, reading one another, and giving awards to each other. The Kapranov brothers, along with Maryna and Serhii Diachenko, are oriented toward blockbuster literature, but they do not want to be associated with postmodernists because, as the Diachenkos say, postmodernism with its aspirations for serious literature in reality is a narrative "about nothing."

In their turn, the authors who comprised the Bu-Ba-Bu group and were definitely more interested in postmodernism than others, clearly demonstrated the transformation of postmodernism in the Ukrainian context. For instance, after 2000 Yuri Andrukhovych, on the one hand, experiments with the techniques of Pop Art while with Serhiy Zhadan and Andriy Bondar (the Poetic concert *Maskult* [Kyiv, 2003]) and practices the poetry of "quoted phrases" ("chuzhi frazy")—the slogans and clichés of mass culture. On the other hand, Andrukhovych tries to analyze the limits of his own postmodernist writing in his novel *Twelve Circles* (2003).

Today, postmodernism seems to be something from the past—a phenomenon that seems to be a function of the cultural, spiritual, emotional, and psychological context of the end of the twentieth century; it appears tinged with the eschatological bent of the time and dissolves into the period. Hence postmodernism, which had grown from a tendency to relativize time, space, and history, is itself perceived now as a historical phenomenon. And it does not matter, as critic-postmodernists themselves said, that postmodernism is itself utterly relative, transient, and devoid of ontological substance. "It is local and fragile and incorporates paradox," as Hans Bertens stated."[14]

It seems that Ukrainian postmodernism, even though it has become a nationwide phenomenon and caused the decentering of the cultural and mental space of the whole of Ukraine's culture, has not turned all literature into the field of a play with reality and has not instilled a new sense of identity based on interaction. As a consequence, it is perceived as a wavy revolution, or, more precisely, a revision of the national culture. Having generated the post-totalitarian "ludic man" and carnivalesque, it has brought a festive mood and has shown how to change the codes of the game itself. It reopened the old, unhealed wounds

---

13 Orest Slyvyns′kyi, "Zapiznila myt′ prozrinnia," *Vitchyzna* 5–6 (2001): 81.
14 Hans Bertens, "The Postmodern Weltanschauung," 51.

of Ukrainian cultural self-awareness, especially the question of the modernity or non-modernity of Ukrainian literature, and deepened the gap between the center and the margin as well the canonical and anti-canonical.

However, postmodernism legalized what can be called a de-totalized, fragmented, and pluralized thinking. It immersed us in the element of the transformation of society, the individual, and, indeed, the very cultural field in which high and popular culture are not just different but codependent. Thus the question of mass culture becomes, after postmodernism, the most important.

Although debates about whether postmodernism is (was) present in Ukraine will not fade, perhaps for a long time, their focus is new. Organically or inorganically, postmodernism came to Ukraine. What is more interesting is that postmodernism is superimposed on the post-totalitarian period of the development of Ukrainian literature—that is, upon the period of change of the paradigms of artistic thinking, modes of writing, and cultural ideologemes. It affects the character of the reception, formation, and "working-over" of the situation of modernity in the last decade of the twentieth century.[15]

---

15 N. Starikova, ed., *Postmodernizm v slavianskikh literaturakh* (Moscow: Institut slavianovedeniia, 2004), 3.

CHAPTER 14

Postmodernism as Ironic Behavior

Postmodernism in Ukrainian literature is remarkable for its forms of stylization, ironic play, and masquerade. First of all, however, it is a new linguistic behavior. To use a statement by one of the theoreticians of American philosophical postmodernism, Richard Rorty, language games, which embody rebellion and undermine socio-cultural stereotypes and clichés by dint of argot, slang, and metaphorical language, can be regarded as a model of linguistic behavior.[1] The time of the ironist comes. The very presence of ironic linguistic behavior gives a further opportunity to develop not only verbal but also socio-cultural forms of behavior, such as public meetings, performances, and protest actions. Hence postmodernism in Ukraine, from the very beginning, undermines totalitarian culture in general and colonial Ukrainian culture in particular. This undermining is implemented first of all linguistically: through the invention of a new language that is alive, easy to understand, and at the same time adequate enough to enable one's self-expression. People who assume ironic postures—and the first postmodernists were ironists—are aware that the final vocabularies we usually use are in reality transient, created socially and culturally, and therefore self-contradictory and fragile; they can be changed, and thereby the ideological and cultural meanings they contain can also be altered.

Language games, along with the polyphony (heteroglossia) of languages, discourses, linguistic hybrids, and marginal vocabularies, become a hallmark of Ukrainian postmodernism. Yuri Andrukhovych's tautological phrase, Yevhen Pashkovsky's phrase-octopus, Bohdan Zholdak's patois, Oksana Zabuzhko's lyrically "suspended" phrase (identical with the held breath)—those are

---

1   Richard Rorty, *Contingency, Irony, and Solidarity* (Cambridge: Cambridge University Press, 1989), 9.

only a few manifestations of the linguistic invasion of the world found among Ukrainian author-postmodernists. Together with the new public behavior of postmodernist writers, their masks and images, along with the mass forms of presentation, such linguistic behavior signified a change in the entire cultural situation: a new literary formation was being born.

A particular variant of the carnivalesque—the post-Chornobyl carnivalesque[2]—becomes a form of postmodern consciousness in Ukraine. As Yuri Andrukhovych testifies, "carnival combines the most contradictory things, juggles with hierarchical values, sets the world upside down, and challenges the most sacred ideas in order to save them from petrification and ossification. Carnival is a war on Death."[3] In this way, the freedom of self-expression coincides with the search for a new communicative space—friendship, a shared field of artistic creation, and a new history, united by carnival. Thus carnival becomes a global metaphor for early Ukrainian postmodernism.

Although Boris Groys notes that "the quotation game, 'polystylisticity,' nostalgia, irony, or 'the carnivalesque,' on their own are not a postmodernist strategy in that the very context of such a strategy remains undetermined, thereby excluding any appropriation of the Other,"[4] the carnivalesque is the modality that is most in accord with post-totalitarian culture in Ukraine. Carnival brought an appropriation of the hitherto alien into Ukrainian literary zones and spheres, in particular those that had been marginal, foreign, or reserved exclusively for mass culture. Sexuality, obscene words, irony toward traditional teachings, a juxtaposition of formerly incomparable objects, and the eclecticism of various metamorphoses are from here on not as much a sign of the cynical rejection of rules and the manifestation of the total relativism as they are ways of mastering a new perception of the world, defined by essence of the post-totalitarian individual. Indeed, as early as in the late 1980s it was important to win a space of personal freedom and to create a full-fledged, rather than reduced, historical, social, and national context of being. It was even important to resort to a game with possible and alternative stories, analyzing mental states and the psychology of those who do not want to be great heroes and

---

2   More about this in Tamara Hundorova, "'Bu-Ba-Bu,' Karnaval i Kich," *Krytyka* 7–8, nos. 33–34 (2000): 13–18.
3   Iurii Andrukhovych, "Apologiia blazenady (Dvanadtsiat' tez do sebe samoho)," in *"Bu-Ba-Bu" (Iurii Andrukhovych, Oleksandr Irvanets´, Viktor Neborak): Vybrani tvory*, 23–24.
4   Boris Groys, "A Style and a Half: Socialist Realism between Modernism and Postmodernism," in *Socialist Realism without Shores*, ed. Thomas Lahusen and Evgeny Dobrenko (Durham: Duke University Press, 1997), 82.

associate themselves with bohemia or with the "gallery of marginalized noblemen or noble outsiders," as Viktor Neborak put it. The winning of autonomy is achieved through language games, collecting the aestheticized remnants of past stories, and keeping up with everyday gossip and socialist realist ideologemes. Corporeality here plays a special role—the postmodern individual is first and foremost corporeal, erotic, and sensual. The body itself undergoes constant changes and metamorphoses, while consciousness lives in various worlds simultaneously, wandering between reality and fantasy.

Carnivalesque ideology is equivocal and multifunctional. On the one hand, postmodern carnival serves as a form of self-assertion for the new generation of Ukrainian authors and is associated with the child, play, eccentricity, mass culture, and familiarity. On the other hand, Dionysian carnival frightens through its buffoonery, breach of the sacred space of national culture, and demonic straining of humor and apocalyptic limits. It opens the door to the rhetorical apocalypse of quotations and repetitions, and even threatens to turn into post-totalitarian kitsch. This tendency of postmodernism to be built on kitsch was first confirmed in the work of the "Bubabists"; later, Volodymyr Tsybulko, experimenting with political and social art in his collection *Mein Kife* (2000), pushed it to its grotesque end.

Ukrainian modernism, however, cannot be considered a merely destructive phenomenon. It contained huge constructive potential. Postmodern carnival ushered in a polymorphic individual, fragmented into various masks, as well as collections of memories and quotations. However, this fragmentariness is not a goal in itself; rather, its aim is, through local micro-narratives, to restore the memories of origins, family chronicles, and private history that are denied the totalitarian individual. The autonomous, subjective worldview undergoes a renewal, and individual consciousness overcomes the power of ideologemes. Such liberation occurs first of all verbally—in ironic linguistic behavior, which materializes in neologisms, linguistic repetition, quotation, the use of foreign languages and words, and a hybridity of names. Language becomes a field of individual freedom. However, character thereby loses its totality and integrity, turns into a fiction, a mask, and a polyphony of various linguistic roles. A character-adventurer, a traveler who goes through various worlds and experiences various adventures in search of his or her own identity, is endowed with special potential.

The fluidity of masks and identities, and the mixing of physics and metaphysics, reality and fantasy, mark a new field in which the postmodern character manifests itself. Language, the body, and space are here heteronomous, at numerous topographical levels, and have various qualities. All of this has an existential meaning: the post-totalitarian and postcolonial Ukrainian subject self-affirms on various levels and in various forms. As a postmodernist ironist, he craves autonomy in order to "get out from under inherited contingencies and make his own contingencies, get out from under an old final vocabulary and fashion one which will be all his own."[5]

In essence, the carnivalesque consciousness of such a postmodern subject is self-parodied and ironic. For instance, orphic and other esoteric mythologemes or psychoanalytic archetypes in the works of Andrukhovych, Zabuzhko, Izdryk, and Pashkovsky point to a deep structure conditioned by cultural history and myth. Further, praxis based on language games refers not to depths and origins but to fluid, dynamic, and synchronous modernity. This is why mythogenic plots from the cultural-historical treasury (for example, from the *opera buffa* in the form of *pasticcio Orpheus in Venice*, in Andrukhovych's *Perverzion*) are mixed up and imitated in order to become a mirror-shield, in which, plot notwithstanding, the modern history of the protagonist is again recreated and reinterpreted. On the whole, Ukrainian postmodernism is not only a post-Chornobyl rhetorical apocalypse of stylization and quotation but also a traumatic experience of post-totalitarian culture, tinged with madness, aggression, love deprivation, loss of identity, the death of the symbolic father who made the Soviet system meaningful, and the dismemberment of the maternal body of the USSR, which everybody was persistently taught to call the "fatherland."

Ukrainian postmodernism has its own topography and map. It legitimizes nationality, manifests regionalism, bears witness to the generation gap, sketches the vectors of repulsion and gravitation, and introduces literary zones and groupings. Such a heterotopical cultural topography encompasses the eccentric Bubabist play with past cultural codes and signs of pop culture, the neo-baroque stylization and adventures of Andrukhovych's characters, the polymorphous decelerated prose of Prokhasko and Izdryk, the teenage bitter loneliness along with quotations of the other's experience in the poetry of Serhiy Zhadan, and the hermetic meditation-collages of Ivan Andrusiak. Not just various stylistic designs but also generations have their meaning on this map. Looking at the literary landscape through the principle of generations is an aim of Volodymyr Danylenko, who describes the generations of the eighties, nineties, and the first generation of the twenty-first century.[6]

---

5 Rorty, *Contingency, Irony, and Solidarity*, 97.
6 Volodymyr Danylenko, *Lisoryb u pustyni. Pysmennyk i literaturnyi protses* (Kyiv: Akademknyha, 2008), 119–34.

Postmodernism in Ukraine is also a geo-cultural project, which reveals the sense of Ukraine's existence both on the margins of Europe and its relation to the imperial metropolis, Moscow. It also functions as a cultural reflection that critically reevaluates the half-colonial national culture and lays bare its tabooed places and topoi: urbanism, sexuality, schizophrenia, alcoholism, bohemia, and teenage homelessness.

As an artistic practice, it is multilingual, public, and a mix of high and low forms of literature. In Andrukhovych, Zabuzhko, and Izdryk, it draws on mass literature, such as detective stories and romances, and uses it to explore high cultural themes and intellectual conflicts. Viktor Neborak even formulated a new ironic canon for postmodernist Bubabist writing: "it is worthwhile to write succinctly, with a developed plot, and without an excess didacticism. Characters should be portrayed in the heyday of their vitality—young, erotic, adventurous, and dynamic. Characters should travel, get in traps, and break free due to their talents and various unexpected occasions."[7] A special role is reserved for "unexpected occasions" or devices, among which are "all kinds of exquisite paintings, verbal suggestions, illusions, toreador fights with stereotypes that are fed on native pastures, self-irony, the trying of all sorts of forbidden fruits with splashes of teasing toward bureaucratic instructions and crowned classics."[8]

In all, a significant consequence of Ukrainian postmodernism is a restructuring of the national literature and the development of mass literature. The competition "Word Coronation" plays a vital role in supporting Ukrainian mass literature. At the end of the twentieth century, on the shelves of bookstores, one could find romances and a detective stories, fantasy fiction and erotica, novels for young adults and gothic tales. It was a significant achievement for Ukrainian postmodernism, which became a laboratory for intellectual reflection and formal experiment—the post-Chornobyl library indeed.

---

7 Neborak, "Iurko i Sashko, Sachko i Iurko," 43.
8 Ibid.

Part Three
THE POSTMODERN CARNIVAL

CHAPTER 15

Bu-Ba-Bu: A New Literary Formation

"In the year of our Lord before-two-thousand, this is literature finally, with its own philosophy (yes!), worldview (yes!), stylistics (yes!), smile (yes!), and fright (yes!). A separate branch of literature and literary tradition, which in the old style we can call Ho-Hei-Go—Hoffmann, Heine, and Gogol"[1]—thus George Shevelov defined the role of the Bu-Ba-Bu group—the most outstanding manifestation of carnivalesque Ukrainian postmodernism.

The Bu-Ba-Bu group appeared in 1985 and signaled a cultural tension between Lviv and Kyiv. The authors comprising the group lived in Lviv and Kyiv, came from Ivano-Frankivsk, Rivne, and Lviv; they arranged "poetic-friendly get-togethers" in various "closed lodgments,"[2] as one of them wrote ironically—apartments, workshops, and attics; and their first public presentation took place on 22 December 1987, in Kyiv Young Theater. For the group's members—Yuri Andrukhovych, Viktor Neborak, and Oleksandr Irvanets—the name referred to Burlesque-Slapstick-Comedy ("Balahan"). Buffoonery was to become a "radical way to overcome depression, an antidepressant" (Neborak) due to the situation in contemporary Ukrainian literature.[3] The group was born out of the friendship and youth of these three men of letters ("it was a mutual love and a mutual understanding");[4] that is, it was a bohemian phenomenon, but most important was that it changed the rules of the game. "Simply put, we turned the game into poetry," writes Irvanets.[5]

---

1 Iurii Sherekh-Sheveliov, "Ho-Hay-Ho. Pro prozu Iuriia Andrukhovycha i z pryvodu," in Iurii Andrukhovych, *Rekreatsiï. Romany* (Kyiv: Chas, 1997), 266.
2 "Iak vono bulo …," in *"Bu-Ba-Bu" (Iurii Andrukhovych, Oleksandr Irvanets´, Viktor Neborak): Vybrani tvory*, 17.
3 Ibid., 18.
4 Ibid., 19.
5 Ibid., 18

Andrukhovych marked the milestones of Bu-Ba-Bu history: "Our association became for us nothing less than a way to survive in the second half of the eighties. At the beginning of the nineties, we seriously considered erecting a monument to ourselves. On 9 May 1994, we were exactly one hundred years old, and so youth had passed."[6] Viktor Neborak announced the advent of the Bu-Ba-Bu era as follows:

> BOO TO DIFECAMBS[7] BOO TO TABOO
> WE'LL GET A DENTURE MADE BY BUBABU
> WE DON'T GIVE A SHIT ABOUT THIS SOVIET BOO
> LET'S DRAW A FEMA LE NU AND BLUE!!![8]

Neborak also emphasized the supra-individuality of Bu-Ba-Bu and started to talk about the "great BUBABU" as a "style of life, characteristic of modern individual, in particular, Ukrainian individual."[9] In fact, it was about the situation of the post-totalitarian existence—"the environment that generated us, in which we exist, and which we fertilize with ourselves, that is, saturate."[10] Thus BUBABU has become an emblem of transitory existence, where

> Our life—BUBABU.
> Our history—BUBABU.
> Our politics—BUBABU.
> Our economy—BUBABU.
> Our denominations—BUBABU.
> Our art—BUBABU.
> Our country—BUBABU.

At the beginning of the 1990s, a sign of the postmodern situation was carnivalesque-spectacular genres, when poets became actors and singers and the music festival Chervona Ruta and the Festival of Modern Poetry seemed almost homogenous. It was novel, states Viktor Neborak, that poets went on

---

6   Iurii Andrukhovych, "Avtobiohrafiia," in *Rekreatsiï*, 31.
7   "DYFIKAMB" in the original—apparently, a hybrid of "dithyramb" and "fecal" (translator's note).
8   Neborak, *Litostroton*, 322.
9   Viktor Neborak, "Dekilka utochnen′ z pryvodu napysannia zvukospoluchennia [bubabu]," in *"Bu-Ba-Bu" (Iurii Andrukhovych, Oleksandr Irvanets′, Viktor Neborak): Vybrani tvory*, 26.
10 Ibid.

stage and began to communicate freely with their spectators, and poetry itself "as a purely internal lyrical phenomenon" got "in contact with outside reality."[11] The atmosphere of a linguistic ironism was forming, in which parody found its object in the official vocabulary, existing as a populist, educational, and Soviet hybrid. What is more, the discourse of totalitarian society is ironically criticized at that time. Verbal games record the inadequacy of existing language for the expression of individual meanings and feelings, while word repetitions, expletives, and the quotation of quotations embody total reflectivity and the tautological quality of speech in a post-totalitarian society. As Irvanets writes, "bubabu is a muttering of an imbecile of a post-Chornobyl paradise on earth."[12]

The rhythmic melody of a phrase, together with its terrifying shifts, sounds almost self-sufficient, while the openness of a phrase to which one can arbitrarily add words based on the principle of assonance and association gives the impression of an automatic language. It turns upside down the totalized, homogenized style of socialist realism. Words are interchangeable, devaluing their meaning; therefore, it is performativity, manifested in ironic language games, that embodies the freedom of self-expression of the speaking subject in the post-totalitarian era. Even the boring ironic rewritings look like vital incantation rather than vulgar repetition.

Ukrainian postmodernism in its Bubabist shape emerges as life adapted to the stage and is a polymorphic phenomenon: it is literary but at the same time national, linguistic, social, conscious of gender, and political. Besides its transformation of language, Bubabism depersonalizes the self, fragmenting it into personas and creating a supra-self—many-sided and multidimensional. The originality of this early form of postmodernism is that it not only presents itself as an artistic practice and aspires to introduce a new artistic style—Bu-Ba-Bu— but it is also a form of cultural critique which reformulates the national cultural paradigm, first of with regard to mass culture. Along with this, the codes of the national culture significantly change—for example, the image of the national poet and his or her role in culture is transformed. After all, in the Ukrainian populist version of the nineteenth century and later in socialist realism, the romantic myth of the national poet gradually loses its sacred character and acquires didactic features, turning the poet into a teacher and an ideologist.

---

11 Viktor Neborak, "Z vysoty Litaiuchoï Holovy, abo Zniaty masku. Rozmova z V. Neborakom," *Suchasnist'* 5 (1994): 57.

12 Oleksandr Irvanets', "Keske bububu?" in *"Bu-Ba-Bu" (Iurii Andrukhovych, Oleksandr Irvanets', Viktor Neborak): Vybrani tvory*, 27.

The generation of the 1980s, liberating themselves from the pressure of totalitarianism, was inclined to ironically undermine not only totalitarian consciousness and language but also the image of a national poet. The object of criticism is the mass image of Ukrainian poet, in particular, the poet as a people's deputy. Scandalizing the Establishment, Viktor Neborak described the Ukrainian man of letters thus: "What is a Ukrainian poet in mass perception? A paunchy fat-face with a people's deputy ID or without it, who is whining about increasing paper prices and the difficulty of publishing a new collection."[13]

The new avant-garde wave that opened literature up was primarily a question of traveling to the Land of Play, but it was also a social action. The point was to ruin the cultural canon: "to write for someone" rather than "to God" ("na Boha"), as Andrukhovych, a master of stylization, aptly remarks. Literature stopped being the logos which underpinned the ontology of national, cultural and political being, and became instead a field of experiment and freedom of self-expression. This self-expression was distinctly personal, and not slave to a foreign version of self. It was impossible to be satisfied with the traditional anymore, for Ukrainian literature's themes—such as "grandpas and grandmas" ("didy z babamy"), "groves and trails" ("haï-plaï"), and the "poor childhood" ("nepovnotsinne dytynstvo"). Furthermore, official language, the party and patriotism, those things that made "characters talk like they're at Communist party meetings or Shevchenko get-togethers," were moribund.[14]

The sacred role of the official poet was undermined by masscult—the name also given to performances by postmodernist authors in the post-Bu-Ba-Bu period,[15] even though masscult had already emerged in the Bubabist period. One feature of this newly born culture is an apology for youth, even for adolescence. The desacralization of the poet-messiah was realized from the position of youth subculture: the bohemian hero—often an alcoholic, a "kitschman" (a creator and consumer of kitsch), a teenager—interested in travelling to the Land of Play (Viktor Neborak). The national literature was tuned into a virtual body and lists-museums of authors and quotes—a museum of wax figures. (It is apt that in the Museum of Literature in Kyiv, a decade later, there was a gallery of wax sculptures.) Literature becomes less public matter than a "private collection"—the title of an anthology of modern prose published in 2002.

---

13 Neborak, "Z vysoty Litaiuchoï Holovy," 62.
14 Neborak, "Iurko i Sashko, Sashko i Iurko," 43.
15 *MASKULT* (Kyiv: Krytyka, 2003).

Postmodernism is generally critical toward the ideology of the organicism of Ukrainian culture. For this reason, Bubabism becomes a world "through the looking glass" of the national tradition. Literature for Bubabists is a virtual thing, which can be reshuffled, appropriated, and supplemented as an alternative reality—for instance, a baroque one. In fact, Ukraine itself is no less virtual. Andrukhovych explains that he is preoccupied with the search for "some *alternative* Ukraine, some 'country of the baroque' ... a frenetic, fantastic Ukraine."[16]

The national culture is perceived not in the light of holy and ideal totality but rather as a grotesque Bakhtinian body, with its bumps and hollows. It "used to direct and lure," states Andrukhovych, "with its unpeopled nooks, uninhabited margins, obsolete taboos, so tempting for a trespasser." The Bu-Ba-Bu's specific postmodernist intention was "to do away with at least some gaps in that torn-into-pieces and rarefied space, whose name is Ukrainian culture." Andrukhovych admits that "it was not we who created something in this culture—the culture created us."[17] The erotization of culture as a feminine (even maternal) body is a main feature of Bubabist cultural revision, in contrast to modernism's interest in breaking the father's symbolic law. Literature is perceived as seduction, as an almost thing incestuous. It lured with "its virginity," Andrukhovych explains, "which so befits a young lady just before her nuptial bloom yet does not suit her in her maturity, before her ageing and fading. Hence we undertook the task of hastening this nuptial bloom."[18]

Among the elements of such postmodern cultural critique is not only the process of de-canonization but also the creation of new canons. Numerous anthologies that appear in the 1990s were attempts to create new canons. Generational anthologies *The Eighties: An Anthology of the New Ukrainian Poetry* [Visimdesiatnyky. Antolohiia novoï ukraïns′koï poeziï] (1990); *Appellative: An Anthology of the Nineties* [Imennyk. Antolohiia devianostykh] (1997); and *The Nineties* [Deviatdesiatnyky] (1998) took on this task. There were also regional anthologies, such as Ivano-Frankivsk's *Pleroma* (1998), personal anthologies (edited by Viacheslav Medvid and Valery Shevchuk, for example, and chronological anthologies like *Good Morning, Millennium!* [Na dobranok, mileniium!] (1999).

Canonization coexists with decentralization. Regionalism, whose most conspicuous manifestation is the acclaimed "Stanislav phenomenon," is revived

---

16 Ola Hnatiuk, "Wirtualna perwersja. Z Jurijem Andruchowyczem rozmawia Ola Hnatiuk," *Dekada Literacka* 19 (1998): 6.
17 Iurii Andrukhovych, "Ave, 'Kraisler'! Poiasnennia ochevydnoho," *Suchasnist′* 5 (1994): 8.
18 Ibid.

in Ukrainian literature of the 1990s. Yuri Andrukhovych, Yurko Izdryk, Taras Prokhasko, Volodymyr Yeshkilev, and Halyna Petrosaniak, belonging to the Lviv-Ivano-Frankivsk school (or *Stanislav* school, to give it its nostalgic tint of the Habsburg empire), are the most important group in terms of regionalism. They declared their fealty to postmodernist discourse and the emphasized visuality in literature. One of the ideologists of the "Stanislav phenomenon," Volodymyr Yeshkilev, defines its genealogy as a hybrid resulting from the superimposition of the "meta-gestalt of the postmodernist discourse," borrowed from the West, on the naturalness of the "cultural provinciality" of Galicia. Indeed, the works of the representatives of this group have a distinct Galician, Carpathian, and Lviv flavor; in addition, they gravitate toward a Central European way of writing.

Beside the general idea of decentralization, which is so significant for postmodernism, the self-proclaimed Stanislav phenomenon aspired to a central place in the literary canon of the end of the twentieth century—irrespective of the emergence, in the course of the 1990s, of numerous get-togethers and groups in Kyiv (such as "New Literature," "The Lost Letter," "Museum Side Street 8," and "Creative Association '500'"), in Kharkiv ("Red Cart" ["Chervona Fira"]), in Ivano-Frankivsk ("New Degeneration"), and in Lviv ["LuHoSad"]).[19] None of these groups was able to achieve such a conceptual self-definition as the so-called "Stanislav phenomenon."

However, on the literary map of the 1990s there was an attempt to juxtapose it with another conceptual platform—the so-called "Zhytomyr school." The Lviv-Stanislav trend is opposed to the Kyiv-Zhytomyr school, represented by Viacheslav Medvid, Yevhen Pashkovsky, Oles Ulianenko, and Volodymyr Danylenko. "The Zhytomyr school is difficult writing, obscure associations, while the Galician one is playful, ironic, and frivolous"[20]—this is how Volodymyr Danylenko defined the governing polarity of the literary situation in the 1990s.

---

19 Comprised of the first two letters of the group members' last names, the word means "meadow-garden" (translator's note).
20 Volodymyr Danylenko, "Zolota zhyla ukraïns'koï prozy," in *Vecheria na dvanadtsiat´ person*, ed. Volodymyr Danylenko (Kyiv: Heneza, 1997), 6.

CHAPTER 16

The Carnivalesque Postmodern

The carnivalesque version of postmodernism juxtaposes variants of urban writing, the auto-thematic, and mass culture with "rustic" and "organic" literature. The carnivalesque games of the beginning of the 1990s were open to the influence of the West and exploited the forms of youth subculture, while the poetic opera *Chrysler Imperial* (1992), which has become the apotheosis of Bubabist performance, emphasized the idea of the advent of the masscult era. After all *Chrysler Imperial* manifested the very idea of carnival. It revealed, as Andrukhovych brilliantly, imprecisely, put it, "the whole nature of Bubabism: its adepts are happy, entirely immersed in the inexplicable joy of being," while carnival itself has been a "great and brilliant failure."[1]

The symbolism of *Chrysler Imperial*, along with the rhetoric that serves it, is hybrid: the relics of Soviet ideology ("the family of peoples," "mind, honor, and conscience"), bohemian-artistic slang, masscult rewritings, baroque melodies along with popular Ukrainian arias, and quotations from Andrukhovych, Irvanets, and Neborak. In practice, the poetic opera has turned into a Bu-Ba-Bu memorial—ironical, bohemian, and burlesque. It combines singing, video, and poetry, foregrounds the carnival of masks ("The Flying Head"), demonstrates favorite Bubabist devices—repetitions of repetitions, the eroticization of ideologemes and cultural topoi, the polyphony of alien voices, and so forth. A new urban kitsch emerges—Bu-Ba-Bu, which echoes the urban tone painting of the famous futurist and hooligan Mykhail Semenko.

The idea of poetic opera was suggested by an automobile Chrysler Imperial, which had stood idle, waiting for its time, for almost half a century somewhere in a garage in Kyiv. As the Bubabists recall, having seen the automobile

---

1   Iurii Andrukhovych, "Ave, 'Kraisler'!," 11, 12.

sometime in the 1980s, they were convinced that the Chrysler Imperial symbolized both a great dream and its absurdity. It had to be imaginatively transformed, however. Andrukhovych notes: "it ought to be a book. It could have been a movie. Or a dance, or rock 'n' roll, or a cake, or a play."[2] The history of the Chrysler Imperial model begins in 1926, and during the twentieth century the vehicle symbolizes power, luxury, and style. It recalls the American mafia of the 1930s, Elvis Presley, and rock-n-roll, and becomes a symbol of fulfilled wishes.

The story of the vehicle becomes a Soviet allegory. This "American monster" turns into a desired phallic object, associated with the absurdity of the Soviets, imperialist chic, the Chornobyl apocalypse, and the apotheosis of happiness. Neborak's lyrical subject proclaims:

> The skyline is ours! Ovation and snot!
> One hundred percent of our dream!
> We are walking as though in Chornobyl,
> But damn it—because we've got you![3]

"Our beautiful brother," "mind, honor, and conscience"—all these mottos, familiar from Soviet times, are transferred, not without irony, to the "Chrysler," which symbolizes the integration of the new world of the market, mass culture, and consumerism. It also embodies an idea of the "second coming." This is a materialization of the deus ex machina, which ushers in a new, Americanized Jerusalem.

Like *Jesus Christ Superstar*, the Bubabist poetic, operatic, puppet show parodied pop culture and at the same time appealed to it. The tokens of the past and the future cultures are demonically hybridized and monstered: Bubabist erotic songs, such as "Babu by!" are solemnly sung by Young Pioneers in yellow-and-blue ties in the high loft of the opera theatre, while *Chrysler Imperial*, as a pop genre, is performed on the stage of the historical opera house in Lviv. Represented on stage are visual clichés of various subcultures: those of the youth, nationalists, Christians, and Satanists, as well as Soviet and the American kitsch. The performance also featured "repetitions of repetitions"—Bubabist images and masks, known from Bubabist works: "the sexy schoolgirl," "the nurse Raia," the demonic Mr. Bazio, the Tamed Classic (William Shakespeare Street), "The Songs of the Eastern Slavs," and so on.

2　Ibid.
3　Neborak, *Litostroton*, 302.

A new—Bubabist—kitsch is being established. The sociocultural ground that allows kitsch to grow also allows replication (of books, things, images) and fashion (which is also based on repetition and begins to dominate over various individual forms of self-expression). The psycho-ideology of kitsch is introduced in order to comfort, satisfy, and enable the collection of impressions that turn into souvenirs and fetishes. There grows a need to eroticize the beautiful and to aestheticize the everyday. Kitsch reveals itself as ostensibly real, and that we, perceiving it, are seduced by it. It is worth noting that the notion of kitsch evolves in the second half of the twentieth century from "bad taste" into the admission that we all are to some extent "kitschmen," or people without whom kitsch cannot exist at all—both consumers and creators of kitsch. *Chrysler Imperial*, then, gathered under roof of the Lviv Opera both the inventors of a new kitsch and its consumers.

This tendency to quote pro-Soviet symbols and official youth mottos, the rhetoric of Soviet holidays, as well as the use of jargon, slang, and self-repetition, defines not only social art but also the semiotics of mass actions in the era of carnival. Literature likewise turns into a scene, performance, and carnivalesque action. Representative and anti-representative at the same time, "the carnivalesque is anti-Christian and anti-rationalist." It is tragic because it is "murderous, cynical, and revolutionary," and it is attractive because it is subject to the power of the unconscious—that is, sex and death.[4]

The images and masks created by Bu-Ba-Bu authors are multifaceted. They are characterized by the deconstruction of the ideal protagonist of the Soviet era, as well as the desublimation of high pathos—the so-called love for Lenin, the Communist Party, the fatherland, and socialism itself. A special role in the de-ideologization of Soviet culture belongs to social art—an emphasis on popular socialist realist mottoes, notions, and images. The creative work of Irvanets gravitated to social art. His "A Deputy's Song," "A Poem to the Mother Tongue," and "The Classics Lessons" exploit motifs familiar from the secondary school and didactic slogans such as "drop by drop to squeeze a slave out of oneself," "a human is born for happiness like a bird for flying," and "guilder-rose language," revealing their artificiality and bias. This rewriting of significant patriotic texts is distressing: for example, the famous banned Volodymyr Sosiura poem "Love Ukraine!," which expressed Ukrainian patriotism in a concentrated and sublime manner. However, in the post-Soviet period the poem became inseparable from the era in which it was born, and its sublimity was associated with

---

4   Kristeva, "Word, Dialogue, and Novel," 50.

the didacticism and absurdity of socialist realist optimism and heroism. For this very reason, Oleksandr Irvanets turns the appeal to "love Ukraine" into an ironic appeal to "love Oklahoma":

> Love Oklahoma! At night and at noon,
>
> As Mom and a dandy exactly.
> Love Indiana. And love no less
> Both South and North Dakota.
> . …. . …. . …. . …. . …. . …. .
> You won't be able to love other states,
> If you cannot love like a brother
> The fields of Arizona and so dear to the heart
> The expanses of Alaska and Nebraska.

By and large, the Bu-Ba-Bu authors create a new, carnivalesque, overman kitsch. Masculine, erotic symbolism becomes an important element of this kitsch. "The very name hints at something salacious, obscene, and ruffian—Bu-Ba-Bu. Exactly thus: "babu by!" ("give me a girl!")," ironizes Andrukhovych.[5] "Maliuite BABU holu Bu" ("Let's draw a fema le nu and blue"),[6] Neborak exclaims in an almost ecstatic and orphic manner. Cult images, created by the Bubabists, became easily recognizable; they offered a new bohemian-narcissistic portrait of both the Ukrainian poet and the Ukrainian as such. Play with Narcissus revealed the stylistics of narcissistic and homosexual autoerotism, with reference to pop stars and rock idols. Khomsky, (hinting at Homsky), from Andrukhovych's *Recreations* openly demonstrates this. Looking at his own image in the mirror, Khomsky states:

> … "and a quick look at myself in the mirror.
>   Just right, Khomsky—a long, loose grey coat, a week's stubble on the chin (Broadway style), hair gathered in a ponytail, sunglasses circa 1965, a hat, just right, the traveler, rock star, poet, and musician Khomsky, Khoma for short, this cool son of a bitch is bestowing upon provincial Chortopil the joy of a visitation by his very own person."[7]

---

5   Andrukhovych, "Ave, 'Kraisler'!," 7.
6   A poetic translation (translator's note).
7   Andrukhovych, *Recreations*, 20.

All Andrukhovych's novels are tautological and auto-fictional. "From one novel to another I simply open myself more and more and still more talk about myself,"[8] Andrukhovych admits:

> *Recreations*: a conscious dispersion of my own features unto four characters—that is, when you cannot not talk about yourself any longer but at the same time do not dare to disclose yourself. *The Moscoviad*: certain fragments are one-on-one autobiographical, but the protagonist—on purpose—not I (in the gestalt understanding). ... *Perverzion*: about myself, but not so much "real" as "desirable for myself." ... *Twelve Circles*: the protagonist is not alone, again, but this is not dispersion anymore; practically almost all of me is concentrated in Artur Pepa, and the rest—that "almost," which is more important—in Zumbrunnen.[9]

The author's narcissism is confirmed by Andrukhovych's auto-quotations in his texts and auto-portrayals of his characters, who unequivocally call themselves poets. "You are a great poet" and "I am proud that I was acquainted with these fellows—they are brilliant poets," his characters repeatedly say about themselves. Thus the bohemian narcissus-poet enters Ukrainian literature of the end of the twentieth century; somewhat ironically, and at the same time quite seriously, he introduces himself as a superstar. In *Recreations*, each of the three Bubabists, easily recognizable behind the main characters, receives the title poet and even "genius poet":

> "Who are you? continued Hryts. "I know everyone here. I know Orest Khomsky, he's a wonderful poet from Lenig ... from Leningrad, I know Rostyk Martofliak, he's a grand poet, he's my friend, I know his beautiful wife Natalia. ..."
> "Marta," Bilynkevych corrected him.
> "Marta," repeated Hryts. "I know Yurko Nemyrych, he's a colossal poet and a wonderful friend, I know myself. But who are you?"[10]

The free indirect speech of Rostyslav Martofliak also ironically plays up and at the same time demonstrates the narcissistic image of Poet:

---
8   Iurii Andrukhovych, "Vona robyt´ mynule zhyvym i nezavershenym," 4.
9   Ibid.
10 Andrukhovych, *Recreations*, 61.

> "You, Martofliak [he is talking to himself], love such moments dearly. Then listen as if you were a prophet, every word worth its weight in gold, and you feel like the minister for external or internal affairs at a briefing: your replies are impressive and you are wonderful, Martofliak, it's been a long time since you've found yourself so attractive."[11]

In a word, the bohemian poet becomes a psycho-ideologist of the "kink" ("vyvykh")—the name of one of Bu-Ba-Bu's public actions. Departing from the normative and practically dead Soviet reality of the 1980s, he did not descend to the underground but came out unto the square. "The time was ours," says Andrukhovych. The Bu-Ba-Bu group was born out of the enthusiasm of youth yet at the same time out of a desire to avoid the official ideology by immersion in the atmosphere of brotherly contact.

The idea of contact is one of the most significant for a carnival. The Bu-Ba-Bu is a carnival of masks, the assignment of a role to each individual (Andrukhovych—"Patriarch," Irvanets—"Bursar," Neborak—"Procurator"), and play with cultural symbols. Literature turns into a public matter, while performance, display, and kitsch become a form of self-presentation. The group members persistently emphasize grotesque hyperbole, membership of the masscult "bottom," and the pseudo-sanctity of the name "Bu-Ba-Bu." All these attributes serve the carnivalization of the literary field. Yet carnival attracts them not only with its playfulness but also with its promise of joy through liberation from taboos and admonitions, as though forever rooted in Ukrainian literature. It seemed that by playing on the border of reality and fiction, and by travelling to the fantasy world of imagination and irony, it was possible to change both life and literature. On the margins of the official mindset and late Soviet culture, the Bubabists attempted to create an island of freedom. For some time it seemed that they had succeeded. Young people followed them; an audience emerged; fans were born. The Bubabists immersed their audience in low culture and turned icononostases into the kitsch-like images of the poet-prophet, poet-demiurge, poet-lover, and poet-bohemian. Irony, theatre, and performance became the forms of the poet's self-expression. The preconditions for irony were already established in the alternative culture of the underground, in the demonic laughter of late Soviet jokes, and in the traditions of Kyiv urban prose of the 1970s and 1980s—in the works of Volodymyr Dibrova, Bohdan Zholdak, and Les Podervianskyi.

---

11 Ibid., 72.

What did carnival mean for the Bubabists? It meant a great game, youth, and an illusion of freedom. It escaped the bounds of totalitarian space and time—that is, "physics"—and ushered in a world of metaphysics, because it was about not only social freedom but also spiritual liberation, which was symbolized by wanderings to other worlds and spaces. The liberation of totalitarian man was promised by the Bakhtinian theory of carnivalization, popular in the 1980s. It endowed the small Bubabist carnival with the character of a "great carnival," which bordered on the "Kink." While laughter, parody, and play revealed the nonidentity and falsehood of the totalitarian person's existence, as well as pointing to its transformations and mutations, they also facilitated an experience of recreation and eradication of complexes. Carnival promised an eternity of holidays. Hence each of the Bubabists experiences his own carnival as an attempt to overstep the limits of solitude. Andrukhovych admits that "Our association became for us nothing less than a possibility to survive in the second half of the eighties"; and he also acknowledges the evolution of the group: "At the beginning of the nineties, we seriously considered erecting a monument to ourselves."[12]

All this is symbolized by Andrukhovych's very first novel *Recreations*. In it, we witness not only the carnivalization of late Soviet reality but also the demystification of both the great carnival and the local, individual carnivals and recreations, experienced by each main character of the novel in his own way: Yurko Nemyrych, Hryts Shtundera, Orest Khomsky, and Rostyslav Martofliak. The great carnival, the Festival of the Resurrecting Spirit, takes place in the symbolic topos—the city of Chortopil.[13] It is the central action of the work, in which the "colorful walkaround pulls the motley public into itself; the divide between carnival, the surreal world, and Soviet reality ceases to exist,"[14] notes Ola Hnatiuk. Ultimately, as the novel suggests, this whole great carnival turns out to be an organized and directed action, which corresponds to the logic of carnival according to Bakhtin. After all, medieval carnivals took place only at certain times and in certain places—that is, carnivals needed permission. Post-Soviet carnival, occurring during the time of Gorbachev's perestroika, also has its limits and its director. The last act of the carnival in the novel is a joke, a diabolical event, organized by the director and stage manager of the festival, Pavlo Matsapura: tanks, a military action, the expectation of a mass shooting—in

---

12 Andrukhovych, "Avtobiohrafiia," in Andrukhovych, *Rekreatsiï*, 31.
13 "Devil city" (translator's note).
14 Olia Hnatiuk, "Avantiurnyi roman i povalennia idoliv," in Andrukhovych, *Rekreatsiï*, 16.

one word, an action resembling an "executed renaissance." All this turned out to be just another section of the festival, a path in the labyrinth. However, in this moment of fear, everybody, it seems, for a moment becomes himself: Martofliak takes Marta's hand, just like when they were courting, and Hryts remembers his last unfinished poem.

Andrukhovych explicitly hints at the illusive quality of great carnival, then. Yet even if illusory, it helps people survive the euphoria of freedom and utilize it. Even more importantly, it creates a field for a therapeutic renewal of the identity of each character of the novel. Each of them has his own planet and "each will go his way," or, as the narrator says, "each of you will lose his way in his own manner."[15] The festival makes ground meat of everyone, rendering them all unrecognizable and undifferentiated and turning them into a grotesque and total body:

> [A]nd now a kobzar is singing about the red winding-sheet or about the red quelder rose, and students are staging a mystery play about our Fatherland, and you can buy a horoscope at a free enterprise stall or eat a kabob, or shoot arrows from a bow at a huge cardboard Stalin, or ogle the next-in-line backside of the next-in-line competitor in the "Supermiss" contest, or drink straight from a bottle, or make up your kisses with blue and yellow paint. ...[16]

However, it is not the dissolution in the crowd that the poets need. The festival attracts, yet also casts away, bewilders, gets people lost in the side streets leading to and from Rynok Square. It is one of these side streets, on the margins of the festival, that the main metamorphoses of the novel's characters take place. All of them take on a strange identity: Rostyslav Martofliak is as big as a bear and at the same time infantile; Khomsky-homsky explicitly flirts with "lady variants"; Hryts Shtundera wears a black strand of sorrow because he belongs to the descendants of Ukrainians deported to Siberia; Yuri Nemyrych lives with a melancholic nostalgia for the distant Habsburg era. All of them become participants in their personal stories in Chortopil: they reveal their complexes. Having met another Marta, a prostitute of his mother's age, and having spent a night with her, Martofliak through incest realizes his infantilism. Hryts Shtundera,

---

15 Andrukhovych, *Recreations*, 70–71.
16 Ibid., 69.

having changed into a UIA[17] sergeant major's uniform, travels at night to the place where Siltse once stood—the place of his grandfather, deported by PCIA[18] agents; and, as though in a trance, he identifies with his grandfather and relives his story. Yurko Nemyrych, having got behind the looking glass of the prewar period in the Villa with Griffins, encounters ghosts of the past; and, despite the colorfulness of the types and characters, becomes a participant in a satanic mass in which his own life is at stake. Khomsky, however, stays with Martofliak's wife, Marta, and, for real or only in imagination, turns out to be a tender and passionate lover, thereby removing the homoerotic mask implied by his name.

All the initiations and labyrinths travelled by Andrukhovych's character are not only created by the great carnival but also overstep its limits. The main thing here is that the carnival cannot remove problems or resolve them; it lays them bare and diagnoses them. Martofliak, frightened and disgusted, flees from Marta-the-prostitute; Shtundera falls into a ditch, without having reached Siltse; Nemyrych jumps out of a window at the Villa with Griffins; and Khomsky does not know if his time with Marta was fact or just a figment of Marta's imagination. It appears that play-acted (repeated) reality—historical, sexual, national—could have become a recreational zone and could have helped restore lost identity; however, the real, as opposed to the imitated, is much stronger and more demonic than all of its possible repetitions and simulacra. What is left for each of the characters is an opportunity to become—albeit for a moment—*other*. "Each of us spent last night in his own way and to his own satisfaction. ... But let it belong to each of us individually," says Martofliak.[19] The carnival reveals the shadows of each of them but does not lead to a merging with the past. This is confirmed by the motifs of falling, escaping, and fever dreams that accompany the characters' actions. There are also individual micronarratives; personal stories; and individuals made out of perversions. However, there is no totalized Soviet individual, no great hero. Indeed, heroism itself is absent.

Micronarratives appear due to carnival, which mixes the real and the fictional, the high and low, individuals and types, and also the sexes. Khomsky's name hints at Chomsky; Martofliak is a male incarnation of Marta; and Marta perceives him as her husband and as her child at the same time. With Martofliak, the carnivalization acquires an erotic-poetical-infantile form; with Hryts

---

17   UIA—Ukrainian Insurgent Army (translator's note).
18   PCIA—People's Commissariat of Internal Affairs (translator's note).
19   Andrukhovych, *Recreations*, 120.

Shtundera, it has apocalyptic-patriotic form; with Yurko Nemyrych, it takes on a noble-diaspora-satanic form; and with Khoma, it has a homoerotic-overman form. Thus, in *Recreations* Andrukhovych, on the one hand, shows through carnival the multiplicity of identities in one person, sketching in this way a field of metamorphoses for the Soviet individual, and, on the other hand, demonstrates that mass actions such as carnivals are first and foremost trips to the Land of Play.

Is carnival able to restore the totalitarian subject's lost identity—that is, counter the subject's alienated family, nation, kin, and his or her own body? The very title of *Recreations* refers to such a restoration of the physical, intellectual, and emotional powers of a person, or a renewal of the nation, and suggests the possibility of a renaissance. But the novel is also irony toward itself and contains a demonic element. Indeed, the plot demonstrates that the novel is not about a universal renaissance but about a zone of recreation—in other words, a specific place where playing is possible, allowed. In addition, play is a transient thing. Nevertheless, *Recreations* affirms the possibility of metamorphoses and recovery from totalitarian traumas. Therefore, in the late totalitarian era, ludic man becomes a new cultural hero because in play, and through play, he delineates a new horizon of expectations and renewals.

CHAPTER 17

Yuri Andrukhovych's Carnival: A History of Self-Destruction

The low carnival of Bu-Ba-Bu bore witness to the revival of the neo-baroque element in Ukrainian literature of the second half of the 1980s and the beginning of the 1990s. At that time, baroque art was perceived as the only aesthetically fruitful and organically autonomous aesthetic form that had developed in the history of Ukrainian culture. Hence, deleting the recent past (socialist realism) and disappointed with the weakness of the national version of modernism, the postmodernists appeal over the heads of the "fathers" to the "forefathers"— to the period of the premodern baroque. Andrukhovych himself confesses that he is preoccupied with the search for the baroque Ukraine. And he is not alone. In the 1980s, a neo-baroque poetic group appears—"LuHoSad" (Ivan Luchuk, Nazar Honchar, and Roman Sadlovsky)—which uses baroque verbal games and combines kitsch and hybrid text with visual poetry, lyrical irony, and futuristic tone painting ("a ryma dveryma hup" ["but rhyme slams the door"]—Nazar Honchar). The characteristics of another poetic group are parody and irony, alcoholic hokku and burlesque ("Duma about a Little Elephant," "nocturne 1"), and the absurd ("The Lost Letter" ["Propala hramota"]—Yurko Pozayak, Viktor Nedostup, Semen Lybon). The work of the poet-palindromists Mykola Miroshnychenko and Anatolii Moiseienko revitalized the baroque form of visual poetry.

It is Bubabist carnival, however, the evolution of which is reflected in Andrukhovych's novels, that has become the most outstanding expression of Ukrainian neo-baroque postmodernism. In his novel *Recreations* (1990), Andrukhovych demystifies both the image of the national poet and carnival itself. At the carnivalesque, phallic Festival of the Resurrecting Spirit in the town of Chortopil, three poet friends, their personalities and masks (the patriotic, the erotic superman, and the nostalgic passéistist), now separating, now joining, distinguish the individual's alienation from biography, history, family, sex, and alienation, which became absolute in Soviet times. It is notable that in *Recreations* the reader can recognize quotations from Bu-Ba-Bu authors, aspects of late Soviet reality, and the typical landscape of a western Ukrainian town in which the action takes place. The rhetoric of phrases in *Recreations* dominates and becomes autonomous. It is a language simulacrum, with no meaning or intention to communicate. Martofliak prophesies in hackneyed slogans, almost like an automaton:

> Love girls—and they will give birth to your very selves. Raise bees and don't trample ants, and it will be returned to you a hundredfold. … Listen to your blood, for blood is the state. Respect each blade of grass, for grass is the nation, it is hope. … Music is moving architecture. … Electrification is communism minus Soviet power. In strength is the unity of the people. This sweet freedom is the Word.[1]

The text fools with the masks of the characters-poets and hints at the recognizability of real biographies and real people. One of the key episodes is the one in which the name of the poet Mykola Nahnybida is mentioned. From a completely real person, he becomes the mystical name of a second-rate, official Ukrainian poet. Here we witness a phenomenon that can be called "carnivalesque homogenization"—the erasing of the differentiated meanings of names, characters, sexes, and languages. The novel's main characters are also themselves a totally carnivalized body because the Festival grinds everything down to the condition of "stuffing [made] by a good cook."[2] For a moment, they become hybrid and fungible, before going back to their solitude again. Everyone experiences his own carnival, his way of reaching a place of spiritual contact and intimacy, but they ultimately end up lonely.

The ideological landscape of the novel lies in the profane transformations of the triune protagonist, Nemyrych-Shtundera-Khomsky, a poet and a superman—transformations that ostensibly symbolize the mystery of the unity of person, mass, and history at the Festival of the Resurrecting Spirit. The key figure here is the poet as a participant and a creator of the Festival:

> and the festival tramples you with its feet, you are minced like stuffing by a good cook, for, as Nemyrych said earlier, you are all alone, so it is doubtful

---

1 Andrukhovych, *Recreations*, 73–75.
2 Ibid., 70.

whether your walking will lead to anything among these tents and stages, among these beautiful cripples, on this square surrounded on all sides by mountains and by Europe, where each of you will lose his way in his own manner, there, it's beginning, they're calling out, whistling, shouting, pulling at your sleeve, begging, demanding. ...[3]

At this carnival, physics transforms into metaphysics and the real city into a labyrinth; a mask becomes a face and an overthrow a carnival. It is well known that carnival is a transient and local action, often organized by the authorities themselves. Indeed, the authorities can insert it into any situation, including a concentration camp. Andrukhovych's novel also demonstrates that the joy of getting together at the carnival is illusive and that the collective body that forms there quells and erases individual desires. Behind the incredible freedom of the carnival lies the demonic power of its director. The revival of the sacred idea of oneness at the Festival of the Resurrecting Spirit resembles the seductive illusion of that totalitarian kitsch in which everybody unites in ecstatic expectation of a celebration.

Andrukhovych's second novel, *The Moscoviad* (1992), categorized by the author himself as a "horror novel," has become a collection of late Soviet ideologemes, discourses, and characters. *The Moscoviad* gives voice to the mythologemes and slogans of the late Soviet era (including the tale about huge rats inhabiting the Moscow underground). Andrukhovych continues to develop his take on the adventure novel, but he also introduces autobiographical material. (At the end of the Soviet Empire Andrukhovych was a student at the Literary Institute in Moscow.) In this way he adds an anti-colonial theme. Having immersed his protagonist in the atmosphere of the metropolis, the capital of the Soviet Empire, Andrukhovych employs a novel-travelogue form, combining on one level the periphery and the center, and the top and bottom of the imperial body. The boundless journey of a Ukrainian poet, a student at the Literary Institute Otto von F., through the circles and tiers of the ghost city acquires symbolic meaning. This postcolonial novel-apocrypha parodies the imperial topos from the position of an outsider—a Ukrainian poet, who not only quotes the famous Venichka[4] but also repeats his alcoholic odyssey in the center of Moscow.

The author presents the environment of Moscow/the metropolis as a travel space for the protagonist, in an emphatically antagonistic manner because

---

3   Ibid., 70–71.
4   "Venichka" is the protagonist of Venedikt Erofeev's novel *Moscow to the End of the Line* (*Moscow-Petushki*), first published in 1973 (translator's note).

Moscow is an anti-world for the colonial subject. However, the identity of the adventurous hero with his surroundings, potentially implicit in his chronotope, practically removes this opposition. The colonial protagonist is identical with his imperial anti-world: his feeling of alienation notwithstanding, he is one of the many Soviet people with an imperial mentality, something especially evident in his statements about so-called "natsmen."[5]

Oksana Zabuzhko associates the idea of *The Moscoviad* with Tadeusz Konwicki's novel *A Minor Apocalypse*, translated from Polish by Andrukhovych. Zabuzhko writes,

> Andrukhovych transposes, or "transplants," into Ukrainian reality not only the plot, which in its essence is Joycean—the one-day odyssey of a character-man-of-letters (who is also the narrator) through the city of his purposeless existence with his ambiguously designed death at the end—like a peculiar "minor apocalypse," called upon to symbolize the "major" one.[6]

As is appropriate for a colonial subject, its consciousness is split between the past and the present, and its whole world is defined by the field of the Lacanian "imaginary," which does not allow it to force its way to the "real." The reason for this is not only the permanent alcoholic intoxication but also the power of symbols, mythologemes, and ideologemes, which invade the consciousness of such a subject in foreign territory, in the capital city of Moscow,

> [t]he city of granite monograms and marble ears of wheat and five-pointed stars as large as the sun. It only knows how to devour, this city of puke-covered courtyards and crooked picket fences in poplar fluff-covered lanes with despotic names: Garden-Forehead-Beatin-sky, Kutuzov-Roach-Mound-sky, New-Executioner-sky, Cudgel-Beat-up-sky, Minor-October-Graveyard-sky. ...[7]

Apparitions of the KGB, grotesque images of men of letters and of the Literary Institute (the cradle of multinational Soviet literature and the arena

---

5   "Natsman" is the vulgar term for a member of a national minority (translator's note).
6   Oksana Zabuzhko, "Pol´s´ka 'kul´tura' i my, abo Malyi apokalipsys moskoviady," in Oksana Zabuzhko, *Khroniky vid Fortinbrasa* (Kyiv: Fakt, 2001), 323.
7   Yuri Andrukhovych, *The Moscoviad*, trans. Vitaly Chernetsky (New York: Spuyten Duyvil, 2008), 76.

of pseudo-discussions), and the alcoholic wanderings of the poet-outsider transform the desired road home into a labyrinth and a parallel anti-world from which he can only escape from with a bullet to his head: "Since tonight I am not running away but coming back. Angry, empty, and with a bullet in my skull to top it all off. Why the hell would anyone need me? I don't know that either. I only know that now almost all of us are like this."[8] This is how Otto von F. concludes the story of his odyssey—by becoming aware that he is just one of many, although up to this point he has considered himself the other.

By inverting the high and low, sinking the corridors of power into the underground, and transforming the famous politicians and figures of the past into dolls and monsters, Andrukhovych creates a strange puppet show—a macabre, simultaneously seductive and repulsive, fantasy image of post-Soviet Moscow. *The Moscoviad* is a tale about a Ukrainian poet-intellectual trying to separate himself from the unified and completely grotesque body that is called the "Soviet people." First of all, he separates himself from the all-encompassing depersonalization created by the empire through a hybrid body that produces "completely new phantom nations and ethnic groups with names so twisted that they will be ashamed of them themselves: Rossiacs, Ukralians, Karelo-Mingrelians, Cherboslovats, Romongolians, Netherbaidzhanis, Sweeks, Gredes, Fruzbeks, Byeloswabians, Kurdofranks, Jerattlers, and Carpatho-Ruthenians."[9] This tale, however, is not only about the empire and a colonial subject but also about the contamination of the character's consciousness by the recent KGB past.

Andrukhovych's third novel, *Perverzion* (1995), is about Europe. This is a view from Ukraine, on the "spit upon margins of Europe," on the border of the postmodern West, a Ukraine which feeds upon its own past and produces parallel discourses and worlds. In this novel, a Ukrainian (who represents the Second World) encounters the West while Third World migrants are entering Europe. These events are summed up in a prayer to the "Lord of the Germanic Gate" and unite the (Eastern) wretched of the earth against a Europe that has imagined them as "barbaric," "wild," "other." In their prayer they had to invent a different god for themselves: "they were buttered with hunger and bombs, epidemics, AIDS, chemicals, the most polluted wells, and the cheapest bordellos were filled with them, they had weapons and patience tested on them, they had their forests burned and their deserts trampled ..."[10]

---

8   Ibid., 185.
9   Ibid., 175.
10 Andrukhovych, *Perverzion*, 27.

By opposing itself to both the East and West, the Ukrainian postcolonial individual is looking for his own way. He perceives Western thinking as a harbor of existentialism: "this woman, this candle, this coffee, this cigarette, this ephemerality, this passage, this terminality, this is only 'here and right now.' In this way I first, apparently, touched upon the greatest of the mysteries of the West ... it is not for us to understand all the depths of this unhappiness, these depressions of sweet sojourn,"[11] admits the protagonist of *Perverzion*, and goes on to say,

> We envy—fiercely and blackly, as though we were lost children, for those Western people, for those inhabitants of blessed evening gardens, their food, their profits, their diversions, their cars, we are ready day and night to gather oranges for them or to wash their spittoons, just to enter, to filter through, to be admitted, to be closer, but every one of them, poor buggers, returning from the office, crouched up beneath the illuminated sign, took the cheapest burger for themselves with ketchup and, washing it down with cola, nearly began to cry: Why this world? ...[12]

This is the way a Ukrainian intellectual-outsider feels the innate irony of the West, where McDonald's—a sign of globalized world and an important actor in the new colonization spreading all over the world—is combined with the intellectual, existential experience of uprootedness and of the missed encounter with the other, reflected in Western philosophy and culture. Having met with the West, the Ukrainian poet—due to the fact that postmodernism there has discovered the power and beauty of différance intrinsic to language itself—realizes that words are perhaps his only weapon.

Thus, for the postcolonial subject, language is a shelter from history, imperialism, violence, devalued official and authoritarian orders. For this reason, he identifies as a "speaking subject" and likes verbal improvisations and verbal games. These improvisations are moments of slowed consciousness, and give him an opportunity to realize that I exist because I speak. The protagonist of *Perverzion* has a special fancy for language games such as "O Italy, what reason do I love you so? Because you blow into the butt of a boat."[13]

> Ada where you who you Ada?
> O Ada. Oasisiada.

---

11 Ibid., 71.
12 Ibid., 71–72.
13 Ibid., 31.

O you Ada.
It's me Ada where you, Ada?[14]

Important here is the spontaneity of speech, which can be captured and recorded by a voice recorder. Language can be a being itself, and Andrukhovych offers a whole collection of speech-beings in his novel, which is a work about the many-sidedness of both ontological and linguistic masks. Such is Mavropule's speech-gibberish, Paul Oshchyrko's blues (the monologue "To listen to reggae, to die beneath the sky, to breathe in the scent of the grass"),[15] and Gaston Dejavu's words-assimilations ("But you can always find at least One property for Assimilation. Everything presages Everything with itself and even Nothing").[16] Language is capable of coming to a standstill, transforming into hybrid twists, flowing like a river, being both elusive and omnipresent, and turning into a mirror that simultaneously reflects a riddle and closes in on itself. However, "Understanding that the Riddle arises before our vision, we deal with It all the more this way and, striving to look into the Mirror of reflection, explaining Visions Alone with the help of Others, laying out the Mirrors One against the Other."[17]

As in Andrukhovych's other novels, the plot of *Perverzion* is set around an adventurous protagonist. This time, though, the author wants to substitute the play of the text itself for the play of the character. He mixes different types of writing and discourses, develops parallel texts and divides them into fragments, introduces inserted and introductory text-teases and text-pastiches (such as the opera of operas, *Orpheus in Venice*), and generally emphasizes that the protagonist is only a fiction—a function, and puzzle, of the text itself.

The plot of *Perverzion* is fairly simple: Andrukhovych's many-sided, carnivalesque protagonist, Stakh Perfetsky, who in the circles of modern bohemia "had countless faces and countless names,"[18] is invited to Venice, to the heart of Western culture, for a seminar on "The Post-carnival Absurdity of the World." The gnostic, mysterious, and bodily transformations of Perfetsky, interlaced with his erotic adventures and nostalgic memories of the past, become a foil for Andrukhovych's cultural investigations into the nature of postmodernism. They echo Andrukhovych's later reflections on Eastern Europe as a zone of the

---

14   Ibid., 90.
15   Ibid., 128.
16   Ibid., 95.
17   Ibid., 96.
18   Ibid., 8.

postmodern. In *Disorientation on Location* (1999), he talks about the postmodern as "underformed, but already felt post-totalitarity," as well as a "constant neo-totalitarian threat"; about the butchered modernism of "any type—Viennese (standard), Prague, Krakow, Lviv, Drohobych." Hence postmodernism for him is "emptiness, a great exhaustion with an endlessly open potentiality, a great promising emptiness."[19]

Perverzion is also a text-emptiness, in which the protagonist himself disappears. This is a metaliterary novel, a novel-anthology of discourses, fragments (of invitations, theatre programs, dispatches), and quotations. And although the fictional publisher claims that the sequence of the presentation of the documents is arbitrary ("It would be interesting, how much different could this sequence be?" he asks the reader, flirting),[20] the text has a strict plot structure, in the center of which there are the adventures of a bohemian poet. These adventures are based on the Orpheus myth (which gives the novel its depth)—they concern memories of a dead lover and an attempt to bring her back from the Kingdom of Shadows, even if by dint of a musical spell. With the help of stylization, quotations, and fragmented writing, the orphic protagonist tries to gather the world from traces, fragments, and images. Somewhere in a parallel reality a cosmogonical-demonic story breaks out. Its traces—contained in messages and dispatches sent to a man called "the Monsignor"—periodically surface. The biography of the poet himself is also a mystery: there is a suggestion that he is a hired gun and simulates his suicide. At the same time, the text literally falls to pieces because of the polyphony of stylization, "linguo-cabalistic expressions,"[21] hybrid voices and consciousnesses, bodily and verbal metamorphoses (for example, those of the guests: "not actors or ministers, and not models, and not bankers, but mostly lemurs and vipers and sirens and eight-eyed dragons and manticores with lacertines. ..."[22]

Despite employing fragmentary writing, Andrukhovych builds events around the theme of love. The romantic adventures of Perfetsky and Ada are central, and there are constant references to Stakh's dead wife. At the same time, as a cultural treatise, the novel discusses the similarities and differences of the fates of Europe and Ukraine. As the writer's mouthpiece, in his presentation at a symposium (where in reality he was sent as a killer) Stakh Perfetsky juxtaposes

---

19 Iurii Andrukhovych, *Dezoriientatsiia na mistsevosti. Sproby* (Ivano-Frankivsk: Lileia-NV, 1999), 120.
20 Andrukhovych, *Perverzion*, 19.
21 Ibid., 128.
22 Ibid. 268.

decadent Europe with the vitality of the nomadic people's that populate the expanses of western Ukraine. He also contrasts cultural and effeminate Europe with the orphic mystery of love between man and woman, the kind of love that is ostensibly still alive on the margins of Europe. "There, where love ends, the 'absurdity of the world' begins," he says.[23]

Perverzion is a perverse novel about perverse love (necrophiliac, polygamist, and androgynous), which is also a thesis about cultural and European identity, and which is organized around the Rilke's orphic myth. As Andrukhovych's protagonist states,

> A hole in the soul—this is the semantic
> Chasm between me and the world,
> When my beloved from the other side
> Calls with decay, with the mortal, with summer.[24]

The trip to the West, to Venice, is not only a real journey but also a symbolic descent into the Kingdom of Decay, as well as an attempt to appropriate the feminine cultural essence of the West that the sinking city harbors. By remembering his dead beloved wife, flirting with Ada, who resembles Salome rather than Eurydice ("The head of John the Baptist didn't make a particularly great impression on Ada, unlike the cherry-colored cape of the young harlot"),[25] becoming familiar with Venice, Perfetsky—as an orphic character—personifies masculine power and even aspires to revive decadent Western culture through the vitality of his young Ukrainian nation.

The orphic myth defines the carnival (masquerade) and metaphysical space of the novel. Thus, on the one hand, we deal with the de-heroization and fragmentation of the modern bohemian hero; on the other, we witness the sacralization of his vital masculine power. By throwing himself into a Venetian canal, Perfetsky, whose totem is Fish, becomes a symbol of cyclical return—the revival of European identity, which, as repeatedly stated in the novel, is a continent that is disintegrating, dying, and lacking in love. The totem Fish is also a symbol of a gnostic author, who equally successfully descends into hell and flies to the heights of intellectual game. Such an author fools his reader: "For Stakh Perfetsky, who nearly everyone in the world considers a suicide-by-drowning and who is the real author of all (not just some!) of this book's excerpts, Stakh

---

23 Ibid., 237.
24 Ibid., 161–62.
25 Ibid., 78.

Perfetsky continues to be among us. He is alive and, I'll say more, he will return. First, just like a book, cunningly lain at my door by him."[26]

Andrukhovych offers a fragmentary, multitiered play that combines the adventures of his protagonist, a cultural study, the orphic myth, and the esoteric mystery of the Fish-Christ. He thereby creates a neo-baroque postmodernist novel. *Perverzion* marks a departure from postmodern carnival, which turns out to be a mere simulacrum. The Venetian carnival is transformed into a performance, and the novel turns wholly into a baroque quotation: "The idea of the total and of course totalitarian mimicry of everything deadly and grand. Quotation, collage, and deconstruction have displaced something more distant in time, more primordial, and more authentic."[27] The play *Orpheus in Venice* becomes the apotheosis of postmodernism (as an internal text). The stage manager, Matthew Kulikoff, intends "to create an opera beyond operas, an opera of operas, where the very elements of *opera-ness*, its inner actuality, its substance, is parodied, rethought and, if You accept this, elevated even higher."[28]

With this in mind, he uses the whole opera experience of the seventeenth and the eighteenth centuries, and resorts to pastiche—baroque "pasticcio," in which new operas were created based on deconstructing and recombining elements of existing operas. It appears, however, that emphatically artificial postmodern imitation—the opera-buffa—becomes the place where reality is really born and the real Orpheus revived. A new Orpheus appears—the persecuted Stakh Perfetsky, who on a tow descends from a balcony to the stage:

> Only this was a *different* Orpheus, a new one. The one who flew down onto the stage from the very top of the arched roof. He entered the reality of my opera just as to his own home. He was easily recognizing the musical phrases and was getting into them without straining. ...
>
> He was the first who introduced *soul* into this mechanical performance. I am nearly sure that he is a real Orpheus. For I tormented myself for too long to summon him from nonexistence. But from my torments *he* appeared at the most difficult moment, my materialized dream, my Venetian madness. Right now it already seems to me sometimes that it had been *my* intention all the same, that from the very beginning I had planned to do this.[29]

---

26 Ibid., 314.
27 Ibid., 36–37.
28 Ibid., 173.
29 Ibid., 189.

—admits the director; and with his help Andrukhovych deploys the myth of vital Ukrainian orphism. In this way, despite numerous hints at the sterility and oversaturated decadence of Western culture, symbolized by Venice, the pastiche *Orpheus in Venice*, created by the director, not only seemingly creates a new genre for performances but also helps the protagonist save himself. Disguised as Orpheus, he escapes his killers.

Andrukhovych's *Recreations*, *The Moscoviad*, and *Perverzion* are simultaneously a laboratory of, and a commentary on, Ukrainian postmodernism. In *Twelve Circles* (2001–2003), he plays with the "already written": he uses and rewrites plots and images from his previous postmodern novels. Among them are an orphic love story, a bohemian poet, the Eastern-European revision, a pseudo-conference, and an author-demiurge. Andrukhovych juxtaposes all these images and plots with the "reality" produced by mass culture—the reality seen in the video commercials of "Vartsabych Balsam," a popular film about Bohdan-Ihor Antonych.

Here lies one of characteristics of the evolution of postmodernism, which more and more integrates into popular culture, creating a phenomenon that later will be called pop-postmodernism. Ukrainian postmodernists widely and successfully employ elements of mass culture and pop culture: for example, Irvanets's social art, Zabuzhko's erotic romance, and Andrukhovych's adventure novel. However, their works were not yet mass media texts, despite media chronotopes such as the conference in Venice or the Festival of the Resurrecting Spirit in Chortopil. Philological language games and mythologism led further, beyond the so-called effect of actual "reality," and enriched the texts with multidimensionality. The mass media pop-postmodern text, in general, cultivates superficiality and the "actual" presentation of here-and-now action. Combining video and music, Andrukhovych resorts also to the media-staging of his novels—as in the play *Absent* by the bands Karbido, Cube, and ArtPole.

In *Twelve Circles*, Andrukhovych even earlier declared his inclination toward media representation. This leads to open automatism and fragmentariness, and the novel itself transforms into a feuilleton. *Twelve Circles* is a highly self-reflexive novel: everything in it looks artificial and contingent, and the author openly plays with reader expectations ("oh no, sorry, this is *Recreations* suddenly spilling out of me";[30] "Wait, again those phantoms of youth gone by, repugnant repetitions and self-repetitions?":[31]

---

30 Andrukhovych, *Twelve Circles*, 31.
31 Ibid.

> It is undoubtedly true, in fact I agree one hundred percent with the pertinent reproach that in at least two thirds of my novels produced up to now, the heroes are *transported* somewhere. In the past I had successfully summoned for them trains, Ikarus buses, a momentously symbolic Chrysler Imperial, and also—the pinnacle of phantasms—an infernal semi-airborne 'Manticora' capsule, designed possibly in Beelzebub's subterranean design bureaus and tested at fire-drenched ranges under a storm of stones, sulfur, and excrement.[32]

He adds theses about the death of the bohemian hero and about the death of the novel. Referring to Izdryk, Andrukhovych intrigues from the very beginning: "In one of the books I know this is called 'The Arrival of the Heroes'. However, I am not sure whether they are really heroes. Or whether this is truly an *arrival*."[33] Andrukhovych sets off the plot about the death of the bohemian hero with sensational image of the poet Antonych, whom he tries to show as the archetypal image of the "cursed poet." This is how Andrukhovych's hero—the poet-bohemian—ought to look. The outlandish and perverse character of Antonych in *Twelve Circles* is an exact mirror of this hero, and the poet-Bubabist associates himself completely with bohemia. At the same time, Andrukhovych develops the very metaphor of the hero's death through the plot about the accidental death of Carl-Joseph Zumbrunnen and the clinical death of Artur Pepa.

The destruction of the novel as a genre is another issue limned in the novel. It displaces the multifaceted nature of the adventurous character and his journeys, which used to be a common plot in Andrukhovych's postmodern novels. Quite noticeable here is the Central European journey of the dead Zumbrunnen, presented as a video sequence played in reverse. As the author says, in *Twelve Circles* he divides himself into several few characters, one of whom is the orphic Zumbrunnen. Ultimately, Antonych is incidental in this text, as are Artur Pepa, Carl-Joseph Zumbrunnen, and every other character, because each one of them fulfils only a partial function and is merely a mirror that reflects the character of the total hero-Bubabist. It is his symbolic death that the text is interested in.

Andrukhovych demonstrates that the total novel, one which would make life and art fully identical—and this is an idea that has concerned the Bubabists for a long time—is impossible in the post-postmodern era. Therefore, the

---

32 Ibid., 32.
33 Ibid., 27.

author begins to talk about a sub-novel, which serves for him as an example of how he cannot write. This is how the idea of a novel that will never be written takes shape: "The story of the Hutsuls' great journey to the Imperial Capital" in 1949 could bring "the tension of an eternal drama of *the artist and power*" and could help survive "bridges across time in all directions" in the middle of the twentieth century and to see the folk dances and the life of the Immortal Joseph's empire.[34]

Andrukhovych lists the possible plots and themes of this unwritten novel and thereby creates a sub-novel. The writing of the work requires too much knowledge, and the protagonist, Artur Pepa, to whom the very idea of the novel belongs, realizes that he can only have a "presentiment of the novel," that he does not want to be either Marques or Flaubert, that he "has not learned how to narrate full-fledged stories," that he knows neither geography nor "the labyrinths of unlimited power," that he cannot reincarnate into a woman in order to feel "what rape is," that he knows nothing "about resistance" or what to do with "Hutsulia," that he does not know many things that are necessary for a novel: how a dying witch groans, the statistics of tuberculosis in the Alpine regions of the Carpathian Mountains, the history of musical instruments, and many other things. He writes of a "sub-novel" that it had to be "extremely fragmentary, only a hundred typewritten pages long, and nothing listed above need be there; but there had to be the knowledge of all of the above—otherwise a novel like this simply couldn't be written."[35] Ultimately, "knowledge of all the listed" without knowledge itself is already a device in Andrukhovych's postmodernist work. His "lists" of names were registers of that which could have been a context of what is said, and only a few separate, unconnected words are what remain.

Thus, rejecting the novel-story, as Pepa does, Andrukhovych resorts to self-reflection. He ironically hints at his own manner of a fragmentary writing and justifies it. However, his adventurous protagonist, whom he used to cement this fragmentary manner, is a thing of the past. His divided, multiplied self resists fusion; the separate narratives of each character are unfinished, interrupted. Dead Zumbrunnen travels to Vienna and returns to his Central European identity; Pepa and his wife escape to once cult "Chrysler Imperial," now broken and rusty; and almost crazy Antonych remains in his own time. At one time, Nemyrych, Shtundera, and Khomsky experienced their adventures in *Recreations* and at least for a moment liberated themselves from their shadows.

---

34   Ibid., 99.
35   Ibid., 104.

They also met in order to continue to be audacious, young, and lonely. Now, crossing the threshold of the middle age, Andrukhovych's protagonists part and isolate themselves in their localities.

Andrukhovych's fragmentary postmodern writing, which evolved through his first novels, in particular *Perverzion*, has become a fashionable style. It has been easily mastered and appropriated by mass culture. Svitlana Pyrkalo's *Green Margarita*, Vasyl Kozhelianko's *A Procession in Moscow*, and Natalka Sniadanko's *A Collection of Passions* actively use the principles of collage and interludes, lists of names and anthologies of styles—techniques that Andrukhovych successfully exploited in his work. Ultimately, mass culture reproduces all forms of representation, from the retro to the underground, from techno to folk art.

Twelve Circles discusses the very mode of the novel writing in the postmodern and post-postmodern era. Not only Artur Pepa, but also the implicit author, explicitly break with the classics, and refuse to write psychological novels à la Flaubert or mythological novels à la Márquez. However, this refusal signals both the strength and the weakness of authors who lack "knowledge." Looking for a new form, Andrukhovych flirts with media mass literature (take, for instance, his mockery of glamorous erotic images in video commercials!), introduces the metaliteral and cultural commentaries regarding his previous plots, and shatters his fictions into fragments and multiple plots. Ultimately, in order not to write a mass cultural and kitsch novel like *Green Margarita*, *A Procession in Moscow*, or *A Collection of Passions*, Andrukhovych writes a novel-feuilleton containing the recognizable realities of contemporary life. Herewith, the rejection of novel-kitsch becomes an aesthetic position opposed to already extant kitsch—the commercial scripts and videos for Vartsabych's Miraculous Balsam and Yarchyk Volshebnik's "apo-video"[36] for the song "The Old Antonych," featuring "this so-called Antonych—a tall old-looking guy in a hat and trench coat, earrings in both ears, looking like some kind of urban phantom,"[37] as well as the numerous commercial slogans scattered throughout novel. One of the main slogans is very expressive: "*Heroes of Business for the Heroes of Culture.*"[38]

The commercials for Vartsabych's Miraculous Balsam interchangeably and eclectically combine the patriotic, and techno- and lesbian kitsch, represented by the two strippers dressed in Hutsul folk costumes fulfilling the wish

---

36 Ibid., 177.
37 Ibid., 178.
38 Ibid., 323.

of a customer who ordered the commercial: "Grab two *chicks*, dress them up, you know, as Hutsuls, take them into an alpine meadow and …"[39] Their names, "Lilia" and "Marlena, "evoke the nostalgic (yet faintly erotic) song that was so popular with soldiers, both German and Allied, during Second World War. (Rainer Werner Fassbinder also used it in his film of the same name from 1980.) Traditional Hutsul style, widely reproduced in cinema, postcards, and the "Hutsul lady" dance, becomes emphatically imitational, mass media-constructed, and monotonous: "[t]his sameness is of the sort that makes all pop stars, tarts, fashion models, high school seniors, vocational school students, in other words, all our *female contemporaries* almost indistinguishable from one another, since this sameness has been created by television, magazine covers, and *our Soviet way of life.*"[40]

Thus, in *Twelve Circles* Andrukhovych links male midlife crisis with the contemporary crisis in the novel (as an artform), and the collapse of the ideal of the bohemian hero. The plot is, at the same time, an act of "public theatre" and intended to scandalize the patriotic Lviv public, primarily through Bohdan-Ihor Antonych, who is a bohemian poet in the novel. The orphic theme leads to the real and unreal death of the two protagonists—orphic personas—only to end up with Pepa's delirium and the mythical Author's appearance in his ravings: Vartsabych, the multifaceted cameraman who, like a deus ex machina, almost joins the action at the end of the novel. The mystification of the author is attained through videos and images common in mass culture, which point to the orphic quality of such an author-qua-director:

> the komsomol activist, the Bull Terrier, the nerd, the maniac inventor, and the circus dwarf, and the old-fashioned lady, and the virtuoso hacker, and a whore, and tramp, and hawk-nosed young witch, a she-wolf, a crow, a snake, and even a dream, for all of them were Vartsabych, or more exactly, that intense yellow substance by the controls.[41]

This is how the image of the author reincarnates—this time not as a Scripter, to use Roland Barthes's term, but as a Cameraman.

In spite of the themes of the death of the author and of the novel, *Twelve Circles* aspires to be a meta-novel, and the author aspires to be a meta-author.

---

39 Ibid., 122.
40 Ibid., 38.
41 Ibid., 319.

In this text, Andrukhovych has gathered together all popular plots and images in his own works, their "repetitions and self-repetitions." They have been reshuffled, and now the author-commentator trumps the reader's irony with his own: "Clearly each of you has a right to an ironic grin here. Wait, again those phantoms of youth gone by, repugnant repetitions and self-repetitions?"[42] The meta-author resorts to double irony and thus restores his right to mystify the reader: "But am I really talking about those? My goal first of all is the truth of *this* story."[43] In the end, having rejected the coherent story, the author has in reality hidden himself behind numerous masks (Antonych, Pepa, Zumbrunnen, Vartsabych, and the Doctor), behind images, and controlled everything and everyone.

He rewrites and comments on the previous texts. Talking about the death of Zumbrunnen as a Central European, the author says that "finally the waters of the Stream accepted a large Danubian fish with its final secret wish never to return."[44] Thus, he supplements (that is, comments on) the conclusion of *Perverzion* and at the same time reveals the indestructible Author, who comments on the description of the hero's death as though he's watching it in a movie:

> Karl-Joseph loved swimming and splashing in greenish mountain waters. And generally speaking, like all my heroes, he loved water. So, should I reserve a bit of hope for him and me? And write that he *felt* good? That his body did not sense pain, but sensed the current? That his hair flowed in the stream like river grass in Tarkovsky's films? That he felt like a fish in water?[45]

The office of meta-author is a privileged one because the author's rights have transferred to the kitschman, and the view from the position of the "already said" may belong not only to the Author but also to the clip(kitsch) master Volshebnyk, who is governed by the effect of the "beautiful":

> Yarchyk Volshebnyk looked at him [the corpse—*T. H.*] from the river-bank and thought: "Quite beautiful." Such an amazing angle, just shoot away: the head thrown back and ceaselessly bobbing in the waters. The trunk picturesquely spread over rocks. An elbow bent in an interesting way; the other arm stretched sideways but with the wrist submerged in water (Yar-

---

42 Ibid., 259.
43 Ibid.
44 Ibid., 297.
45 Ibid., 239.

chyk did not know that the bones were broken in three places, but this is not essential—we've agreed that *here* there is no pain), so Karl-Joseph was stroking with his hand a languid river-dog.[46]

Twelve Circles, then, is also a narrative about history and the passage of time. A man and his wife have found shelter from a blizzard in an old Chrysler Imperial they found in a scrapyard. The Chrysler Imperial is a cult object for the Bubabists. Rather than a phallic, overman image, as in the times of Bu-Ba-Bu, it becomes an almost maternal body, which protects the lovers:

> ... *this* wounded iron body, which some ten years ago had been a veritable fortress on wheels (as it was seen then in the midst of fireworks on the streets of Chortopil), so this body, or rather shell, right now after unknown road adventures and plain adventures over the last ten years, found itself in *this* ravine, hiding two people simultaneously close to *and* distant from one another.[47]

Thus Andrukhovych sums up his postmodern experience, using a new topos of the post-postmodern age: the topos of maternity.

The novel-feuilleton is a mix of auto-quotations, reflections, repetitions, and analogies. Ultimately, *Twelve Circles* fully reveals the tautological quality that is characteristic of Andrukhovych—after all, all his novels are about one and the same situation and about one and the same protagonist. In this post-postmodern work, his romantic mystery of love between Man and Woman also remains unchanged. Only the carnival bids farewell.

---

46 Ibid., 239–40.
47 Ibid., 259–60.

CHAPTER 18

After the Carnival: Bu-Ba-Bu Postmortem

> Our youth went by concurrently with the totalitarian era. And wasn't our Bubabism, which we also considered a challenge for totalitarianism, something that at the same time could only exist due to that totalitarianism?
> —Yuri Andrukhovych

One of the main questions in Oleh Ilnytzkyj and Marko Pavlyshyn's discussion about the essence of Ukrainian postmodernism concerns the "transplantation" of postmodernism as a product of Western civilization and its "organicity" in Ukraine.[1] However, the development of postmodernism in Ukraine was not predicated on the logic of late capitalism, as it was in the West, but on the deformation of the totalitarian mindset after communism. Hence the Ukrainian case essentially changes the very idea of postmodernism as a solely Western phenomenon; instead, postmodernism is found to have regional varieties.[2]

At the end of the twentieth century, postmodernism fulfills the function of a new ironism, which reevaluates all existing cultural discourses. This is not the same, however, as subversion or revenge; rather, it is a *re*-description of things which, according to Richard Rorty, lies in the discovery of the model of a linguistic behavior that would seduce the next generation and would urge it to look for new forms of a non-linguistic behavior.[3] As though to confirm this formula,

---

1   See Il′nyts′kyi, "Transplantatsiia postmodernizmu," 111–19.
2   Mihai Sehed-Masak correctly points out that "Looking at the literature of the former communist world, we stay reassured that even the most perceptive theoreticians of postmodernity pay little attention to Eastern Europe" ("Postmodernizm i postkomunizm," *Krytyka* 5 [1998]: 18).
3   Rorty, *Contingency, Irony, and Solidarity*, 9.

at the end of the twentieth century, Ukrainian carnivalized ironism had first and foremost the form of a linguistic behavior. Yuri Andrukhovych, for instance, declares himself as a primarily "verbal person." Linguistic, ironic behavior defines the image, function, and nature of Bu-Ba-Bu and, becoming a thing of the past, gives way to new models of behavior, this time not of a linguistic but of an institutional and even social content, such as the Orange Revolution.

In the 2000s, it is confirmed by the involvement of "Smoloskypers," the critical strategy behind the reevaluation of values that was elaborated in the periodical *Literatura plius* and the "demiurgic personal creativity" proclaimed in *Pleroma*, and is interpreted not only as an aesthetics but also as an ideological challenge. Further, it reveals the postmodernists' opposition to so-called "testamental, rustic discourse."[4] We can add also the programmatic masculinism of the Psy Sviatoho Yura literary workshop (which quite naturally expressed itself in the somewhat overripe "that which is underneath" ["te, shcho na spodi"] of Yuri Pokalchuk); and, ultimately, the appearance of new "canons" in the form of various secessional anthologies. In this way, we will have a full picture of the postmodern Bubabist ironic attitude toward the ideological interpretation of literature.

Thus, Olha Sedakova's idea about the powerless pacifism behind the postmodern attempt to "play in the world in which there is no power at all" looks inappropriate or even erroneous when we take into consideration, for example, Viktor Pelevin's strategy of engaging the reader. Likewise, Bubabist postmodernism does not lack power because the negativity and ironism of Bu-Ba-Bu—most clearly demonstrated in the faux ritual of the Second Coming (with the Imperial at the head of the procession) and the metaphor of Christ (and Anti-Christ) hidden in the Chrysler—undermines the cult of the bohemian Poets.

It might even be that the birth of postmodern irony proper happened thanks to the late totalitarian era. When the totalitarian mindset collapses, we lose an ironic stance itself, with the discourse of power behind it. Selectiveness and aggressiveness often accompany the acceptance or rejection of authors, supporters, critics, sympathizers, and sponsors. Here mind rituals, such as "the party organization" and "the party literature" come to mind—it is they with which modern Ukrainian literature associates itself more and more. Some critics even publish new strategic "May theses."[5]

---

4   Ol´ha Sedakova, "Postmodernizm: zasvoiennia vidchuzhennia," *Dukh i Litera*, 1–2 (1997): 375.
5   "Ukrainian literature at the border of the millennia is a fundamental and global topic. Therefore, I will try to outline a few problems that, in my opinion, are the most salient for discussion at this specific stage. For a clearer presentation of these problems I choose the form of

Viktor Neborak also confirmed the phenomenon of the "partinization" of literature—this time "paragraphs" rather than "theses."[6] As a sign of the return of irony he proposed to the Association of Ukrainian Writers an end to aesthetic hostilities: "we should not be insincere when discussing each other's works and deeds related to creative activity."[7]

Additionally, if we take into account Volodymyr Yeshkilev's clarification regarding the hierarchical nature of the so-called "Stanislav phenomenon" ("variably salon [metasalon] must be reflected in the "shadows" by the search for the theme of "power" through the alternatives of the lower, cheaper kind")[8], the "party," or "command," structure of the actual Ukrainian literature becomes even more pronounced. This "party-ness" seems to be very different from the ironism-carnivalism, which the Bubabists brought to Ukrainian literary life at the end of the 1980s. Does it testify to the coming of post-postmodernism, or to the crisis or end of Ukrainian postmodernism? Without making a fetish of postmodernism, I would answer both yes and no. Yes, because to a great extent Bubabist carnivalesque and ironic energy have become either sublimated by official culture or have developed into mass culture and kitsch, leading to writer's block in individual Bubabists. No, because a certain stage and form of postmodernism, identical with post-totalitarianism, has come to an end along with the Bubabism-carnivalism.

Thus, the creative work of the literary group Bu-Ba-Bu has become the first most expressive form of the "postmodernist turn" in Ukrainian literature. Moreover, in tandem with a whole cluster of contextual metaphors, ideologies, and models of ironic behavior, Bu-Ba-Bu is also an entire socio-cultural phenomenon called Bubabism.

1. BUBABISM AS CULTURAL CRITIQUE

The "vocabularies" of culture that were the rhetorical and social target of Bu-Ba-Bu's subversive energy were taken from the official lexicon of a hybrid of national enlightenment and social kitsch. Soviet mythology and socialist concepts proper,

---

      theses and call them, by analogy, the May theses" (Ievhen Baran, "Literaturni devianosti: pidsumky i perspektyvy," *Kurier Kryvbasu* 116 [1999]: 3).
6   Viktor Neborak, "Literaturna orhanizatsiia i literatura (Prynahidni mirkuvannia kolyshn´oho 'zvil´nenoho' sekretaria z ideino-vykhovnoï roboty komitetu komsomolu L´vivs´koho medychnoho instytutu)," *Literatura plius* 5–6 (1999): 16.
7   Volodymyr Ieshkiliev, "Tin´ stanislavs´koho fenomenu," *Literatura plius* 9–10 (1999): 4–5.
8   Ibid.

it seems, interested the Bubabists to a considerably lesser degree. Only Oleksandr Irvanets openly resorts to parody, subversion, and stylization of kitsch-like socialist realist motifs and melodies ("A Deputy Song," "A Poem to the Mother Tongue," "A Song of Eastern Slavs," and especially the famous anti-populist poem "Love! ..."—more widely known under the title "Love Oklahoma"). In an ironic-grotesque way, Irvanets also comments on the famous "classics lessons," which from elementary school level onward defined the codex of the Soviet education. Andrukhovych rewrites the discourse (or, as Kostetsky and Shevelov call it, "balaka") of totalitarian society on a semiotic level. In particular, Andrukhovych's linguistic experiment traces the communicative slippage of language, and his famous "vocabularies" of repetitions create a mirrored, tautological message. By integrating into their writing clichés, obscene language, semantic lacunae (which can be considered Freudian slips), by using maxims and phrases (such as Irvanets's "classics lessons"), postmodernists emphasize the devaluation of any meaning but the official one, as well as the fungibility of any word and any language in the homogenized discourse of the Soviet Empire.

Andrukhovych fills *The Moscoviad* with verbal blocks borrowed from the archive of the classics, profaned by the school curricula (from Shevchenko to Lesia Ukrainka), and with vocabulary that is automatically reproduced thanks to Soviet propaganda. He also adds some masscult frames based on phrases that have been popularized by pop songs, such as "nesypmniesol'naranu" or "mertvi bdzholy ne hudut'."[9]

Bubabist linguistic irony, not without narcissism, was a peculiar form of the carnivalesque that could combine, in an almost gnostic synthesis, language and matter, spirit and body. In short, those things that are falsely and absolutely divided in totalitarian society. This was due to the fact that totalitarianism, according to Terry Eagleton, is an amalgam of romantic idealism (more precisely, the undiluted kitsch of bombastic idealism) and cynical materialism in which individual bodies and events become undifferentiated and fungible.[10]

Bubabism in its essence was a public event, first of all, a performance, overcrowded with voices (both the authors' and the masquerade's) and designed for reading aloud. Voice is a certain, virtually angelic aspect of being, which was the only one to fix the autonomy of both the colonial and the totalitarian subject

---

9 Literally: "Don't-put-salt-on-my-wound" (Russian); "dead bees do not buzz" (Ukrainian) (translator's note).
10 Terry Eagleton, "Estrangement and Irony in the Fiction of Milan Kundera," in *The Eagleton Reader*, ed. Stephen Regan (Oxford: Blackwell, 1998), 93.

in the late Soviet era. This was the final stage of depersonalization. In fiction, the body was virtually annihilated (despite the emphatic biographical reality of names, the protagonists of Russian postmodernist writers such as Venedikt Erofeev (*Moscow to the End of the Line* [*Moscow-Petushki*]), Sasha Sokolov (*Palisandriia*), Sergei Gandlevsky (*Trepanation of the Skull*), and others, are almost entirely disembodied subjects, most commonly drunks whose appearances and ages are indeterminate. Spiritual being, instead (on the level of the individual as well the level of the masses), has been boiled down to an "inner voice," which metonymically replaces the personality of the subject.

Bubabists took notice of these disturbances of physics and metaphysics and called them a masquerade (or, more precisely, a carnival), in which a mask ("maskhara) took the place of "person" and "face", while language, spirit, and the self retreated into the voice. Pirandello drama, which proposed that it was possible to open a chink between the face and the mask, is nowhere to be found here. There is no face anymore. Only the voice remains, which is the field of the freedom, action, irony, and narcissism of the Bubabists.

In this regard, it is worthwhile attending to the words of Andrukhovych, Bu-Ba-Bu's founding father. He is interested in "joints"—that is, in areas where "the passage of spirit into the material" occurs.[11] "Everything I do in literature," he notes, "can ultimately be boiled down to a mysterious and almost maniacal groping for these painful and sweet joints."[12] Hence the topography of the "inner voice" (the place of alienation and loneliness) in the field of the bodily, often lost in the collective social-cultural background, changeable and interchangeable, and very often female, is the moral and philosophical (however bombastic it may sound) basis of Bubabist carnivalesque. A game with Narcissus is its special feature.

Bubabism is a phenomenon that neatly fits the context of late totalitarianism. Formally and ideologically, it is closely related to the jokes so common in the informal culture of totalitarian society. Indeed, Alexandra Hrycak very convincingly demonstrates that the Bubabist's subversion of both pro-Soviet symbols of official youth organizations and official Soviet celebrations inspired the mass performances at the beginning of the nineties. For example, Bubabists were the main authors and characters of Vyvykh–92 ("Kink–92").[13]

---

11 Andrukhovych, "Avtobiohrafiia," 31.
12 Ibid.
13 Alexandra Hrycak, "The Coming of 'Chrysler Imperial': Ukrainian Youth and Rituals of Resistance," *Harvard Ukrainian Studies* 21, nos. 1–2 (June 1997): 63–91.

The Bubabists are not interested in rewriting, or more accurately "suspending"-imitating, the stylistic matrixes of traditional Ukrainian literature (although Shevchenko, Sosiura, and Tychyna, for example, are minor exceptions). They are largely indifferent to the vocabularies of Nechui-Levytsky, Hrinchenko, and Honchar, and other writers central to the hybrid Soviet-populist canon. The Bubabists may occasionally use them metonymically, however, as in their employment of Mykola Nahnybida as an allegory of noncreativity. The following aside in *The Moscoviad* is typical: "'Well I never!' you want to exclaim, like a character from Soviet Ukrainian prose."[14]

Devoid of an avant-garde aspiration to supplement or subvert Ukrainian literary tradition, Bubabism has not become a retrospective apocalypse. The object of the Bubabists' interest is a play with the "voids" and taboos of the national culture. The Bubabists' inversions and transgressions (the annihilation of borders and the transfer of meaning), however, are free from what Jean Baudrillard calls "repentance." (Recall Tengiz Abuladze's famous film *Repentance*, popular during the time of perestroika.) They are not dominated, either, by another element of postmodernism at the time of its inception in the West—by what John Barth called "resentment" (meaning both indignation and offence). Repentance, along with offence, combined in ritualized and carnivalized laughter, and was seen by the Bubabists as the only way to escape totally prostituted Soviet (indeed, totalitarian in general) officialdom.

Thus, Bubabism as a version of post-Soviet postmodernism is a sociocultural critique rather than a radical avant-garde movement. Bu-Ba-Bu as a whole was not an experimental avant-garde group such as OBERIU, with which it is often compared. Oberiuts, as is well known, were united around the idea of "real," or "concrete," art: art based on the ideas of "estrangement" and "nonsense" ("zaum") developed by Russian formalism. It was supposed to open up the hidden impulses of art itself. Bubabists, instead, wanted mass action and resonance, blurring the limits of poetic language as such. Bakhtin's theory of carnival fulfilled their requirements much better. First and foremost, Bu-Ba-Bu was a challenge to the literary Establishment in Ukraine by the youngest literary generation. It reflected the need for "gangsterism" and decapsulation of official literature, along with a radical reimagining of the idea of the Ukrainian man of letters as almost the primary actor in modern Ukrainian history. "What did

---

14 Andrukhovych, *The Moscoviad*, 35.

Bu-Ba-Bu do with respect to this context?" Neborak reflected in this regard.[15] "It named itself in a quite provocative manner, began to practice on the most responsible way of presenting poems (first of all, recitation by heart, which you cannot expect from Kordun or, let's say, [Ihor] Malenky), and about the general structural integrity of a party. In one word: about the image."[16] As Yuri Andrukhovych said later, it all grew out of a single: "to melt … the lump of the dreary undereducated gravity set on everything Ukrainian."[17] In the final analysis, Andrukhovych states, having constructed the ideology and legend of Bu-Ba-Bu, "it was not we who created something in the culture; it was the culture that created us."[18] If, then, we approach Bubabism as a cultural and historical fact that, unfortunately, has already passed, it seems that it was not only an inversion but also a negation of the semiotic space of the national culture.

2. BU-BA-BU AS CARNIVAL

The ideological motives of the Bubabists, however, should not be overstated. In the case of Bu-Ba-Bu, we are dealing with something like a group of young poets' initiation ritual—a group whose ally was the buildup of revolutionary tension forming among the lower classes and feared by the upper classes. A socio-cultural desire *and* need for play, coincided.

Hand in hand with the Soviet Union's drift towards collapse, the Bubabists start their game by drawing on the Bakhtinian carnivalesque. Let us take into account that this theory, although created in the late 1930s, becomes popular and generally accessible in the 1980s and early 1990s (for example, in the 1990 edition of *Rabelais and His World*, as the English translation calls it, and the publication in 1975 and 1979 of Bakhtin's aesthetic-phenomenological work). Essentially, the discovery of Bakhtin is an event of the late Brezhnev era. After Bakhtin's death in 1975, indeed, his vogue emerged in academic circles.

"Ours was the time," Andrukhovych writes, and this statement is worth remembering. It was a period of youthful enthusiasm and of a desire for something other than official culture—a fraternal or even familial worldview.[19] A

---

15 Neborak, "Z vysoty Litaiuchoï Holovy," 57.
16 Ibid.
17 Andrukhovych, "Ave, 'Kraisler'!," 6.
18 Ibid., 8.
19 The idea of contact—one of the most significant themes in Bakhtin's work on carnival—also occasioned Bu-Ba-Bu's choice of ideology: an aesthetic worldview associated with the liminal condition of being and the play between life and nonlife. This exploration of borders

moment of inversion occurred: an aesthetically programmed form of life becomes dominant. As Bakhtin emphasizes, carnival resembles a real (if transient) form of life: "Thus carnival is the people's second life, organized on the basis of laughter. It is a festive life"; in addition, "the utopian ideal and the realistic merged in this carnival experience, unique of its kind."[20]

The Bakhtinian carnivalesque also appealed to the Bubabists because of its subversive attitude toward official culture. As the reassessment of culture at the beginning of perestroika was limited to filling in "blank spots," the cultural mindset itself remained largely unchanged: irony still had no place. The lower-class revolution that was Bubabism, then, was directed toward the subversion of official culture's sacred idols and semiotic codes.

As is well known, for the Bubabists carnival became the formula for bringing together the socio-cultural with linguistic play in the late 1980s and early 1990s—that is, during perestroika. It made no difference that carnival had become kitsch long ago and, in addition, often had a role in official culture. Carnival is also characterized by transience: it is a concentrated time-space with an author and a director behind it. As anthropologists maintain, every public celebration amounts to a rope bridge hung across a ravine; or, perhaps, in Andrukhovych, the rope upon which Stakh Perfetsky teeters over a theatre pit.

The idea of carnival, which the Bubabists took from Bakhtin, at the end of the 1980s was associated with the national revival and was interpreted as the quintessence of the mass "People's Movement" ("Narodnyi Rukh"). Further, at its heart was the energy of the young. The apotheosis of the Bubabist carnival, Vyvykh–92, however, confirmed that despite the Bubabists' euphoric confidence in freedom, there were special police detachments ready to catch anyone who went too far. According to Andrukhovych, the director of Vyvykh, Serhii Proskurnia, was nearly arrested.

Carnival is, to a great degree, a verbal phenomenon. Although Bakhtin differentiates verbal, jocular, and public works from ritual-spectacular forms (that is, celebrations of the carnivalesque kind), linguistic games are the basis of carnival. As Neborak's Pliashkosmoktach maintains, "if people care about the word ((the art of the word, not just words)), they should center their lives

---

    lead to the notion that reality itself might be different. Ideas of transience and illusion pointed toward an alternate reality. As Bakhtin wrote, "a certain form of life, which was real and ideal at the same time" (Mikhail Bakhtin, *Rabelais and His World*, trans. Helene Iswolsky (Bloomington: Indiana University Press, 1984), 8.
20 Ibid., 8, 10.

around it."²¹ This gap between word and matter became the place where the Bubabist linguistic-ironic—almost cabbalistic—game unfolded. Together with other Satanic, masonic, and orphic pseudo-rituals, it gave Bubabism a mysterious character. The underground origin of the Bubabists is worth noting. Andrukhovych states: "What were we? A sect? A secret club? An informal association of snobs? A gathering of juvenile Vodka-Philosophen?"²²

It would be a great simplification, however, to perceive the subversion of the official, or high, culture (which legalizes Bakhtinian carnival) as an unusually organic phenomenon that is caused by both the ambivalence and creative energy of low culture, as the theorist maintained. It is well known that mass culture in the postmodern, technological era employs the forms of high culture, without rejecting or subverting them, but rather transforming them into a series of utilitarian "vocabularies." This is a process that can be categorized not as a form of organic reflection, characteristic of the modern era, but as a mode of a "non-naive" postmodern thinking. The Bubabists' works, which have become "readable" and have shaped the phenomenon of the mass Ukrainian-speaking reader, combined the constructions of high culture and the forms of mass literature.

3. BUBABISM AS KITSCH

Bakhtinian carnivalization contained a huge charge of radical-proletarian (low) ideology—"this conviction that a radical change and renewal of all that exists became 'necessary and possible'"²³ After all, the theory of the carnivalesque, which Bakhtin developed during the Stalinist period, was born out of the socio-linguistic practice of the totalitarian era and, in its essence, was an inversion of the socialist utopia. It was caused by an educated, neoromantic voluntarism, rational idealism, and the politicization of aesthetics. In other words, Bakhtinian carnival, with its advocacy of low culture, contained an inverted, grotesque-demonic topos of plebeian revenge. Having adopted the Bakhtinian carnivalesque, the Bubabists unconsciously also appropriated this plebeian revanchism. Andrukhovych wrote: "It seems that at first it was not as much aesthetics as a mode of survival. Or aesthetics as a mode of survival. Or vice versa. In other words, it was an attempt to be as free as possible in a situation that was

---

21 Viktor Neborak, "(Pliashkosmoktach)—(Voda). Rozdil iz romanu *Pan Bazio ta reshta*," *Suchasnist'* 5 (1994): 55.
22 Andrukhovych, *Dezorientatsiia na mistsevosti*, 110.
23 Bakhtin, *Rabelais and His World*, 274.

generally devoid of freedom."[24] Neborak agreed with him: "It seems to me that we reproduced the spirit of the time when people were slowly coming out of their masks. For that we were rewarded with attention, although often one mask was just replaced with another."[25]

Aestheticized, carnivalesque kitsch became the form through which, and the limits within, the Bubabists were winning back for themselves (as well as for the whole youth society of the late totalitarian era) space for freedom. Bubabism joined in (along with other mass actions of the time) with the creation of the phenomenon of the Ukrainian kitschmen—those individuals who started to read, listen, and sing "bubs." ("Bubs" is an abbreviation of Bubabist, refers to the word for German boys—"buben"—and is a kitsch nickname for the Bubabists.)

Bubabist carnival was deployed ritualistically: as a series of symbols and codes for ironic behavior and assessment. It marked the transition from semi-official Soviet thought to individualistic, liberal thought. This explains the ritual function of the role of masks/characters in the poetic opera *Chrysler Imperial* and displays the nature of the Bubabist metaphors: *Chrysler Imperial* (the coming of the New Spirit, the ideas of gangsterism and subversion in Ukrainian society); *The Queen of Imbeciles* (an erotic transcription of the "bewitched" ("prychynna"); *Cossack Jamaica* (a postmodern version of *Cossack Mamai*—"on this side there is Bahamas-mama; on the other side there are the palms of Haiti"); *Recreations* (the tripartite personality of Poet, Lover, and Narcissus as the main subject of the national literary revival); *Flying Head* (an actualized tradition—a baroque metaphor for the mask and square); *Love Oklahoma* (a treatise on the erotic-patriotic complex); and Turbation of Masses (on the erotic-civic role of the man of letters).

Carnival becomes the principal idea of Bubabism (from aesthetics to the "practical, quasi-philosophy of life," according to Andrukhovych), which created the fatal and sacral, universal and carnivalesque nature of Bu-Ba-Bu. The masculine bent of Bu-Ba-Bu is obvious. What did the carnivalesque mean for the Bubabists? It meant the Great Game and an illusion of freedom, and also the victory of life over death in the most general sense. In other words, it meant the revival of the soul (or, more precisely, of "little" local narratives). As Neborak asked: "can the soul die with its blood and laughter, poetry, cursing, wine, music, farce, love, audacity, buffoonery, ritual, magic, theatre, once again

---

24 Andrukhovych, "Ave, 'Kraisler'!," 5.
25 Neborak, "Z vysoty Litaiuchoï Holovy," 58.

buffoonery, laughter, crying, high, taste, jazz, rock, jazz-rock?"[26] The history of Bu-Ba-Bu shows that yes, it can—when youth and the game end, when you admit, like Naborak in his "Taking off from Akademichna Street in our Lord's Year 1997", "The Akademichna, sister, you are leading nowhere!"[27] "V do**bu** / **ba**nknotnu / ya / vid**buv**!" ["I have left for the bank-note era!"][28] confesses Neborak's lyrical subject and he graphically inscribes the phenomenon of Bu-Ba-Bu into the "banknotna" ("bank-note") era. It seems, as Neborak implies with his ironic-gnostic attitude toward the rhythmic foundation of being, as well as his great disappointment with the "ironic negativity" of carnival, that carnival is slowly turning into a fetish and topos of self-will."

The experience of Bubabist experiment with the carnivalesque supports the thesis that modern carnival culturally and socially can only be organized action. It is, therefore, transient and even fake. Oriented toward the domination of the "bottom" and an illusion of unity (a collective, grotesque body), it links, for instance, street celebration and late imperial Moscow, since in both cases everything is boiled down to hypertrophied corporeality—that is, permanent performance.

"Each of us breathes, drinks, loves, stinks the same way," proclaims the representative of "the great nation" Yura Golitsyn in Andrukhovych's *The Moscoviad*,[29] a novel-feuilleton about empire and post-imperial discourse. This shared condition apparently provides a foundation for "being together" and for feeling like a "big family." Brotherhood, freedom, and equality, perverted by Soviet ideology, are ripe for carnival and kitsch, themselves predicated upon an illusion of unity as well as upon an imitation of satisfaction and beauty. In its seductive, counterfeit world, with its characteristic false hovering over being, and with its comfortable oblivion, any heterogeneity and ambivalence of knowledge is eliminated, while the image dominates over the object and reality.

Kitsch also uses automatism of speech and the narcissistic reflexivity of language seen in Andrukhovych's prose fiction when he plays with "lists" of names as their simulacra: "Perhaps you could ask Dzerzhinsky. The Iron Felix. No, the Iron Sigfried. Ask Konrad Klaus Erich Dzerzhinsky. Or Rainer Anselm Willibald Kirov. Or Wolfgang Theodore Amadeus Lenin."[30]

---

26 Ibid., 14.
27 Neborak, *Litostroton*, 417. "Akademichna Street" in Neborak is one of the streets of the Bubabist carnival.
28 Ibid., 416.
29 Andrukhovych, *The Moscoviad*, 46.
30 Ibid., 103.

Language turns into imitation. This, by the way, becomes the ground for the Bubabists' linguistic-ironic carnival, which is equivalent to postmodernist kitsch, where quotes become interchangeable and anonymous, and culture is reduced to a library collector and analogous with sewage. However, it does not surprise because, as Otto von F. states in his pseudo-notes for his Noble Prize speech, "a library ... is nothing else but a large (bigger or smaller) sewer of human spirit."[31]

Entirely in accord with carnival, the author also becomes serpentine, changeable; after all, "[w]hat difference does it make what the name of this frail body is? The main thing is the immortal soul. But it does not have an earthly name, any human name is too small for it. In some sense you are no Arthur either. ...";[32] this is the way KGB agent "Sachko," Otto von F.'s double, ironizes, reflecting the "postmodern situation." Hence it is not a coincidence that Stakh Perfetsky also "had countless faces and countless names":

> They called him Stakh Perfetsky and Carp Loverboysky and Sheatfish Saintlymansky and Pierre Fukinsky and High-as-a-Kite Birdsky. But they also called him Gluck, Bloom, Vrubl, Strudl and Schnabl. In addition he was Jonah of the Fish and Siura of the Balls and Glory of the Days. But he was also Sargent Pepper, Juan Perez, Petey Peppa, Pepperonimon, and Ertz-Hertz-Pertz. Some knew him as Persiansky, Parthiansky, Personsky, Profansky, and Perfavorsky.[33]

Narcissism is another attribute of Bubabist kitsch-carnival. All the inversions, doubles and masks in, let us say, Andrukhovych's *Recreations*, can be interpreted as the unfolding of a global metaphor about Ukraine's poetic narcissism: the "external" and "internal" voices of his bohemian poets are divided among the "beautiful poet," the "colossal poet," and the "tremendous poet." One could also add to this list the "great sleeping prophet." The Bubabists' play with Narcissus developed as they saturated their work, especially the prose ones, with biographical elements and with echoes of real names and recognizable faces. This is how the narcissistic image of the Bubabists was created, and through it also the image of Ukrainian bohemia in the nineties.

---

31 Ibid., 116.
32 Ibid. 127.
33 Andrukhovych, *Perverzion*, 154.

4. CARNIVAL AS A COMPLETE FACT

Carnival naturally subsists on carnivalization; the latter, however, is not reduced to carnival alone. Carnivalization contains a form of irony and a double *ludus modus* (a double mode of play), which takes socio-cultural action outside carnival itself. Carnivalization, as it were, pulsates with carnivals, but only in order to reject them and to affirm the infinity of being itself. At the same time, the double *ludus modus* of carnivalization (among other modes) emphasizes the variability of life and the contingency of game itself. Carnival also has a demonic nature: it attracts and bewitches only to later reject. It takes off its mask.

The bluff of carnival was already apparent in Andrukhovych's *Recreations*. There, carnivalization permeates various levels of the social, cultural, and erotic behavior of the characters and transports the Festival from the Market Square to the margins of side streets and outskirts. The jocular nature of the carnival, with its inversions and disguises, turns out to be a merely superficial imitation. In reality, the deeper ironic perspective of the action lies in the fact that a military coup can take a form of a masquerade. The metaphysics of the carnival—free and intimate human contact—also turns out to be an illusion (such as the touching of Martofliak and Marta's hands at the end of the novel).

As a socio-cultural action, carnival counterfeited an inversion of authoritative roles and masks and asserted the illusory nature of seemingly invincible institutions (from the nation and history to the family). It required bloody sacrifices: for instance, Khomsky's initiations are linked to the beating of a street junky. Each character of *Recreations*, having experienced his or her own bodily incarnated carnival, remains lonely and solitary. (Yet, as we know, the sense of carnival, according to Bakhtin, is an experience of unity.) All attempts at repeating patrimonial (as well as Austro-"European") history for Hryts Shtundera and Yurko Nemyrych end in literal and symbolic failure. It seems that the pseudo-incestuous and pseudo-transgressive eroticisms of Martofliak and Khomsky do not change their characters.

Overall, carnival appears to be only a legalized form of socio-cultural deviation allowed by authorities. The figure of the "greatest" director ever and anywhere, discernible behind the scene of the carnival, if we depart a little from the prototype Serhii Proskurnia, transforms into a demonic image of evil, in which the baroque rogue Matsapura and the aestheticized image of Beria merge. (The demonic subtext of a "great conspiracy" is clearly seen both in *Perverzion* and in *The Moscoviad*.)

Discovering the demonic side of carnival-as-imitation, Bubabism becomes an apocalyptic performance rather than an ironic one. It also shows itself as a phenomenon inscribed in the structures of official culture. Carnival—an illusion of freedom, a freedom test—clears the way for an alternative (or pseudo-alternative) official culture. Evidence of this is the destruction of the privileged position of the author, something only possible in the post-Soviet period, and the creation of a new official culture that includes also such "younger laureates" as Oles Ulianenko, Yevhen Pashkovsky, Ihor Rymaruk, Vasyl Herasymyuk, and Viacheslav Medvid. Worth noticing here is a proposal to nominate Andrukhovych for the Shevchenko Literary Award.[34] Against the background of the post-Soviet literature becoming the new official culture, the Bu-Ba-Bu carnivalesque flirts with the masscult and kitsch (all Bu-Ba-Bu's performances and presentations were nothing but kitsch).[35]

In the most general sense, through the carnivalesque the Bubabists showed the transformation of traditional feudal Ukrainian society into a society of performance, as described by Guy Debord. In other words, they metaphorically modelled (and made more pointed) the idea of democratic society, which, according to Debord, is first and foremost a consumer society in which everything is bought, advertised, and sold, and hence a complete spectacle. The desire to salvage the "book" world and to fathom the power of the word in the face of the onslaught of mass culture, telecommunication, and technology is based, for example in Pashkovsky, not on a game but on a revelation: in the post-Chornobyl era, he wants, in writing, sincerity to run out of steam, and associates his authorship with "unnamed graphite rods" in the hell of the Chornobyl reactor. The apocalyptic world of Pashkovsky poignantly reflects the irony carnival, the utopian and ideal functions of which break down when encountering the existential torture of Pashkovsky's lyrical "I." In the ironic-grotesque world, carnival revives with the help of parody. We can see that in *Perverzion*, in which the opera of operas *Orpheus in Venice* is staged, this truly postmodern simulacrum and pseudo-carnival consisting of repetition and quotation all of a sudden becomes a field of freedom—the scene of the birth of the "real Orpheus." Perfetsky, the novel's protagonist, escapes pursuit by fleeing from the audience to the stage and ending up, thereby, within a new play.

---

34 As is well known, Andrukhovych refused to take part in the competition.
35 See Iurii Andrukhovych, Andrii Bondar, and Serhii Zhadan, *Maskul´t. Eseï ta poeziï z novykh knyzhok* (Kyiv: Krytyka, 2003), 8–17.

In this way, Andrukhovych obtrusively and ironically entangles us in the unending transformations of his "eternal" hero. Only in the sense of such a "suspended" irony-carnivalism we can interpret Bubabism as a phenomenon of postmodernism. Hence "the end of the carnival," proclaimed by the "bubs" themselves, was not as much a token of carnivalesque postmodernism's coming to a close as it was a culmination of carnival. Having created it, some of the Bubabists (like Neborak) experience the "spoiling," which occurred "according to all the rules of the ritual," when "the Son of Man" is turned into a Cult Thing.

> I left behind—myself, as a stupid TV show,
> As a chain of metamorphoses, as a photograph album, as a hall,
> Filled with the wax of figures or the gypsum of sculptures or skins
> And stuffed animals. I have got outside. I have liberated my sight.[36]

Others, like Andrukhovych, hint at the Eternal Return, disappearing, together with his totem Perfetsky-Fish, in the waters of a Venetian canal.

5. A POSTHUMOUS ENCYCLOPEDIA ARTICLE ON BU-BA-BU

The period of the greatest activity of Bu-Ba-Bu (thirty-two concerts and poetic performances) occurred in the years 1987–91. Bu-Ba-Bu's apotheosis was the festival Vyvykh–92 ("Kink–2"), when the main festival action comprised of four performances (from 1–4 October 1992) of Bu-Ba-Bu's poetic opera *Chrysler Imperial* (directed by Serhii Proskurnia). In 1996, the print version of *Chrysler Imperial* (*Chetver-6*), illustrated by Yurko and Olha Koch, marked the end of the "dynamic period" of Bu-Ba-Bu's existence.[37]

6. BU-BA-BU'S COMEBACK

The escape from utopia, which marked the period of postcommunism, influenced the understanding of postmodernism's possibilities. Mihai Sehed-Masak has a point when he says "that the passage from communism to post-communism probably caused the decline of postmodernism."[38] Postmodernism inevitably parts with

---

36 Neborak, *Litostroton*, 367.
37 "Bu-Ba-Bu," in Ieshkiliev and Andrukhovych, eds., *Pleroma. Mala ukraïns´ka entsyklopediia aktual´noï literatury*, 35.
38 Sehed´-Masak, "Postmodernizm i postkomunizm," 20.

the carnivalesque because it is grounded, as it has been noted already, on the idea of a socio-cultural utopia and the aggressiveness of the "bottom." Postmodernism itself, in spite of its flippancy and play, has restored for us the value not of "great illusions" but of "small narratives," such as human contact, understanding, the postponement of death, the intrigue of knowledge, the evanescence of a moment, and the indestructibility of vocabulary. Indeed, this was what the Bubabists also did.

The nature of carnival is ironic yet unheroic; it is full of life and eroticism yet not masculine. In time, however, a common factor in Bubabist carnival-kitsch became apparent: the superhero who behaved like a superman. In a sense, Bubabism bore the narcissism of the superhero and criticized the national poet-messiah from the position of the bohemian poet. Carnival's masculinity, declared by the inversion of the name (Ba-Bu-By!),[39] was later confirmed by the emergence of the literary group "Psy Sviatoho Yura" ("The Dogs of St. George"), created by analogy with an order of chivalry. All the Bubabists became members. Yuri Pokalchuk, one of the group, associates the expectation of the renaissance with the image of St. George slaying the dragon. "The taming of the dragon is first and foremost taming yourself, your own frenzy; it is a manifestation of the eternal need of the male element to conquer itself," proclaims the program performance of the group.[40] "After all, what is the dragon if not a woman who twines around the stem of manhood."[41] Neborak's anti-feminist definition of the "pussy-postmodernist" has become notorious: "Pussy is a fluffy system, self-sufficient and independent, / like an empress," who, however, "at least once a year, o poor darling, requires / a cat!"[42]

As though referring to Bakhtinian carnival, with its "low" and hence a "feminine" basis, Marta in *Recreations* sums up the Bubabist superman-narcissistic complex: "all you know about is yourselves, all of you."[43]

7. MUTATIONS OF THE GREAT CARNIVAL

The prose trilogy of Yuri Andrukhovych reaffirmed the metamorphoses of the idea of carnival and the unfolding of postmodernist thinking. The first novel, *Recreations* (1990), demystified the ideal of a national poet, immersing in the

---

39 Approximate translation: "Give me a girl!" (translator's note).
40 Iurii Pokal´chuk, "Vershnyk letyt´ nad svitom," in *Psy sviatoho Iura. Literaturnyi al´manakh* (Lviv: Prosvita, 1997), 10.
41 Ibid.
42 Neborak, *Litostroton*, 379–80.
43 Andrukhovych, *Recreations*, 113.

vortex of the carnivalized, phallic Festival of the Resurrecting Spirit the Bubabists' personas and masks. (In the author's afterword, "Life as Dream" (1997), Andrukhovych once more confirms the "biographism" rather than "literariness" of the characters.) The author combines erotic, patriotic, and superheroic motifs and grinds each character's identity in an inversion game. This is an auto-parody of Bu-Ba-Bu, which at the same time testifies to an overcoming of alienation from one's own nature, history, and true love, and hints at the complete bankruptcy of any revival or change achieved through carnival.

Another participant of the Bu-Ba-Bu group, Viktor Neborak, later exercises not an autobiographical but, this time, an archetypal analysis of the carnival phenomenon. Neborak uses Ivan Kotliarevsky's *Eneida* (1798)—a burlesque travesty and a cult work for the Bubabists—as the carnival's prototype. The work attracts the authors by its baroque play, stylization, irony, and the author's freedom, which perhaps for the first time was asserted in Ukrainian literature.[44]

Neborak's archetypal analysis of carnival reminds one of Victor Turner's theory, which regards carnival as a state of liminality and rite of passage. Maidan (meaning "square") is a liminal place, the "center" of the play's action. During the play-passage, the players periodically find themselves on the periphery of the stage; for this reason, play marginalizes. The demonism of the play shows itself exactly in this process of marginalization.

In time, play—the element of the bubs, including Neborak as one of the carnivalists—suddenly started to scare people away with its demonism, while the freedom offered by carnival turned out to be the stubbornness of an outsider. Neborak ties the game to "uprootedness," like, for instance, in a running race, when your legs appear to be tied together and tangled as though in a sack. Beside the physical uprootedness, there is always a danger in stopping during the game, not finishing, and therefore not completing the cycle of the playing-transformation-initiation-spiritual consecration.

Such a metaphysics of the game opens itself to Viktor Neborak during his rereading of Kotliarevsky's comic *Eneida*. After all, the wanderings of the bohemian Aeneas can be considered an archetypal image for the Bubabists, while *Eneida* can be regarded an autobiographical work in a symbolic sense. Starting from the vortex of transformations and the masquerade mutations, perceived through *Eneida*, Neborak has analyzed the "Bubabist" as an archetype of a hero who gets "uprooted."

---

44   It is worth noting that in 1998 the Bubabists celebrated the 200th anniversary of the publication of Kotliarevsky's *Eneida*.

Overall, Neborak only generalized the metaphysics of carnival that time and again was expressed in Andrukhovych's novels. Andruhovych's "horror novel" *The Moscoviad* immerses the reader in an atmosphere of Moscow as a metropolis and the capital of a dead empire. The wanderings of a Ukrainian through the circles and streets of the ghost city are accompanied by irony and by parodies of the signs and symbols of the empire. Its genre, in more precise terms, would be novel-apocrypha. However, at the same time, the colonial subject here gets reduced to a shadow-victim.

Venedikt Erofeev's *Moscow to the End of the Line* demonstrates that carnivalization does not need an inversion of the center from the perspective of the margins or the bottom.[45] Along with Venichka, Erofeev does not want to overcome the opposition of the center and the periphery, corporeality and spirituality, power and subjectivity. However, he apophatically—almost in accordance with St. Augustine—states that the bottom is the truest elevation, or the top, and that they do not need any reversal. This is another version of carnivalization as irony; in Ukrainian literature it has been left unused, breaking up into the linguistic-cultural (even cult!) and subversive play of Andrukhovych and the marginal-apocalyptic anti-play (seriousness) of Ulianenko, Pashkovsky, and Medvid.

Andrukhovych's novel *Perverzion* (1995) is an anthology of postmodern heteroglossia—it is a materialization of all the masks of the "supermanhood" of Bubabism (the bohemian, lover, colonial subject, and superhero), and also an acknowledgement of his bankruptcy. Additionally, it is an attempt to turn the socio-cultural, ironic myth of Bu-Ba-Bu into an esoteric, metaphysical one (the eternal non-return of Orpheus, or the mystery of the King-Fish). On the other hand, if we consider this encounter with the West from a postmodern perspective, it echoes a quotation from the past. In the seventeenth-century manuscript *Relation about the Origins and Customs of the Cossacks* by Alberto Vimina, the author notes that the Cossacks were not short of a "pleasure and wit in a conversation."[46] To corroborate his statement, he provides the following story:

---

45 For this reason, Erofeev's anti-carnivalism is understandable. Mikhail Epstein interprets it as a polemics with Mikhail Bakhtin ("The phenomenon of Venichka, which grows out of pantagruelism, overgrows it," he says. "The carnival itself becomes an object of the carnival, which leads to the sphere of a new, peculiar seriousness." He goes on: Bakhtin "sensed in Erofeev his own that was becoming alien. He sensed a carnival that ceases to be a carnival."). See Mikhail Epshtein, "Posle karnavala, ili Vechnyi Venichka," in Venedikt Erofeev, *Ostav′te moiu dushu v pokoe: Pochti vse* (Moscow: Izd-vo HGS, 1995), 18–19.

46 Alberto Vimina, "Relatsiia pro pokhodzhennia ta zvychaï kozakiv," *Kyïvs′ka starovyna* 5 (1999).

> My assistant told about the grandeur and wonders of Venice, which those in attendance were interested in, so, having narrated sufficiently about dishes, crafts, and riches, he added more about the enormous number of streets, saying that there are so many side streets that even the Venetians can get lost in the city. "No," the Cossack says, "you won't astonish me with Venice because the same happened to me in this small room: after I have sat a few hours at that table, I cannot find the door to go home."[47]

Thus, the modern opposition between the West and Ukraine gets suspended. The Venetian *Relation* looks like a mirror of *Perverzion*, while *Perverzion* itself turns into a "pleasant and witty conversation" from *Relation*.

The hyperspace of the stylization of *Perverzion* bears witness to the fact that kitsch is an important form (and "place") for the desires and intentions of the post-totalitarian Ukrainian subject to be realized in newly discovered Europe. There, inside the kitsch called "Europeanist," the Ukrainian, provincial individual—that is, the subject Central Eastern Europe—who has a predominantly narcissistic attitude, can articulate himself. His sphere, as Andrukhovych puts it ironically, consists of "conversations about Europe, Europa, Epropa."[48]

The universal form of art in totalitarian society is kitsch. Totalitarian kitsch, as Terry Eagleton notes, is a discourse that does not tolerate any doubt or irony. It avoids the dismal grimaces of mutilated life and thought, and smooths them away, reaffirming itself through laughter and sincerity, radiance and euphoria, happiness and unity, all of which are accompanied by a merry march toward a bright future. Romantic idealization, sentimentality, and collective imagery are the basis for this type of discourse.[49]

However, in a totalitarian society there also forms a different kitsch, which lays bare the demonic side of totalitarianism. These forms are based on a carnivalesque worldview. The carnivalesque opposes the world of official festivals and generates a collective energy that undermines ideological power. Its domain is the folk culture of laughter.

Bubabism, having emerged in late totalitarian society in the form of carnival, utilized the de-romantization and irony inherent in demonizing kitsch. Bubabism was a demonizing kitsch of the post-totalitarian era. However, it was also a postmodernism, which rejected carnival as a transient joy and a "bottom" revolution, as well as exposing the seductiveness of the illusion of totalitarian festivals.

---

47 Ibid., 69.
48 Andrukhovych, *Dezoriientatsiia na mistsevosti*, 119.
49 Eagleton, "Estrangement and Irony in the Fiction of Milan Kundera," 95.

Part Four

FACES AND TOPOI OF UKRAINIAN POSTMODERNISM

CHAPTER 19

Narrative Apocalypse: Taras Prokhasko's Topographic Writing

The work of Taras Prokhasko and Yuri Izdryk represents a new wave in Ukrainian postmodernism—its post-carnival stage. This type of postmodernism frees itself from the playful adventure hero as well as from the romantic conception of the bohemian superman. Instead, at the center of this variety of postmodernism is the fragile consciousness and the fractured existential world of an intellectual-outsider who is no longer a sovereign subject. His self-awareness becomes broken up, his body (hybrid and thinking) polymorphous. Such a character is not hermetically sealed; he does not close himself in his inner world and is not preoccupied with metaphysical issues. On the contrary, the consciousness of this kind of postmodern hero is a resonator of sounds and a mirror of fragmented images. Perception is often comprised of informational noises and traces that are left by a reality mediated by multimedia. The self-consciousness of this character is neither organic nor autonomous because its thinking is a medium among multimedia, mass culture, microhistory, and the individual psyche.

A characteristic of this kind of postmodernism is its expressive structuralization, or heteronomy, of writing. This textual polymorphism (from the Greek "morphe"—"form") means that the use of recurring formal microstructures becomes fundamental to the process of image creation and narrative construction. The text seems to generate and, at the same time, fragment itself. This rhetorical structure is often underlain by a topographical and ideographical image. Such images define the unfolding of a fictional narrative that refers not to reality but to a constructed tale about this reality. A distrust of a linear story finds its expression in the fact that reality is replaced by a series of plots: historiography becomes topography, and topoi turn into a source for building stories.

In Taras Prokhasko's novellas (from the book *Anna's Other Days* [1998]), the new model of writing with an expressive mythological core is revealed perhaps for the first time. The narrative deploys mental images-constructs that become virtual projections and then again superimpose themselves on reality—like forms of landscape, for example. Such images can be perceived as a phenomenological topography. In this regard, it would be worthwhile to remember Husserl and Wittgenstein, who maintained that real knowledge is only achieved when we turn from objects to notions; and it is notions that express "possibility" and "meaning"—that is, "reality"—while language formulates such a "reality."[1] We can also recall Gilles Deleuze, who suggests that the novelist

> extract[s] the non-actualizable part of the pure event from symptoms (or, as Blanchot says, to raise the visible to the invisible), to raise everyday actions and passions (like eating, shitting, loving, speaking, or dying) to their noematic attribute and their corresponding pure Event, to go from the physical surface on which symptoms are played out. ..."[2]

Prokhasko's topoi are such noematic attributes, which make it possible for a pure Event and the symptoms of reality to meet.

The genre structure of Prokhasko's work is also worth attention: he writes novella-essays, in which a philosophical idea plays not a small part and the narrative follows a topos, or an image-idea. These topoi (image-ideas) are geometrical schemes, maps, series-collections: an architectural project, herbarium, a tomogram of the brain. As mental images, they deconstruct reality because they superimpose upon it and render it virtual, a likeness of dream-delirium. At the same time, by dint of these topoi, Prokhasko's characters combine various fragments, events, and plots of reality, and thus put together (construct) their hallucinations-dreams, revealing their "pre-sence." They also create a special mythology of the Carpathian region, which grows out of the "tension of mood" and resists an ethnographic description or a fragmentary travelogue.

Prokhasko's novella-essays demonstrate how the flow of time is caught, how traces of life become collections, "how space changes due to the empti-

---

1 See Edmund Husserl, *Logicheskie issledovaniia. Prolegomeny k chistoi logike* (Kyiv: Venturi, 1995), 235.
2 Gilles Deleuze, *The Logic of Sense*, trans. Mark Lester and Charles Stivale, ed. Constantin V. Boundas (London: The Athlon Press, 1990), 238.

ness, nakedness, and the erasure of boundaries" between consciousness and reality. In this way, the protagonist of "The Sense of Pre-sence" ("Vid chuttia pry sutnosti"), for example, collects the traces of metaphysical presence while getting ready to write a filmscript and realizing that what fascinates him is the moment of detachment from life and hence the moment of suspension, materialized in the very process of memorizing:

> Pamva long ago stopped trying to remember how space changes due to the emptiness, nakedness, and erasure of boundaries, how the best correlation between coldness and the brightness of sunshine establishes itself, how leaves forfeit their own smells and begin to all smell the same, how a special plasticity of resistance against constraints appears. It seemed to him that this way he deprives the world of its last properties, that he is not supposed to take anything by memorizing.[3]

The bursting through to the "pure Event" resembles Eastern meditation, while the attention to the "symptoms" (flowers, trees, and air) holds imagination on the border between memory and forgetting.

That which Prokhasko's protagonist is able to achieve is tantamount to a peculiar destructuralization of reality and a reversal of time. The depicted can also be a called necro- and morpho-logical vivisection of bios (life), with a simultaneous collecting of its traces in human memory. There is an impression that it is taking life onto the surface and playing with the sensations of being itself, a mimicry of human existence. Prokhasko's morphological writing reproduces the process of the disintegration of memory, of consciousness, into tiny microstructures: sensations, moments of time, flashes of memory, and plots. All this is carried out in order to recombine anew such elements and compile new structures from them, in a reversed order: from the periphery to the center, from the end to the beginning, from the inanimate to the living, and from the imagined to the real. These surreal action-transitions and the reflections of the surfaces of various things could be illustrated by the lithographs of the well-known Dutch postmodernist graphic artist Maurits Cornelis Escher (1898–1998).

This way, in Prokhasko's novellas bio-genesis and morpho-logy interweave in order to liken the natural and the manmade, life and death, the living life and its formal capture in sensations and words. An unusual combinatory

---

3   Taras Prokhas′ko, "Vid chuttia pry sutnosti," in Taras Prokhas′ko, *Leksykon taiemnykh znan′* (Lviv: Kal′variia, 2003), 78.

phenomenology is established here, something resembling a meditative solitaire from the remnants of things and ideas—the "basic structures of life." His character feels the same:

> Pamva remembered how when he was twenty-five he thought that he had known already all the basic structures of life and that further on they would only repeat themselves, obviously—alternating a little every time, with nothing principally new to occur.[4]

These are maps of the imagined which resemble the rhizoid topography of Gilles Deleuze. At the same time, it is a literary eschatology, because what matters here is not the merging of sensations but their disruptiveness, the gap between an idea and event and between the man-made and the natural. This chasm has yawned between the real and the imagined. Thus, paying special attention to rhetorical and topographical microstructures, Prokhasko creates a catalogue of the end of the world (of the sensations of reality). At the same time, resorting to descriptions and lists and superimposing a map on reality, he does not renew or revive the naturalness of reality but reinforces its fragmentariness, disruptiveness, formality, that is, the artificiality of the real. The symbol of this writing is not "eternal recurrence" but an unceasing restructuration, an "eternal combinatoriness," which brings an understanding that "culture is knowledge about a few people whose properties are rewritten in your genetic code. ..."[5]

Such a rhetorical apocalypse does not expose a quotation, as in Andrukhovych, but a topogram that lies at the basis of symbolic reality. This type of postmodernism is associated with an apocalyptic text, which is not only a metaphysical message about the Day of Judgment but also a peculiar discourse about salvation. Its semantic unity is not hermetic and is not oriented toward truth-seeking; neither is it exhausted by an optimistic promise of the birth of a "new human." As a consequence of its fluidity and unending restructuration and fragmentariness, the apocalyptic message goes astray from the final judgement about the end (death). This is the cause of the expressive morphological structure of the postmodernist writing built on repetitions and combinations of formal structures. The rhetoric appears to resist entropy and arrests the meaning that is slipping out. The apocalypticity of this postmodernism is predicated

---

4   Ibid., 87.
5   Taras Prokhas′ko, "... botake," *Kur′ier Kryvbasu* 119–21 (1999): 338.

on playing with the "already said," on rewriting the texts of the past, and even on the stylization of the un-styled.

The postmodernist apocalypse is connected with a hypertrophy of discourse and autonomy of rhetoric. Distinct structural properties of form, mono-forms that repeat themselves in the text (for example, dreams in Izdryk's novella *Wozzeck*, Prokhasko's mental landscapes, Zholdak's patois, Volodymyr Yeshkilev's wandering symbols, and the phantasms of Yuri Vynnychuk's *Malva Landa*) create an impression of such a rhetorical apocalypse. Andrukhovych's *Perverzion*, in its turn, is a text of texts, entirely dominated by "quotations and verbal games." Andrukhovych's and Izdryk's prose fiction gives many examples of such a rhetorical apocalypse. Mass literary genres from Svitlana Pyrkalo's youth novella "Green Margarita" to Vasyl Kozhelianko's alternative historiography (*A Procession in Moscow*) use the tradition of rhetorical repetition, fragmentation, and the collection of discourses, in their hierarchical montage. Repetition and fragmentation also become part of erotic prose (Yuri Pokalchuk's *The Underclothes* and Yuri Vynnychuk's *The Harem Life*).

Prokhasko's later works also grow out from postmodernist rhetorical apocalypse. In *The UnSimple* (*NeprOsti* [2002]), the author continues to create his phenomenological-rhizoid topography, this time engaging with Carpathian and ancestral myth. "If there is a place, there is history (if there is history, it means there should be a corresponding place). To find a place is to start history," says the narrator of *The UnSimple*.[6] Toponymy crisscrosses here with genetics, and this intersection becomes the place of the creation of the Central European myth of Galicia: the myth about Ialivets and the Galician person of the interwar period.

The mode of historiography Prokhasko uses is telling ("baiennia"), that is, a fairy tale, narrative, fiction, because as the text says, "When you don't remember how you told, how you are told, then there is no one."[7] The writer confirms his mistrust of Grand History and Grand Narratives, and instead affirms small narratives—stories and fables. "Our life and what will remain after depends of what we say …,"[8] Prokhasko says in one of his interviews. This storytelling, in which the real and the fictional intertwine, becomes a way of catching the existential flow of life. Stories (small narratives) have a semiotic nature and are

---

6   Taras Prokhas′ko, *NeprOsti* (Ivano-Frankivsk: Lileia-NV, 2002), 6.
7   Ibid., 12.
8   Taras Prokhas′ko, "Vid toho, shcho i iak my hovorymo, zalezhyt′ nashe zhyttia i to, shcho zalyshyt′sia pislia n′oho …," rozmovliav Oleh Kryshtopa, *Knyzhnyk review* 23, no. 56 (2002): 1.

a material for culture. Photographs, dreams-stories, an animated film, a house built according to the principle of vegetal aesthetics, pysanka which becomes a moving picture, and even the "mobile daguerreotypes of the silhouettes of a germ moving on the rock"[9]—all these forms-morphemes fortify the traces of life. Stories comprise a cultural reservoir; they can be combined into chains, coiled, and spun as though in a centrifuge. Stories create myth and reveal the tree of life. Prokhasko's rhetorical topography revives the stylistics of the Hutsul secession at the beginning of the twentieth century, with its attention to floral ornament, geometrical figures, and embroidery projected into the mountain landscape.

The reservoir of stories is the basis of not only culture but also mythology—for example, the myth of Central European belonging found in Carpathian Ukraine. For instance, Sebastian, the protagonist of *The UnSimple*, believes in the special country Carpatia ("It mattered for him that the Ukrainian question be brought up exactly in Central Europe"[10]) so he "invented a beautiful country around Ialivets, in which there would be no garbage, everybody would speak each other's language, and the highest institution would be the office of scripts, where everyone could submit something interesting, and the government would be directed by these stories."[11]

The UnSimple fulfil the function of guards and gatherers of stories. "When someone is born somewhere, they sit exactly under those windows and invent his or her fable, like the earth gods."[12] Thus, inventing a life for somebody, the UnSimple tell a story, and the life becomes a consequence of this tale. Actually, these Hutsul spirits-magicians collect as many as possible fables from human lives and analyze "records with stories and ravings, epiphanies and crazy ideas."[13] Consequently, there even appears an assumption that "uneducated Hutsul pseudo-magicians-contrivers manipulate Europe and the world by dint of these stories."[14]

The life in Ialivets could be a life "with no stories"—that is, something that is unchangeable and is here, in this place. However, place, landscape, geography, and even the body become sensible things—stories. "[T]he basis of every private epos is a list of ideas about places in which family history takes place—

---

9   Prokhas'ko, *NeprOsti*, 37.
10 Ibid., 117.
11 Ibid., 123.
12 Ibid., 79.
13 Ibid., 115
14 Ibid., 116.

some sort of a domestic geography of plants,"[15] at the foundation of which is the joy of germination, wandering, entering, passing, staying, and returning. Botany, geography, fables, sex—everything becomes a pretext for myth creation. According to the logic of myth, repetition becomes authentication and narration turns into life creation. Notable are repetitions that embody the heroic fate of the "main man," Sebastian, as well as his cult character. He revives every time he encounters the females of his kin while the stories of these women end with their deaths. The three Annas—daughter, mother, and granddaughter—with whom Sebastian lives as though with his wives, testify less to incestuous love than the joy of Sebastian's male germination, his "unrestrained power to be instilled in it" (the blood of the females of one family).[16] The idea of such a unity is patrimonial—namely, the insularity of the clan, "its un-smearing, restraining it in some limits, boundaries."[17] With this, the author implies that such a patrimonial living or coexistence reaches the point where "it cannot be widened any further than in some sexual, bodily love."[18]

In *The UnSimple*, narrative is condensed on the surface of events and the stories themselves become a wandering-travelling on the surface: on the streets of Ialivets and in the Carpathian mountains. People there do not know their origins and invent their history by reading it from the environment or projecting their dreams-deliriums onto the landscape. For instance, Francis saw

> how this or that thought runs short of space in the head," and hence thoughts emanate beyond the limits of the body and accommodate themselves along the fragments of the landscape. But the main thing is that a story is created by a "bai": bai is not a word. Bai is many ordered words. Bai is already a story. For different causes there are different bais. Bais are plots. Baian' is a narrative, the narration of a story or a plot.[19]

Prokhasko's late postmodern combinatory character of narratives, which resembles Hutsul seccessional modernism, in *The UnSimple* is predicated upon a few ideas: "the concept of place (native or not native, but where you become ingrained), the concept of kin, and the concepts of narrative, conversation, and talk."[20]

---

15 Ibid., 99.
16 Prokhas′ko, "Vid toho, shcho i iak my hovorymo …," 15.
17 Ibid.
18 Ibid.
19 Prokhas′ko, *NeprOsti*, 86.
20 Prokhas′ko, "Vid toho, shcho i iak my hovorymo …," 15.

CHAPTER 20

The Virtual Apocalypse: The Post-Verbal Writing of Yurko Izdryk

The principal interest of Ukrainian postmodernists is the marginal character and their existence on the border of the real and the surreal, their mental "normality" and illness, their integrity and depersonalization. The deficiency of intimate space and close contact is experienced acutely. Earlier, carnival created an illusion of intimacy; now, the absence of one's "own" body and "own" personal story is deeply felt in any encounter with the "other," in any attempt to reinforce one's own presence. The postmodernist character turns not to the inside but to the outside. It does not look for solitude but, instead, breaks through to reality, and tries to force its way into it through ravings and dreams. This character keenly feels the emptiness of the world, because, for example, he has lost of a beloved, and lives in a world of simulations, mirrored reflections, and dreams.

Such is the protagonist of Izdryk's prose fiction. *The Double Leon. The History of Illness*—the last part of a trilogy whose first two parts are the novella *Island KRK* (1994) and the novel *Wozzeck* (1996–97)—presents a character and event of the same type: a love story and the technique of weaving text-collages. In essence, it is about a schizophrenic character, whose main diagnosis is the loss of sociability and individuality.

In *Wozzeck*, Izdryk implements some kind of a visualization of the reflection of "I" in the process of the perception of reality, splitting the subject into "I," "You," and "He." "I" is the thinking substance; "You" is the body's surface, the projection of consciousness, and the boundary between the living and the dead; and "He" is someone I sees (the other). All of these parts make up Wozzeck. Thus, the author takes care not only to close his protagonist in his own thoughts but also to show the very process of the materialization of perception, of consciousness exiting the limits of the body and corporeality, as in Georges Perec's *A Man Asleep*. The protagonist admits, "suspended at the edge of life, you felt that you were neither body, nor your accustomed self, nor a person, but a kind of pulsating substance that might be inside or outside you and could fit inside a poppy seed or expand to the dimensions of the room."[1]

The work records the overcoming of the limit of the corporeality of "I," the exteriorization of the inner self that is breaking through to reality, wandering through multimedia plots and images (of video, film drama). The fragile consciousness is not capable of escaping the limits of recaptures-repetitions and fetishes-things—this is the peculiarity of Izdryk's postmodern character, which lives first and foremost by video images. "In your quest for your own 'I', in your attempts to crystallize out this elusive substance (a prerequisite for any encounter between the first and second persons) you always come up against the purely technical limitations of internal optical magnification—that is, of good old neurophysiological blowup."[2] This is how this character comments on his manner, demonstrating his unpredictability, amorphous fluidity, capacity for unending transformation, and elusiveness. Izdryk consistently and thoroughly studies the process of the postmodernist death of his hero.[3]

Such a character is ostensibly different from the narrator who, with a sharp pencil in his hand, is reflecting of the idea of a new text and the next steps of his protagonist; yet, at the same time, the character is identical to the narrator and himself the stage director of the video about himself.

The director admits, that "[t]he pulsation of this imaginary bubble, this new and sovereign 'I' of yours, was in no way dependent on your will or subject to any regularity. Such uncertainty drained and exhausted you appallingly."[4]

Izdryk builds a virtual postmodern reality, and reproduces the psychedelics of the pain and dreams of the character who escapes into his dreams. After all, reality is invariable; dreams, on the other hand, allow space for freedom and fantasy. That said, "somebody wants to direct the dreams."[5] This awareness of an internal censor testifies to the fact that virtual reality is also under someone's control. Wozzeck (this ideal other) also wants to control the

---

1 Izdryk, *Wozzeck*, 11.
2 Ibid., 9.
3 About this in more detail see Marko Pavlyshyn, "Peredmova," in Izdryk, *Votsek & votsekurgia* (Lviv: Kal'variia, 2002), 7–30.
4 Izdryk, *Wozzeck*, 11.
5 Ibid.

consciousness of his character, named the letter A. The challenge of this name is to find out what A is: somebody else, a primal human, the antichrist, an anti-I? Thus, the postmodernist novel retains the split between the author and hero, essentially modifying their relationship. The power of the author is demonized, as in Andrukhovych's works, or is interlaced in a complex manner, as in Izdryk, with the freedom of the protagonist—the alter ego.

The permanent state of the postmodern antihero is pain, constant penance, and autism. Pain (primordial, ancestral, Cain's) is the sense of existence; it prevents freedom of action and does not admit the other. In fact, an encounter with the other becomes unnecessary because, as the protagonist comments, "There is an increasing danger of catching sight of the other's face—and it might prove to be familiar; there is an increasing danger of revealing one's own face—and it might be noticed. From then on it will be marked. Movement at this stage must be precise. It takes more practice than merely not stepping on the black floor tiles or avoiding the cracks in the asphalt."[6]

Apocalyptic time is divided into "Day" and "Night" (the titles of the two main chapters of *Wozzeck*); it is about the Big Water; it is recorded by the disintegration of the words of a prayer; and signified by an image of falling letters.

Further partitions of self and dematerializations of reality unfold in subsequent Izdryk works. For example, the love story that is one of the leitmotifs of *The Double Leon* grammatically is built in a way that ontologically confirms the repetitiveness of the story and the encounter of "I" and "you." The protagonist of *The Double Leon* lives in the Kafkaesque false world of the Medical Institution of the Closed Type, perceiving reality simultaneously from the outside and from the inside of his self—"I (and hence? You as well) know the story beforehand."[7] The virtual world becomes total and ruins the difference between the real and the fictional:

> thus he was walking, still walking, and I was he, in him, as always evasive, unseen, omnipresent, and nonexistent. He seldom looked at the layouts of the hypnotizing emergency exits, at the room numbers, at the EXIT signs. But the layouts were spurious. And the numbers only imitated the esoteric order of the administrative cabbalistic. And the exits did not lead anywhere.[8]

---

6   Ibid., 14.
7   Izdryk, *Podviinyi Leon. Istoriia khvoroby* (Ivano-Frankivsk: Lileia-NV, 2000), 12.
8   Ibid.

In *Wozzeck*, mental distress shows itself through the "poly-, mega-, trans-, and multi-individualization of the protagonist. The history of illness in *The Double Leon* lies in the totally opposite direction—in the fragmentary inferiority of the protagonist, his invincible but unfulfilled desire to merge with the object of his love," comments a contemporary critic.[9] Rostyslav Semkiv, on the contrary, notes that Izdryk's works are "a craving for an androgynous idyll on the spacious meadows of the sweet virtual Arcady—the good old Austro-Hungary."[10]

Izdryk's multimedia, schizophrenically split text in reality is supported by only one thing: the story of a love affair. The skillfully built inverted monologues of Him and Her, the discreet units of time ("fifteen minutes to the end," "seven minutes to the end," "three and a half minutes to the end," "eleven seconds to the end," "one second to the end"), the disintegrated sounds of life, the repeated experiments, roles, and dialogues played in the subconscious—all these reflect the polyphony and falsity of the postmodern world. The head of the postmodern hero is full of a "mix of utopian visions, communist slogans, hippie dogma, and post-opium mantras," along with the archetypal images of mass culture, both visual and audio. Illness becomes a peculiar refusal to live in a world in which struggle and power dominate and in which there is only one way out:

> In various directions—some into the internal mongolia, some into a man-made madness, some into unending travels; some hide behind masks and roles of various kinds, some escape into alcohol or narcotics, some fall into anabiosis, turning into a house plant; the most talented simply deceive themselves all the time.[11]

For many years, Izdryk has been the editor of the original conceptual journal *Chetver* (*Thursday*), in which peculiar visual collages and clips supplement literary texts. His own prose fiction is similarly multimedia, mosaic, visual, and fragmentary. Izdryk makes videograms of internal monologues, portraying his character on the verge of madness, illness, and hallucinatory consciousness, reproducing his novels' unending experiments, fantasies, bifurcations, and imagined voices. The author enjoys verbal and topographic metamorphoses, horror, the flow of images and pictures, introducing the hallucinatory tech-

---

9   Leonid Kosovych, "Postskript," in Izdryk, *Podviinyi Leon. Istoriia khvoroby*, 176.
10   Rostyslav Semkiv, " Ironiia nepokirnoï struktury," *Krytyka* 5 (2001): 29.
11   Izdryk, *Podviinyi Leon*, 94.

nique into everyday consciousness. Flights of fancy are expressed in a morbid underground—memory. Descriptions become videograms, in which thoughts and senses materialize through fantasy and flow into hermetically sealed internal monologues.

The protagonist is an alcoholic. The fear of a crime and of the past, the flowing of "I" into things, the identification of the self with a fluid body, the biological foundation of life, and the mythologizing of his own horrors accompany his "struggle with the endlessly slowed time." Language is not capable of rendering this state adequately, and the writing becomes hieroglyphic. In this way, the postmodern character is identified with fluid verbal and psychological substance. This morphology of being, as in Prokhasko, testifies to an experience of the end (or the origin) of time—that is, it has an apocalyptic nature. Water, worms, turtles, the fluid and amorphous body—in all these signs of the transgression and disintegration of the self the author demonstrates in his "history of illness."

The apocalyptic end ("The End Proper," as the chapter is titled) is transformed into the sensations of a human-worm, whose way is a fall into darkness ("My inside is like a worm: nothing is left there—only the empty and smelly darkness.")[12] The alternative to such nonexistence is a fictional virtual story, verbal contacts-bridges, and androgynous fantasy—a meeting of two lovers, two close and mutually desiring people. This virtual reality is a conditional mood of being: "We (would) meet one hot summer day at a neglected bus stop, where you (would) wearily drink water right from the bottle, and I (would) all of a sudden approach and kiss you, and you are not at all (would not be) surprised and (would) offer me some water."[13]

Virtual worlds exist concurrently. The conditional mood is opposed to a negative form of being—a negative reality if recorded by "Her" speech, which denotes wounds and losses in the new postapocalyptic world that starts right there. After all, in this world it was so important to meet and it was difficult to reach understanding: "We met (not) right on a desolate street. Divided by a flow of cars, we saw (not) one another, and we understood (not) each other right away."[14]

The negative entity changes the conditional, modal, anagrammatic form of writing, while the subjective flow of speech is structured around the word

---

12 Ibid., 25.
13 Ibid., 26.
14 Ibid., 29.

"fear." The conditional modality records the threshold of consciousness, the fluctuation between hope and fear, and the denial points to the possibility of an encounter with the virtual, which exists in reality itself. Ultimately, the doors of paradise open and the visible and essential meet, He/She meet and happily unite, and He now resembles a little boy while She resembles a little girl. Totally in the spirit of mystical symbolism, there comes an understanding that Will is "a little-adult-girl-woman-bride."[15] However, the moment of the divine beauty passes because it occurs after the apocalypse, when "the turtles have already left their nests, the hummingbirds have pecked out the turtles' eyes, and void and darkness are coming near."[16]

Thus Izdryk's novella modally and ironically realizes the mythologeme of the revival of an individual's integrity, played against the backdrop of the apocalyptic symbolism of water, childhood, the world ocean, love, primeval chaos, and the disintegration of language. Even more peculiar is that Izdryk creates a virtual reality, which for the postmodern schizophrenic subject is even more real than reality itself, and which rejects the protagonist-outsider, causes his illness, and condemns him to solitude.

"Illness" becomes a metaphor of being. The bifurcated, fragmented individual lives in various worlds, with various consciousnesses, and in various texts. This state corresponds with morphological writing, in which the old formulas-morphemes, reiterating, refer to something already known but slightly altered, as in, for example, "the arriving of heroes" or "the departing of heroes," a "prayer" or a "list." At the same time, language itself is also subject to full alienation: words stop naming things and become things themselves: "Words have stopped being the names of things and have turned to things themselves, having fallen into the deadened pages of books, newspapers, and notebooks."[17]

Izdryk's postmodern writing is characterized by the destruction of the verbal subject as a bohemian Bubabist. Idzdryk's protagonist is torn between the mind-psyche-soul, and his dream is to return to primeval words, to language-as-essence. This, as it were, is his metaphysics. There are "salvaged words that are comprised from complete, unscathed meanings";[18] however, the words he predominantly deals with consist "not of letters and sounds but of the debris of the shattered entities. With most of the words, this debris is put together

---

15 Ibid., 35.
16 Ibid., 33.
17 Ibid., 35.
18 Izdryk, "Teoriia vidmovy," in Izdryk, *Take* (Kharkiv: Vydavnytstvo "Klub simeinoho dozvillia," 2009), 267.

carelessly, impermanently. The meaning peels off of it like chalk with which, evidently, it was written down."[19] Thus in one of his later interviews Izdryk formulates a phenomenology of post-verbality, and it is this that he implements in his novellas.

The novella *The Double Leon. The History of Illness* records a transformation of the world into a primeval place of unnamed things; and into an opposite state—names appear to be dead, alien. Thereby the verbal (semantic) line turns into a simulacrum. Words turn out to be aliens coming from some other, parallel world. The unavailability of the meanings of words, the loss of words' meanings, or the infinity of new pronunciations every time a word is used—all this becomes the basis for the total uncommunicativeness of Izdryk's postmodern hero, and even the "repetitions of repetitions" do not mean anything anymore. This is a prototype of the postmodern apocalypse, in which referentiality is absent and what exists is only a mirrored, rhizomatic superficiality of stylizations and repetitions—that is, a world of simulacra. Ultimately, "since things were devoid of names, the world as though has lost a balance, having slipped from the mirrored surface of the obvious into the chasm of self-will and madness."[20]

Normal language is not capable of an adequate expression of these postapocalyptic feelings, so Izdryk invents a new quasi-language. He resorts to graphic signs, idioms, syntagmas, reprises, fragments—the "elementary particles" of text, which repeat themselves, are disrupted, and create the fractured rhythm of the text. The schizophrenic entropy of the consciousness of self is reflected by means of these fragments and "graphs," in which meaning is deployed through the shifting, repeating, and disintegrating of the "graphs" rather than in a linear mode. This is the world of the post-structural subject, which exists on the screen rather than in a series of words capturing the unity of "I"-consciousness. Cinematographic dreams, hallucinatory collages, hybrid monsters-images, auto-parody and ironic crypto-quotations (for instance, found in Andrukhovych), lists-classifications (of assholes), as well as numerous pop-symbols borrowed from movies, literature, music, and even cartoons capture the entropy of consciousness and the birth of a new creolized textuality consisting of the symbiosis of visual and verbal signs.

The origin of Izdryk's postmodernist writing clearly is media. This writing is also polymorphous. Marko Pavlyshyn says that the language game in

---

19 Ibid.
20 Izdryk, *Podviinyi Leon*, 38.

Wozzeck "as a rule, is based on an unexpected discovery of additional meanings and effects, which appear as a consequence of the structures of language itself rather than of the intentions of the speaker."[21] Irrespective of that, *Wozzeck* clearly demonstrates the exhaustiveness of language itself, its dumbness for the post-verbal subject. According to the author's philosophy, one can talk not only with words but also in many other ways: "with scents, evaporations, emanations, fluids, gaze, bumpy and not so much muscles, epithelium, will efforts, and fantasies."[22] When it comes to words, they can be an object of play rather than a means of speech because their meanings have faded. Hence play with words is totally arbitrary and devoid of material form. Such a play "can end in anything: a phrase, a sonnet, a novel, a slogan, a shortcut, an anagram, and a cipher."[23]

In Izdryk, words are often cinematographically unfolded dreams. Not only do they become self-willing and create their own landscapes, filling out the whole consciousness, but they also live their own lives and develop their own plots and hybrid images. They are of an expressive masscult mettle and resemble a cartoon. The fantasy imagination feeds on self-parodies and allusions to real colleagues, who figure in the text under somewhat altered names, like in the "arrival" and "departure" of the heroes: Lysheha, Protsiuk, Tsyperdiuk, Andrusiak, Prozayak, Liubansky, Irpinets, Borakne, Rymaruk, LuHoSad, and so forth. One can also consider Izdryk's novels as an encyclopedia of mass culture. It is not a coincidence that Volodymyr Yeshkilev notes that Izdryk's quotes appear to be mainly "a reflection on the secondary phenomena of the textual space, formalized by the late imperial masscult."[24]

The polymorphous structure of writing and polymorphous language reflect the sick consciousness of the post-verbal protagonist. It is filled with acoustic, infernal glitches, images-mythologemes and the symbols of the Jungian collective unconscious. Both psychoanalysis and orientalism supply this sick consciousness with associations and fantasy images. For instance, one element of a dreamlike hallucinatory collage is an East-West hybrid (the chapter "Acupuncture"), and its ideogram is something like "СОLАянь in PEPSI." After all, Izdryk unequivocally associates his writing with pop culture, which from the very beginning has grown on the images of Zen Buddhism and Eastern

---

21 Marko Pavlyshyn, "Peredmova," 14.
22 Izdryk, "Teoriia vidmovy," 262.
23 Ibid.
24 Volodymyr Ieshkiliev, *Votstsekurhiia bet. Komentari do "vnutrishn´oï entsyklopediï" romanu Izdryka "Votstsek"* (Ivano-Frankivsk: Vydavnytstvo Unikomus, 1998), 5.

philosophy, combining them with the psychology and sensualism of the Western hippie. Thus acoustic fantasies, memories, pop symbols and names, pseudo-Chinese fables, sessions, glitches, and the hypnotic power of Mister Sulfazin immerse "the double Leon" in Zen: he himself wants to stay in places, "about which you would never say 'here' or even 'somewhere'—only 'nowhere' and 'never.'"[25] Eastern philosophy in its masscult version, therefore, is an index of mental collapse in the postmodern subject.

Sinking into abstinent anabiosis, the character periodically returns to reality, but these are only interim moments. He lives in a fantasy-whimsical, hybrid world, dominated by "the so-called syndrome of Lubansky." This is the name for the moment of the unceasingly fluid metamorphoses of consciousness at the time of awakening—that is, attempts to escape the dreams-landscapes in Izdryk's novella *Wozzeck*. An endless play of various parodies, auto-parodies, and play with the twins-masks of the self, cram the novella so that the text swims about, and fluidity, or water, is probably the main substance of the work. To become water, to spill with water, "to be only myself," "my own imprint on the limestone of my shellfish-turtle," to feel like a suspended balloon—the subject perceives all these as states of freezing.

By reproducing a series of constant reincarnations, bifurcations, and trifurcations in *Wozzeck*, and by showing the character's inability to escape the world of mirrored signs in the novella *The Double Leon*, Izdryk turns the postmodern subject into an autist and represents him as an interface of foreign voices. The subject of his postmodernist work is an existential consciousness, mediated by media images: movies, computer games, commercials, and digital television. In this situation, the book, its linear sequence of words, its semantic unity, disintegrates into letters, signifying the gaps in postapocalyptic time, which coincide with the post-Soviet period. As Izdryk ironically-philosophically notes: "researchers of my work perhaps already know that I use the word 'existence' not only in the sense of being or existence as such but also to suggest certain mood, the sweet and sour aftertaste of nostalgia, the ineffability of the eternal solitude, etc."[26]

Beside virtual, schizophrenic narrative, Ukrainian postmodernism gives birth to a virtual-historical one. It shows itself as an attempt to reformulate the linear sequence of history—in particular, through supplanting it with a synchronized rather than hierarchical collage of historical discourses. For example,

---

25 Izdryk, *Podviinyi Leon*, 168.
26 Izdryk, "Biohrafiia artysty," in Izdryk, *Fleshka-2GB* (Kyiv: Hrani-T, 2009), 207.

in his *A Procession in Moscow* (1998), Vasyl Kozhelianko ironizes the utopian cultural myths and ideals of Ukrainian renaissance and offers a virtual alternative history of the Second World War.

A Procession in Moscow (published in *Suchasnist'* in 1998) fantasizes and provocatively materializes the utopian historiosophical ideals that, in various forms, supported Ukrainian self-awareness in the twentieth century. First and foremost, it is a dream about an independent and great Ukraine, whose parody counterpart becomes a dream of the Ukrainian Empire of the Three Seas. The author ironically and playfully ties this historiosophical utopia with a historical provocation, rewriting the history of the Second World War, and making down-to-earth—in a burlesque manner—its main figures: from Adolf Aloisovych Hitler to Beria; from the Head of the Committee of State Rescue Sakartvelo (Georgia) to Ianuarii Vyshkinidze; and General Chantariia, Whiskered Cavalryman Voroshylo, and Khrushchev, who "operated under the auspices of Ukrainian nationalists in the environment of the Soviet government with a pseudonym 'Cornman.'"[27] The collected jokes, allusions to the coup of the State Committee on the State of Emergency, fantasy stylistics, as in Kvitka-Osnovianenko's "The Konotop Witch," allusions to the Chechen War, contemporary pop stars such as the "VV" band, whose hit "The Spring" ("Vesna") has become an anthem of Ukrainian paratroopers—there is everything in this virtual history.

The novella offers a collage of various discourses: a reshaped map of Europe; a chapter from the textbook *History of Ukraine*, a parody of the style of the Soviet historiography whose mimicry is the Ukrainian independence version of historiography; the newspaper style of the Bukovinian press, collected on the pages of the newspaper *Dzygarok*; a gourmand's report about a banquet in the restaurant Prut to honor the newly appointed governor of Bukovina; geopolitical discourse; a sample of popular 1990s political rhetoric; spontaneous-colloquial, epistolary style; and a metacritical self-parody—the essay "The Pathos of Ukrainian Patriotism in V. Kozhelianko's novel *A Procession in Moscow*," written by a student of the seventh grade of the Gymnasium No. 58 in Chernivtsi.

The novella scatters particular styles and jokes, weakening or even destroying linear narrative, shifting chronology and jumping time periods. Even the myth of the winning soccer match of Dynamo Kyiv in occupied Kyiv is dismantled. Among other elements that are entwined in the novella are the inevitable West-East hybrids, such as the Permanent Presidium of the Council

---

27 Vasyl' Kozhelianko, *Defiliada v Moskvi* (Lviv: Kal'variia, 2000), 58.

of the Great Mahatmas of Shambhala, which raffles off the fate of the Earth's history—emblems of the fashionable postmodernist fascination with Buddhism, interstellar wars, and so forth. The most scandalous event in the plot—Victory Parade on Red Square—has the subtitle "A Parade of Complexes."

CHAPTER 21

The Grotesques of the Kyiv Underground: Dibrova—Zholdak—Podervianskyi

The predecessor of literary postmodernism in Ukraine is the Kyiv underground of the 1970s–80s (Volodymyr Dibrova, Bohdan Zholdak, and Les Podervianskyi). Zholdak's images-collages, Podervianskyi's hybrid monsters, and Dibrova's puppet characters-mannequins reflect—unlike the bohemian decentralization of space and the decentralization of the character in the postmodernists—the monstrosity of the Soviet masses' mindset. The Ukrainian postmodernism of the 1990s produces two kinds of *decentered bohemian subject*. One of them is an externally "ex-centric" character, which does not have its own, authentic place, and therefore travels through countries and times, plays and put on masks. This is the character-Bubabist. However, despite his eccentricity, internally it is quite stable, which is confirmed by his narcissism and constantly reproduced oppositions to self and other. The second kind is an internally eccentric character, constantly changeable, fragmentary, and nonidentical with itself. It creates its own virtual reality, like Wozzeck, dissolving and transforming in language and language games. In other words, it is a rhizomatic character.

The general features of the underground creative work are urbanism, anti-totalitarianism, the grotesque, and the absurd, which appeared as a consequence of juxtaposing different worlds—Soviet and not Soviet, official and unofficial, as well as pastiche based on rewritings of classic works. A special role is played by the undermining of totalitarian language, which usually excludes from the codex of linguistic behavior the "low" colloquial style, oriented toward spontaneous and individual speech. An attention to the character of the "small person" and a wide use of popular cultural forms also become common.

In the 1980s, a representative of the Kyiv ironic prose school, the author and translator Volodymyr Dibrova created a multifaceted character Peltse as an embodiment of the total Soviet person, who at all times stays with "the people." Dibrova employs a grotesquely transformed image of the "small person," some devices of the puppet theatre, and the ideology of the Theatre of the Absurd. His character Peltse lives in the atmosphere of double standards and triple truths, at the crossroads of party ideology and mass clichés. For his own life he chooses the line of least resistance, avoiding conflicts and alternatives. However, he gets constantly sidetracked into other worlds and into trying out other forms of behavior. His roles run from leader to tramp. His face also changes: "Sideburns lengthened by a micron turned Peltse into a philanderer, a line 'accidentally' left on the face resembled a bruise, and a slightly pared-down jawbone revealed the signs of Down's syndrome."[1]

Peltse is a character-puppet, comprised of various possibilities and styles of life. As an alpinist falling from a rock, he sees "[m]ore than a single life—a host of lives that glittered with endless possibilities, depending on point of view and, moreover, on choices made.[2]" Despite his variability, he is childish and idealistic, and sees life as a cartoon or a journey to the land of fairy tales. The world, however, sharply contrasts with a fairy tale: "Mother smiles gently, parts her lips to answer, and Peltse sees two lives simultaneously. In one of them his mother says: 'Don't worry, a fairy tale is make-believe.' In another, she says: 'Don't worry, a fairy tale can't lie.'"[3]

Dibrova's "total," grotesque character Peltse is changeable. His element is not as much a vertical infinity of growth as it is a horizontal space of behavioral variability, in which he multiplies, becomes other, sees himself from the side, and depends on circumstances, even though at the beginning he enthusiastically defends himself and his name, for which he gets constantly taunted—"'Peltser-Skeltser, Alka-Seltzer,' chant the heartless brats. 'Shit his panties, couldn't help it.'"[4] Life resembles a cartoon that shows perpetual transformations-reincarnations of the eternal "small, cuddly" and infantile self, which is afraid of its name and body, and of becoming adult.

> Am I really Peltse? It can't be true. I am so small, so cuddly, and Peltse is huge, like that wardrobe, and as mysterious as those things my parents

---

[1] Volodymyr Dibrova, *Peltse and Pentameron*, trans. Halyna Hryn, foreword Askold Melnyczuk (Evanston: Northwestern University Press 1994), 30.
[2] Ibid., 4.
[3] Ibid., 5.
[4] Ibid., 8.

keep hiding in the drawers. Will I really pass the fourth, the sixth, the twelfth grade, grow up to be as big as this name and eventually become a wardrobe myself? Maybe I am not Peltse at all but simply me?[5]

Little Peltse does not stop wondering, compensating his fears with the visions of his own victories and reprisals.

In the totalitarian world, like a world created by a demiurge, a "small person" treasures not only the top and the bottom but also parallel worlds, such as the underground, dissidence, and semi-officialism. Thus in his visions,

> One road leads Peltse into a basement where he sits longhaired, armed with a guitar, full of passionate words, in front of a bottle of wine and a tin of canned fish. The basement has two exits: the first to a cold damp room without furniture or wine; the second to a similar basement, but with wine and a slovenly cohabitant, who is about to bestow upon him a quite curable but long-neglected communicable disease. The other road, having wound its way around the prickly exclamation "Well?" carries him out onto a small town square. ...[6]

He imagines himself as a boss, personifies some leader, whose selected works in a leather case are given as gifts. There are also other variants—the people's hero, a grandpa leader loved by children, "a noble knight-giant," ultimately, a bandit. ... In this way, Dibrova delivers the many-voiced, fantasy tales of the late Soviet era, which combine jokes, kitsch, totalitarian myths and legends, along with folk songs about the "heroes."

One more embodiment of the "small person in Dibrova's work," in this case that of an "unwanted" person, is Burdyk from the novella of the same name, written in 1992–95. The author gives him a recognizable and understandable identity: Burdyk, like the narrator, belongs to the "suppressed generation," "informal culture," and "the victims of the system." "What kind of times do I live in? What for? Why exactly so? Exactly here? Exactly I?" Burdyk does not stop wondering, declining to climb higher and higher on the stairs of life, as do others. Dibrova writes about the opposition and this underground person ironically rather than pompously. He develops his character at the crossroads of different rather than ideal notions. Having fallen behind the lining of his

---

5   Ibid., 12.
6   Ibid., 5–6.

own pocket, Burdyk finds there "a list of names and things," which determined the image of his suppressed generation. Among these, as it were, attributes of underground behavior, there are, for example,

> Carl Jung and Friedrich Nietzsche (a sample of a political joke)
> "sprat in Baltic herring" (can, cheap snack)
> "Bolshevik" (a bottle of hard wine of 0.8 liters.
> Aka "bomb," "faust-cartridge" or simply "faust"),
> Hermann-Hesse, Deep-Purple, Zen-Dao (attributes of spirituality and an expression of a craving for "real" life).[7]

Ultimately, doing an autopsy and revision of his generation, the narrator, who has been given the task of restoring Burdyk's biography, creates a probable portrait of his life, balancing tragedy and buffoonery, knowledge and ignorance, seriousness and irony: "What Burdyk thought about we will never know. But she could get overcome with regret (young Burdyk was an Orpheus!), sadness, emotional tremor (as an echo of an unquenched feeling), empathy (like 'what has life done to you!), or even joy (with Hurskyi I feel safer)."[8]

Dibrova's irony testifies first and foremost to the artificiality of "heroism," "tragicality," and the myth of the "suppressed generation," yet at the same time it conceals the pain of defenselessness, of loss, and of the euphoria of the nineties. Now his friend is up to write the heroic biography of one of the representatives of that time, who has become an alcoholic and whose last moment of life did not differ much from the death of a rat run over by a car. There are a few versions of the heroic biography: "Burdyk was a genius," "he was gifted to a greater degree than any one of us!," "it's a pity that he was born in the Sovietdom rather than in Paris" (as a variant—in Munich, California, at least in the "sister Bulgaria"), "eh, he would have written so much at that time! …"

Dibrova tells Burdyk's life story from his teenage vandalism to student radicalism, from his politicization and interest in literature to his nationalism, and from his attendance at "salvation of the noble language" rallies to raves in the 1980s, thereby chronicling late Soviet times and the period of perestroika. Ultimately, Dibrova refers also to the new literary modes and practices of that period, for example, in the form of a "novel-document of the life of the village Khotsky," plots about a cache in the Carpathian Mountains, and so

---

7 Volodymyr Dibrova, *Vybhane* (Kyiv: Krytyka, 2002), 228.
8 Ibid., 292.

forth—genres popular in Ukrainian literature during perestroika. He retells possible parodies about Lenin: "Lenin at Cyprus," "Lenin at High Water," "Mamin-Sibiriaka Has Come!," "Lenin and a Sentry," and "Why Lenin Cannot Catch Up with a Turtle." The paradox, as Dibrova describes it, is that totalitarianism cannot exist without its underground antipode and vice versa. Therefore, it is normal that "when communism was in power, its opponent Burdyk, although he did not thrive materially, was constantly moving up. And struggled with the authorities to the best of his abilities. As soon as the progressive order declined, however, Burdyk died."[9]

Dibrova creates a world in which reality exists at the border of the surreal; this world is a maze of situations and references to late Soviet mode of life. This sur-reality, in which the whimsical and the alternative stand behind every real situation, is a quintessence of Dibrova's absurd. Burdyk finds out that the visible world in which he lives is built in such a way that

> everything here had a staircase that led either down or up. Therefore, every step, anyone anywhere was taking, was either a descent or an ascent. Even that top where Burdyk was sitting appeared to be just one of the stair steps in one of the innumerable possible and impossible ladders.[10]

In his collection of short stories *Songs of the Beatles* (1982–87), Dibrova offers pictures of underground behavior in the stagnant Brezhnev times using his characteristic irony, in which life borders on absurd. The multifaceted nature of Dibrova's characters corresponds to the absurdity of the totalitarian system at large, in which irony becomes a manifestation of the relative freedom of the underground person. Dibrova's characters exist in several worlds at the same time. The late Soviet underground held a love for the songs of the Beatles, which symbolized hippie freedom and a road to desired worlds. Thus in the time of perestroika the expectations of "geological shifts" are based on the fact that "power is taken by the generation brought up on the Beatles." However, this world coexists with, or even inside, the other, the Soviet Union in which lives "a neighboring teacher with thirty-five-years of experience," in which the "Sevastopol Waltz" competes with the Beatles.[11] There, a free, wandering student, who received from someone a book by Dostoevsky in which the word "God"

---

9   Ibid., 417.
10 Ibid., 399.
11 Ibid., 113.

started with a capital letter, meets a policeman. "It was too far to the border; I wore ironed pants, so when the sergeant told me to sit in the car, I didn't resist and only asked him to drop me off on the highway."[12] However, his stay in the precinct, interviews with the policemen, expropriation of a few draft copies of Mandelshtam's book of prose, in which Kautsky was mentioned, transform the free student—from the point of view of the "officials"—into an "emissary" and deprive the free student of his rosy dreams and sense of freedom, which was so consonant with The Beatles' "Why Don't We Do It in the Road."

However, at the beginning his sense of freedom was full-fledged:

> "Nobody is shadowing us; then why," I cried, "why don't we do it in the road? Three years of school are behind us and two in front of us; the practicum is done; I am going to the seashore and care about nothing. The desired freedom, transparent and clear, was fed on the morning air and the colors of haystacks by the road. Why don't we right away and forever live like that?"[13]

The hilariously quick transformation of freedom into bodily fear and of the student into an emissary, a gentlemanly attempt to translate the ribaldry of the police into normative vocabulary, existence within the brutal reality of the police and the sublime happiness generated in the soul by the songs of the Beatles—all this testified to the total absurdity of the Soviet world.

While Dibrova, a representative of Kyiv ironic prose, preserved a character of the post-totalitarian type, having transformed it into a puppet, Bohdan Zholdak created a verbal monument for him—patois. It is worth noting that language games become an important element for Ukrainian postmodernist authors. In the late Soviet era, particularly popular were the mono-dialogues of Les Podervianskyi, who created a hybrid form of the "new language," gathering together, in paradox, demonic black humor, literary and cultural clichés, and parodies-collages from Soviet mythology. In this way, he recorded the situation of "our time," whose hero has become foul language. With the help of his dramatic parodies-collages from the Soviet mythology (at the verge of paranoia), in his own way he described the apocalyptic situation of disbelief. "The whole population of *The Hero of Our Time,*" as Dibrova says, "is standing at the edge of the abyss, is looking into the abyss, and the abyss is looking back at it."[14]

---

12 Ibid., 115.
13 Ibid., 114–15.
14 Volodymyr Dibrova, "Prynts' Hamlet khams'koho povitu," *Krytyka* 5 (2001): 27.

Podervianskyi reproduced an absolutely heterogeneous and uncensored world, extracted from the underground of the unconscious world of human horrors, desires, and perversions. Such a chthonic world is hybrid, amorphous, fluid, entirely weaved from intertextuality, which encompasses folk fairy tales, classic works, and ideological slogans. In this way he reproduced the anti-utopian state of broken communication in totalitarian society. By inverting the "high" and the "low" in a burlesque manner, he created a linguistic hybrid—a "language of imbeciles and degenerates" that combined colloquial language (patois) and argot. In short, Podervianskyi gave voice to the collective unconscious. However, Podervianskyi's plays, which he himself masterfully staged and soundtracked, were not carnivalesque. The laughter did not purify and did not bring release; it only bore witness to the fact that transgression brings enjoyment.

Serhiy Zhadan maintains that Podervianskyi is a conceptualist: "his thoroughly designed and conceptually shaped scenic space denies the very possibility of the presence of the elements of absurd in this space, transferring the secret mechanisms of the scenic life into the dimension of another—mythological and sacral—logic and subordination."[15] "Hamlet, Or the Phenomenon of Danish Katsapism,"[16] "Fairy Tale About a Turnip," "Danko," "Pavlik Morozov," "The Meeting Place Cannot Be Changed, Fuck!"[17] and other Podervianskyi sketches reflect the black humor of late Soviet times. This is a grotesque and monstrous reality, in which foul language rules and anti-culture parasitizes. The demonic culture of the underground not only created a subculture, being the other side of the official culture, but was also a laboratory of grotesque and polymorphous imagery. Beside linguistic hybrids, Podervianskyi's stage heroes include, in particular, Anthrax, Gangrene, Gonorrhea, and Uterus Prolapse, while other characters are "human-crocodile," "human-net," "human-bird," "human-bench," "human-door," "human-chalk," and so forth.

More representative of the Kyiv Ironic School, Bohdan Zholdak is the author of prose fiction written in patois ("surzhyk")—an anti-language. Patois is a variant of colloquial Ukrainian, a mix of Ukrainian and Russian words and grammatical forms, and distorted by a peculiar pronunciation. It is worth

---

15 Serhii Zhadan, "Pavlik Morozov: mizh pobutovym heroïzmom i pobutovym tryperom," www.samvydav.net.
16 From "Katsap," a derogatory appellation of a person of the Russian nationality, used in Ukraine (translator's note).
17 This and other titles in the original contain various untranslatable intertextual references and elements of a Russian-Ukrainian hybrid patois (translator's note).

noting that in the 1990s in Ukrainian society there were heated discussions regarding the orthography and the appropriateness of patois.

What Zholdak does can be called a realization of Nechui-Levytsky's wish to make the language of the grandmother Paraska the basis for the Ukrainian literary language. Borys Hrinchenko in his time analyzed what the people take from the Ukrainian classics, and suggested ways to develop a "common people's" Ukrainian literature. He made a point about the inappropriateness of all kinds of hybridity in literature. The language of the plays should be comprehensible, he noted, and it is not worthwhile to overuse a "mixed jargon," such as that seen in Voznyi's *Natalka Poltavka* or Holokhvastov's *Chasing Two Hares*. Hrinchenko, on purifying Ukrainian literature, wrote,

> The overuse of those types cannot be tolerated. It cannot be tolerated also because the introduction of the distorted jargon or jargons to the great extent on the stage ultimately spoils the language of theatre, renders it rough and vulgar. After all, it is not advantageous for the authors themselves: jargon, as disgusting phenomenon, is not able to develop regularly; it only changes according to the vogue. Therefore, plays that are written in jargon almost entirely (for example, *Chasing Two Hares*) will soon become impossible for reading and staging, when the jargon of the characters, such as Pronia, Holokhvastov, and others, will disappear.[18]

Thus, Hrinchenko prophesied the disappearance of patois, which he calls jargon (evidently, an urban one). He could not envision either the success of *Chasing Two Hares* in the second half of the twentieth century or the possibility of the emergence of a literature written in patois. However, in the postmodern world, the contamination of languages, or linguistic hybridity, gains special value. We can at least recall the popularity in the Caribbean of pidgin as a colloquial version of English, and how it coalesced with the original dialects of that region. Caribbean pidgin emerged as a result of colonization and the intermixing of various languages of that region. The following words are good examples: baby—pikinini; believe—bilip, bilipim; Bible—Baibel; boss—masta, bos; cake—kek; chair—sea, sia; cheese—sis. There is also literature written in pidgin—poetry, in particular.

Therefore, in addition to being relevant in the context of national language policies, patois also becomes interesting as a phenomenon of the postmodern

---

18 Borys Hrinchenko, *Narodnyie spektakli* (Chernihiv: Tipografiia Gubernskogo Zemstva, 1900), 41.

situation in Ukraine. It is postmodernism that pays special attention to the phenomena of linguistic hybridity, the polymorphism of linguistic forms, and the heteroglossia of discourses. Ukrainian surzhyk and Belarusian trasianka are forms of linguistic hybridity that with their lack of official status, their vulgarity, and their colloquialism undermine the authority of official language. In a sense, this secondary, or conscious, patois is also a discovery or an invention of postmodernism. As Zholdak points out, the "intelligentsia enjoys my patois short stories the most because it has already conquered in itself that stupid patois and knows how to struggle with it. The patois-speaking population has no reaction to it because it does not perceive this patois as a device—it just truly speaks this language."

Zholdak's collection of short stories *Beef* (*Macabresque* [1991]) offers fragmented scenes from life, something resembling urban folklore narrative miniatures, which merge the surreal and real, ideological and everyday consciousness, soldier jokes and parodies of Soviet folklore ("Ivan and Stalin," "Let's go" ["Paiekhali"], "Mania or Tania" ["Mania il' Tania"]). Zholdak's narrator is a homo Sovieticus. His interests in "low" passions and "high" ideals are equal, as, for instance, in the story of the "gal Sveta" ("devochka Sveta"), who "eradicated abroad slave trade, extremism, drug addiction, and lechery as such, thereby advancing them along the way of the social progress toward the higher forms of development, namely—ours."[19]

It is worth noting that irony and laughter are common and are born out of mass consciousness and commonplace situations that suddenly break through their usual limits. A train conductor, on the one hand, evokes an association with someone wearing a "crown of thorns"; an invalid without a leg, on the other hand, appears to be not a dignified veteran but a rat. The narrator, instead of working on a report about the heroes of sport, is for whatever reason more attracted to talking about the stoker who in the boiler house is peeping through a hole into a female locker room ("Just now").

Zholdak's collages, Podervianskyi's hybrids, and Dibrova's character-puppets reflect the monstrosity of the mass Soviet mindset. This type of irony could be called carnivalesque if not for one peculiarity. This is not the inversion of "top" and "bottom" but an almost masochist deferral of such an inversion: the "bottom" and the "top" appear to be identical—that is, they are parts of one common total world space. Tales, gossip, and jokes from Soviet times,

---

19 Bohdan Zholdak, *Ialovychyna (Makabreska)* (Kyiv: Ros', 1991), 61.
  The fragment is written in patois, parodying the style of the Soviet ideological language (translator's note).

narratives-traps and hints, peeping and eavesdropping—all of these capture the gaze from the bottom and from the top, from the inside and from the outside simultaneously. One more step, and patois grows into a newspeak, and the narrator becomes a central figure of the new time—a "new Ukrainian."

Totalitarian culture has tendency toward the expansion of mass communication. Language becomes normative, ideologized; it turns into a stereotype, cliché; individuals' speech, instead, becomes passive. Totalitarian language expropriates from the codex of linguistic behavior a colloquial style because the latter is characterized by spontaneity and individuality. For this reason, it is not a coincidence that in the post-totalitarian period patois revives: "the linguistic tactics of the assurance of significance, of the amalgamation of the partner, confidentiality, forcedly withdrawn from the language use, all of a sudden appear to be pleasurable against the background of the communication norms of the totalitarian type."[20]

Patois is a certain focus that sets off the so-called norm of language, along with the model of ironic behavior built upon analogical alogism. Homogenous situations and characters produced by totalitarian society turn out to be shifted in relation to each other rather than identical. This shifting of the norm, alogism of analogy is a kind of danse macabre, a dance of death, through which the transience and animosity of the everyday mass mindset shines.

Zholdak's element is the gibe. To explain what "pervert" means, his narrator deploys a whole array of analogies and coincidences that stand rather far away from the discussed subject, in order to emphasize the obvious difference:

> People are walking along the street and suddenly see how a child, of three and a half let's say, is falling down from the seventeenth floor. It fell down the way even its parents didn't believe. Especially in that, although so-so tiny, it ended up being completely alive. ... So, our story is even more incredible. Because it's about adults. Not in the sense of their falling onto a tree or bush but in the sense of fallings of a very different kind, even more incredible. Say, for instance, one played harmonium. Well? But he crazed up not because of the harmonium, or cello, or any other music but due to a totally foreign thing. Which every normal woman has ("On Perverts" ["Pro izvrashchentsiov"]).[21]

---

20 N. A. Kuprina, *Iazykovoe soprotivlenie v kontekste totalitarnoi kul´tury* (Ekaterinburg: Izdatel´stvo Ural´skogo universiteta, 1999), 20–21.
21 Zholdak, *Ialovychyna*, 22. The original is written in patois (translator's note).

This somewhat infantile and disintegrating language records not so much the ideality as the grotesque quality of the image created by the marginal consciousness. The Gogol-esque narrative tradition acquires a naturalistic rather than a playful modus. It is not the narrator who speaks now but patois itself, turning into some kind of a talking "snout." Language itself undergoes demonization, behind which there is an impersonal "nobody." Later, in Zholdak's collection of short stories *Anticlimax* (2001), eroticisms, political jokes, parody, and argot, combined with patois, become a mode of self-stimulation, consolidating the orgasms of the everyday, devoid of heroism and the post-totalitarian mindset.

CHAPTER 22

Feminist Postmodernism: Oksana Zabuzhko

Ukrainian postmodernism has an expressive gender orientation. Masculine and overman symbolism is actively used by the Bubabists, the group "Psy Sviatoho Yura" ("The Dogs of St. George"), and a representative of the former New York group Yuriy Tarnawsky. They are opposed by the prose fiction of Yevheniya Kononenko, Sofia Maidanska, Halyna Pahutiak, Nadiia Tubaltseva, and Svitlana Iovenko, while the feminist prose of Oksana Zabuzhko has become an apology for the female literature of the 1990s. The success of the novel *Field Work in Ukrainian Sex* (1996) lies first and foremost in the appearance of a "new heroine"—utterly subjective, sexual, and intellectual, who with her absurd rebellion testifies that the time of the national revival is in its essence an apocalyptic time, which threatens maternity, love, and family, in which the process of growing up degenerates, creative activities are demonized, and the body gets sick.

Zabuzhko's novel is a tale about an unfortunate love affair and at the same time about a fraternal dual between two Ukrainian intellectuals—a female poet and a male artist—who painfully protect their rights to autonomy and freedom. This conflict acquires a broad cultural meaning and becomes an analysis of the postcolonial situation in Ukraine. The motifs that are central to the novel—the "weakness" of males, developed in the course of long years of Ukrainian independence, along with double female marginality, as well as socio-cultural analysis of the nation—intertwine and become explosive. Zabuzhko shows a particular love drama as a deep colonial trauma which has affected the whole nation and has ruined the intimate space of kin (the people).

Zabuzhko's postcolonial prose fiction stands close to the contemporary postcolonial prose of Toni Morrison, Salman Rushdie, and Jamaica Kincaid. Zabuzhko uses gender conflict to reproduce the universal aspects of Ukrainian national history and suggests various interpretations of these collisions:

the victim mindset ("in general, all that Ukrainians can say about themselves is how, and how much, and by which manner they were beaten"); the sexual ("poor sexual victim of the national idea"); the suicidal ("why not now? Immediately? Why wait?); the infernal ("He jumped back like someone stabbed him with a knife; that malevolent flame in his eyes was strange … one the edge of a bared grin with sharply protruding incisors from under the upper lip, like it was *something else* that peered out for a moment through his narrow eyelids, red-rimmed and swollen from lack of sleep"); the crisis-authorial ("All that we're given—like children toys—are ready-cut slivers of reality, fragments, details, colored pieces of some large, unsolvable puzzle").[1]

The postmodernist texture of the novel *Field Work in Ukrainian Sex* is expressively polyphonic: it is created on the basis of open fragmentariness, quotation and auto-quotation, the paraphrasing of the others' speech, the putting in quotation marks of the others' words, and an intermixture of English phrases. In this way, the author manipulates styles and twitches the reader: she takes the mask off her own self and puts it back on, or expressively plays with the erotic, patriotic, or intertextual expectations of the reader. Sharp stylistic shifts are also worth noting: from a "cynical, with in-your-face street-smart mannerisms picked up 'on the inside' somewhere" to the intellectual who reflects on the themes of authorship and suicide in literature.[2] It is notable, however, that the narrative is designed for communication and comprehension and is not autistic.

The author combines "high" ideology and "low" passion; perceptive lyricism and foul language; the demonic and the infantile; autobiography and foreign cultural codes; fictional details and recognizable biographical features. And she creates in the novel multidirectional associations by referring both to fiction and to the real story of the author's life. The prose is interlaced with poetic stanzas, quotations, English phrases, cultural digressions, italics, Latin and Cyrillic letters. This creolization of the text creates the effect of diversity, shaped and maintained by the passionate tale of the heroine:

> What can I tell you, Donna-dearest. That we were raised by men fucked from all ends every which way. That later guys of the same kind were screwing us, and that in both cases they were doing to us what others, *the*

---

[1] Oksana Zabuzhko, *Fieldwork in Ukrainian Sex*, trans. Halyna Hryn (Las Vegas: AmazonCrossing, 2011), 115, 85, 125, 137.
[2] Ibid., 4.

others, had done to them? And that we accepted them and loved them as they were, because not to accept them was to go over to the others, the other side? And that our only choice, therefore, was and still remains between victim and executioner: between nonexistence and an existence that kills you.³

Existentially pinpointed, national and gender collisions bear a distinct genitive idea—the survival of the nation. At the same time, a relationship between the two people turns out to be an arduous search for understanding between the two sexes. The psychoanalytical structure of the novel translates into a therapeutic discussion of the whole traumatic history of the relationships between a man and a woman. "Literature as a form of national therapy," as Zabuzhko ironizes, becomes truly a psychoanalytical study, which echoes the archetypal situations of rape, subjugation, violence, demonic sensuality, powerlessness, and retaliation.⁴ By removing, layer after layer, soil from the past, by analyzing the symptoms of the illness, which the author calls "survival" and which substitutes for "life," the heroine gives a diagnosis to herself and to the whole era: autism, schizophrenia, infantilism. ...

Field Work emphasizes the difference of the gender-existential orientations of men and women. Zabuzhko's woman wants to procreate, to salvage the racial beauty of her nation, to revive its elite ("how much of that Ukrainian intelligentsia is there among us anyway, pitiful, forcibly *dragged back* against the current of history—a tiny group and even that scattered)," while the man pursues his own self-realization—to be a "winner."⁵ He is "a dying species, almost-extinct clan, we should be breeding like crazy, and all the time, making love where and when we can, uniting in orgiastic insatiability into one, yelping and moaning mass of arms and legs, extending ourselves and populating this radioactive land anew!"⁶

In this way, he acknowledges the female protagonist of *Field Work*, who, seeking her own Jerusalem, wants to cure her genus. The instinct of genus protection, related to the maternal instinct ("my darling, my dear, dear boy, come to me, come into me, I will embrace you from all sides, hide you with my body, let you be born anew") becomes a component of her love for the man.⁷ The idea of genus reveals itself also in the almost Electra-esque desire to save him as

---

3   Ibid., 158. Translation modified (translator's note).
4   Ibid.
5   Ibid., 77.
6   Ibid.
7   Ibid., 76.

an artist—brother of "the same blood": "it's okay, bro, no worries, we'll break out, I'll pull, drag you out on my back, I'll have enough strength for all of it: bring half of Ukraine to its feet, bring half of America over to Ukraine to take a look."[8]

The asexual, imbued with frigidity and fear mother, who survived both Holodomor and the detentions of the seventies, is contrasted with the overt sexuality of the daughter. Zabuzhko's protagonist is preoccupied with an existential rebellion against superstitions, trials, betrayal, and her own bodily weakness. Her excessive efforts, ultimately, awaken the hidden in a woman, or as the surrealists put it, the vixen-castrator: "You have taught my body to castrate the perpetrator: all my feminine strength, accumulated for generations, which has thus far been directed toward the light … with you has turned itself inside out, black lining outward, has become destructive," as the female protagonist reveals.[9] The amplitude of the heroine's character runs from a harpy to a little girl, who, like an orphan, cries, "'I'm your daddy', except who's going to say that to a thirty-year-old broad."[10]

In spite of its radically feminist commentary, *Field Work* reflects a liberal vision of female self-affirmation—a need for love, and a longing for motherhood. Particularly important is the generation gap, which sublimates into a resentment of the weakness of the protagonist's father (a former sixtier who is broken by fear and lack of self-realization), as well as into the theme of her mother's frigidity, portrayed as a generalization of the history of sexual upbringing in Soviet times. Overall, the conflict of generations described in the novel—in particular, reflections on the place of the sixtiers in the cultural history of Ukraine—has become one of the most important motifs of Ukrainian postmodernism.

Another important theme in Zabuzhko's novel—that of the West—is based on a perception of America and American reality; in particular, it arises from the opposition of Ukrainian life (passionate, liminal-intensive, existential, experiential) and Western (sterile and fed on mass culture):

> [I]n your culture tragedy is of an exclusively personal character, loneliness, love dramas, those clinical incests that forty-year-old matrons supposedly dig out from their childhood memories in psychotherapeutic séances, and

---

8   Ibid., 107.
9   Ibid., 97.
10 Ibid., 6–7.

> which I, truth be told, don't particularly believe—after a year or two of psychiatric sessions you'll start recalling a lot more than that—however, you are unfamiliar with subjugation to limitless, metaphysical evil, where there's absolutely nothing in hell you can do.[11]

The observations of "Ukrainian sex," which Zabuzhko's heroine unfolds before the American audience, also oppose soulless disembodied communist kitsch, which is dominated by the "collective body" of the total Soviet person. Sexuality here is a mode of revealing the "inner" person, and sexual history presents the archetypes of gender behavior of both men and women. What is uncovered thereby is the individual subconscious, dominated by violence, revenge, contagious aggressiveness, suicidal wishes, and deep narcissism. All these moods are expressed in the heroine's monologue, which is a mode of self-analysis and at the same time a mirror of roles and images.

Zabuzhko, perhaps for the first time in Ukrainian literature, has explored the theme of sexual violence. Her scandalized narrative also refers to the ideal image of childhood:

> truth is found only in childhood, only through it can we find the true measure of our lives, and if you have managed not to trample to death that little girl inside of you (or that little boy standing in the pasture with a stick in his hand, awestruck by the terrifying—because so beyond the human capacity to render—immensity of the multicolored symphony of fire of the setting sun)—then your life has not faltered.[12]

As though accepting Simone de Beauvoir's thesis that every woman hides a little girl and that infantilism is a shelter for women in the patriarchal world, Zabuzhko endows this idea with an existential meaning. The infantilization of a character (male and female) becomes a notable feature of Ukrainian literature of the 1990s (Valery Shevchuk's novels and novellas; the novellas of Yevhen Pashkovsky, Yurko Izdryk, Viacheslav Medvid, and Halyna Pahutiak; and Serhiy Zhadan's poetry).

The search for identity and the heroine's road to herself in the image-archetype of a little girl waiting for a miracle become in Zabuzko's novel a tale about the life in a "world without love." The novel, which is practically a

---

11 Ibid., 110–11.
12 Ibid., 72.

monologue-lecture delivered before a mirror, looks like a rehearsal for a future university lecture. The text revives feminist, narcissistic discourse in Ukrainian literature, which began at the end of the nineteenth century with another Ukrainian female author, Olha Kobylianska.[13]

The genre of mass literature—the female novel—merges with the intellectual essay, intertwining prose and poetic language, patriotic discourse and rhetoric. Zabuzhko does not hide her interest in the emergence of mass literature in Ukraine. She is convinced that "unless there is at least some 'soil' of Ukrainian mass culture," high literature is not possible.[14] Using the forms of mass culture—bestsellers, romances, gothic fairy tales—and immersing the reader in an atmosphere of love intrigue, demonism, and mysticism, Zabuzhko creates genres and forms of readable literature.

Other Zabuzhko's novellas ("An Alien Woman," "I, Milena," "Girls," "The Fairy Tale About the Snowball Reed Pipe") are also dedicated to the image of the "new heroine," her personal and creative self-realization in the "strange," mostly male world. What does it mean to "possess" and what are the forms of female power?; What is the essence of the happiness of creativity for a woman if the imagined separates her from life like a glass wall?; How does one catch and—most importantly—not destroy the fragile balance of life itself? All of these questions are the object of scrutiny in Zabuzhko's texts. The heroine of *An Alien Woman* is a writer who in real life feels like an alien and, observing everyday, usual life, experiences an existential nausea. She instead wants to write, and the process of the birth of writing and writing itself become an object of observation on the part of the mediators—almost a computerized infernal messenger, a known man of letters Velentyn Stepanovych, and the author herself, for whom writing and corporeality are indivisible.

Oksana Zabuzhko's oeuvre contains short stories that look like the fragments of an unwritten novel: "Girls" and "Sister, Sister." The protagonists share a name—Dartsia—and both deal with the author's central themes of sisterhood, moral conformism, social violence, and the corporeality of being. In these short stories Zabuzhko cultivates a peculiar female writing that is predicated upon a delicate and intense groping for some primeval, ancient existential foundation of being—the authenticity related to femininity. In this female writing the soul

---

13 See Tamara Hundorova, *Femina melancholica. Stat′ i kul′tura v gendernii utopiï Ol′hy Kobylians′koï* (Kyiv: Krytyka, 2002), 86–125.
14 Oksana Zabuzhko, "'Meni poshchastylo na starti …,' Rozmova z Oksanoiu Zabuzhko," in *Zhinka iak tekst. Emma Andiievs′ka, Solomiia Pavlychko, Oksana Zabuzhko: fragmenty tvorchosty i konteksty*, ed. Liudmyla Taran (Kyiv: Fakt, 2002), 198.

is not separated from the body, and the innocent fragility of life can only be approached by touches, senses, and tactile probing. The stylistic atmosphere of the stories grows from the approximate groping of words with which one can attempt to express the authenticity of the newly emerging world, at the heart of which there happens to be an innocent golden-haired fawn—Lenka-Lentsia. With her there is a not-yet-perceptible, but already-awakened, love.

"Girls" and "Sister, Sister" depict the existential rebellion of a woman against alienated authenticity—authenticity which has been expropriated from her by society. This authenticity can be associated with the fragility of love, and with the childish defenselessness which is born and lives in a person until it is destroyed, and which is inseparable from the tactile-bodily matter (mystery) of life.

Zabuzhko's novella "Girls" is about a friendship-love between two school girlfriends, in which the central theme is conformism, and in which sapphic games are only an episode in the history of love, betrayal, and power over the other. The theme of sisterhood often attracts Zabuzhko, suggesting an intensive search for the intimate space of life destroyed by the totalitarian past. This theme is embodied in the "brother-magician" of the heroine from *Field Work*, the sisters in "The Fairy Tale," and the girlfriends in "Girls."

"Girls" is a story about the maturation of the Nabokovian Lolita, who experiences sapphic games with her girlfriend, thereby discovering her own body; it is also about an honors student who is adjusting to life, entering the world of adults. Thrown into this world, Dartsia has to struggle for leadership in her class, school contests, and life races. She already has a premonition that life itself has placed her in a situation in which to love the other means to have the other in your possession, although she cannot but feel that this possession destroys the other and falsifies her own life. Dartsia simply vomits this false life at the end of the story, reminding us about the existential nausea of this life, related to the devouring and appropriating of the world that cannot be fathomed and of the people whom we cannot let go. Only having passed the period of initiation, through a betrayal, the girls receive the right for a life of their own, albeit with an aftertaste of permanent guilt.

Oksana Zabuzhko's prose features biologism; it is not a naturalistic biologism, as it may seem at the first sight, but an existential one. She talks about life not in abstract terms but from the inside, and thinks this inside-ness not only socially but also biologically. She is not afraid of talking about the mysterious, usually closed for outsiders—about the moment of the inception of life, which swells in body and increases in soul inside the humid warmth of the maternal womb, multiplies by the will of the cells, and pulsates in the blood rhythm. It is there, in the soul of the not-yet-born Ivanna ("Sister, Sister"), that fear settles—the first and the last one—a fear of not being born. Thus biologism becomes socially meaningful when it comes to dead, aborted children, to life purposefully disrupted by people infected with fear. As usual, Zabuzhko emphatically accuses the inhumanity and monstrosity of the Soviet system which spreads fear.

Zabuzhko's characters often confess a desire to have a sister. Feminist criticism emphasizes that the search for a sister is a female rebellion against the power of gray uncertainty, patriarchy, and the brutality of human relationships. It is also the maternal instinct, which unites all females into one generic being. Caring about an unborn sister is akin to a maternal, female responsibility for the authenticity of the world. It is also one of the ways of searching for one's own authenticity as, in Simone de Beauvoir words, in every woman lives a little girl. The opposite of this ideal female world is a social, people's world where "strange" people kill the unborn sister while "familiar" people kill the already born and felt female love. The conceptually and stylistically coherent version of female writing that Zabuzhko established in her stories reveals the heretofore unused potential of existential female prose.

The characteristic features of Zabuzhko's prose are the subjectivity of the narrative, internal dialogues and monologues, and quotations and allusions, which are organically integrated into the author's speech. Zabuzhko's work, as it were, forms a peculiar image of body-language. Language is my home, the heroine repeats after Heidegger. Zabuzhko's prose features its own multileveled lyrical phrase—not hermetic but utterly open and emotional. Her stylistic periods accumulate several emotional elevations and falls, and reflect the almost bodily pulsation of the linguistic stream.

In "The Fairy Tale About the Snowball Reed Pipe," Zabuzhko changes the associative context in which her heroine lives from intellectual to folkloristic and from urban to rural. By resorting to folklore imagery, restoring the authenticity of retro names, song and ballad motifs, the gothic romanticism of curses and witchcraft, and immersing the reader in the atmosphere of the fictional village, Zabuzhko-the-urbanist rediscovers the world of the folk fairy tale. She is interested in the intensity of the mythopoetic world, which is usually balanced in fairy tales yet erupts under the pressure of love passion and tragic cataclysms in a gothic novella.

In "The Fairy Tale" Zabuzhko is trying to reconstruct the world of the Ukrainian erotic, about which she talks in one of her interviews. Ultimately, the "unique, inimitable total culture with a characteristic philosophy of corporeality and a dialectic of the male and the female" dwells, in her opinion, in folklore. From there, folk ethnoculture, erotica transfers into a new urban culture. Zabuzhko notes that

> The older patriarchal ethnoculture does not exist any longer—it has been destroyed by force, dispossessed, forced to a collective farm, seated in the number of "100 thousand girlfriends" on a tractor, whose vibrations are absolutely intolerable for a female organism, sent to Siberia, where it gave itself to an escort for a dish of skilly, brought by famine and fear to the mark of the highest in the world male impotence and female frigidity.[15]

In "The Fairy Tale" Zabuzhko practically creates a folklore gothic novella, which combines pleasure and horror, love and hatred, uses the mother-daughter archetype. (After all, the gothic often features the secret of the birth of a daughter and the correspondence of her fate with her mother's, which often has a mystic tinge to it.) Thus the fairy tale plot about "Dido's" daughter and "Baba's" daughter gains in Zabuzhko a modern, catastrophic meaning. It is about a horrible female revenge generated by the despondency and pride of the extraordinarily talented Hanna-panna (Miss Hanna). She crosses the border of compassionate life and gives herself to the power of the demonic, which was supposedly a cure for her "resentment toward the people who insisted on measuring her cloth by their own yard, and when she refused to go along with these measurements they, in their immeasurable meagerness, came to despise her."[16]

The daughter retaliates on behalf of herself and her mother, who was forced to marry a person she did not love. This time, not a devil-tempter, as in Lesia Ukrainka, but a sukkumb, a Satan-prince, visits Hanna at night. An openness to the demonic, to witchcraft, seems to be the only alternative to the fate that awaits women who do not fit into patriarchal society. Two versions of such a fate—those of Olenka and of Hanna—become antagonistic. The horrible disharmony of the two sisters repeats the ancient story of Cain and Abel, and out of this conflict there emerges Hanna's disagreement with God: "Why did the Lord accept Abel's sacrifice and reject Cain's one?"[17]

The novella is filled with corporeality—self-studying, self-admiring, savoring, smells, blood. Ultimately, the scene in which Hanna murders her

---

15 Ibid., 183.
16 Oksana Zabuzhko, *Kazka pro kalynovu sopilku* (Kyiv: Fakt, 2000), 70.
17 Ibid., 64.

sister Olenka is perhaps a female version of the Honta scene from Shevchenko's "Haidamaky":

> And it was a luxury probably as piercing as during the nightly frenzies to thrust the knife directly into the breathing with warmth, dense yet at the same time pliable before the knife body, which shuddered along with a short spasm (and this moment of resistance was also luxuriant!)—and softened; instead, an oily, vivifying, hot, red odor rushed into the nostrils, causing voluptuous dizziness—drink it or bathe in it, as though the whole world, finally, popped open altogether and rushed toward Hanna-panna to quench her thirst, and shaking the bloody knife above the head, drunk and wobbling from this unexpectedly rich—could not be richer!—wedding.[18]

Deprived of simple female happiness, rejected by people, proud, talented, and beautiful, Hanna-panna gives in to the power of the Devil. Only he is able to "return her to herself" by giving her love, and his words can throw to her feet "the whole world, with all the people who were, are, and will be in it."[19]

The woman's duel with a life governed by the patriarchal norm and with the female fate transitions from the mother to the daughter and ends in a sororicide. The daughter, as such, avenges her mother, Maria, who was forced by her father to marry a man she did not love. At the same time, and in accordance with Freud, the daughter wants to become the embodiment of the other's—the father's—desire: the father's song about a girl, an inaccessible princess, her mother, whose love her father was not able to win, awakens in Hanna female jealousy for the first time. Having heard the father's song, "the girl lay in a stupor, as though having overheard something shameful about her father, which caused pity and was really hurting, yet at the same time a different, even more cruel wound was waking up and growing—that voice did not address *her*."[20] Thus love intersects with sin and transforms—in a gothic-romantic way—into a core of devilish evil, turning a person into a beast. Zabuzhko shows how in a discordant world an extraordinary personality strays onto a satanic path.

Distortion and straying symbolize sinful blasphemy—the pride of the girl gifted by God, her love toward her own body, and her wish for immortality, at least in a song "composed only for her." In line with gothic imagery, talent and

---

18 Ibid., 77–78.
19 Ibid., 70.
20 Ibid., 13.

longing intertwine with creativity, and it does not matter if this is some sort of knowledge, which Hanna wore as a smile on her face when she was a teenager, or her delving into the secret power of water. Ultimately, the song of a young chumak, played on a reed pipe from the snowball tree the murdered sister has turned into, personifies the myth of "embodied" creativity. The reed pipe, popular in folk songs and ballads, cut out of a snowball tree, which now is the bodily dwelling of the woman-girl murdered by her mother-in-law (jealous sister-in-law), symbolizes the "corporeality" of the sung song. Thus the song that grows out of the body becomes a metaphor of the female author's creative work.

Within the framework of the traditional plot, Zabuzhko objectifies a totally postmodern ideology: crossing the line, the transgression of characters, social norms, and the Christian law—all this can be found in "The Fairy Tale." Its gothic imagery contrasts with imitated depictions of the infernal sabbath in mass versions of the fantasy genre—for example, in Maryna and Serhii Diachenko's *The Witch Era*, in which in one place there coexist aliens, witches, people, inquisitors, and spirits.

The demonic, frightfully wishful crossing of the line happens for Zabuzhko's heroine when she "with a sweet horror saw the Chalice with the Eucharist in the shapes of her own body."[21] The unearthly lover offers her the whole world ("with all the people who were, are, and will be in it"), and this feeling is akin to her own immortality in the song she wants to become. The demonic nature of creativity and the Nietzschean idea that the Christian God is the God of the weak form a frame for the duel between good and evil in Hanna's soul. She asks, "why would this world belong to Olenka?"[22] Zabuzhko's intellectual fairy-tale, dressed as a gothic novella, serves as an example of how high postmodernist literature integrates the genres of mass culture.

Another author, Halyna Pahutiak (*The Notes of a White Bird* [1999]), writes mostly in a fantasy genre that employs fantastic metamorphoses, maternal-female visions, and infantile dreams. At the center of Pahutiak's work is an image of an ideal place or paradise—in one example, the Wondrous Country, which is populated by people who have been freed from a mental facility (Pereviznyk, Kolos, Sheptun, Diohen, Basil). They are natural and easy to communicate with, and have their own superstitions, mythologemes, fears, and dreams about a kind King. A little boy, who comes to the country by chance, only shows childishness that is innate to this world. Depicting the marginal

---

21 Ibid., 68.
22 Ibid., 74.

world on the border of a trash dump and virgin land, Pahutiak explains that she sees the "Other world" (or the real world) "from the inside of the virgin land." Fairy tales, after all, appear to be a "rescue from the inconsistencies of the Other world," from the absurdity that is growing there.

The search for paradise, for virginal land, can be considered a post-Chornobyl syndrome in Ukrainian literature. The prototype of virgin land (paradise) in Pahutiak's work is often the countryside and rustic culture, as well as the experience of the ideal unity of mother and child. It is very like Castaneda's fantasy images and the idea of the destroyed human world as a trash dump. The gap between adulthood and childhood is central to Pahutiak's prose, as is a feminine image of the world. Her texts grow out of meditations and dreams, and reflect a neat and filigree play of prayers, dreams, and fears of the woman-mother; in other words, they grow out of the life of a woman as a small person that has experienced the angst of existence, loneliness, poverty, and weakness. Pahutiak writes about the personality hung on a "crook above the abyss" of words, about the impossibility disentangling from words, and about infantilism as an escape from the adult world.[23] In her works, adults want to become boys and girls in order to feel little, wander through the house, recall the past, feel safe in the posthuman and apocalyptic world:

> Childhood, bloom, old age—those are not the action of time but a state of the soul, its different faces. Like a play. A child plays pretending to be an adult or an infant, trying on to see what fits better: infantilism or responsibility. Afterwards, adults teach it the thing they have mastered themselves: automatism. … How good it is, after all, not to be a human. It is indeed hard to find a creature less natural.[24]

---

23 Halyna Pahutiak, *Zakhid sontsia v Urozhi* (Lviv: Piramida, 2003), 351.
24 Ibid.

CHAPTER 23

Postmodern Europe: Revision, Nostalgia, and Revenge

As sociocultural reflection, Ukrainian postmodernism plays the role of a topographic guide at the border of one's own, and the other's, intimate and social spaces: one's own body and the outside world, biography and literary fiction. It also explores the space between two geopolitical zones: the former Soviet Union and modern Europe. Coined in the 1920s, the slogan about drawing near to the "psychological Europe" offered a new vector in Ukrainian national self-identification in opposition to the Moscovian. In the course of the twentieth century, Ukrainian modernism looked up to the ideal of cultural Europe, and the attempt to catch up with it was the momentum behind the process of modernization. Only at the end of the century did the image of the West undergo fragmentation and different images of Europe itself emerge.

Starting from the 1980s, Eastern European intellectuals (Milan Kundera, Czesław Miłosz, Danilo Kiš) refer to the idea of their "own" Europe first and foremost in their search for a single tradition of Central Europe as a multicultural society—a society which has been destined to be a victim of greater European history and a marginal space of greater European culture. One of the components of Central European identity is link to the Austro-Hungarian Empire. Due to the historical-topographical repartition of the geopolitical map and the romantization of the past, Ukrainian postmodernists in the 1990s join this project of modernization.

"My Europe," for the Pole Andrzej Stasiuk, is neither a virtual nor a linear phenomenon; it is not completely covered either by "postmodernist ... free choice" or by "the modernist desire of limitations."[1] Migration, travel, "is always

---

1   Stasiuk and Andrukhovych, *Moia Europa. Dva eseï pro naidyvnishu chastynu svitu*, 46.

an escape," as the Polish author says. Trips from Warsaw to the East and West for him complement each other and correlate with each other: "In the first case, we were conquered by space; in the second, we were defeated by time."² In addition, and in a globalized sense, united Europe is a phenomenon of mass culture, which knows no obstacle in crossing all boundaries and unifying European space in its entirety.

Therefore, Stasiuk appeals to a "circling around" of his own place, around the border of transgressions; and such a circling, perhaps, "is an awkward form of paying homage," "an ideal geography," which consists "of details, trifles, momentary events, resembling cinematic sketches, of flickering fragments, which flutter in my head like leaves in the wind, and through this snowdrift of episodes there shines a landscape and there shines a map."³

For the Ukrainian author Andrukhovych, on the contrary, "my Europe" is a melancholy revision of history—a Central European revision. The place from which the reconstruction of history begins, as well as the image into which Andrukhovych's vision of Europe shapes itself, is Manichean incompleteness—the ruins of the past and the debris of everyday life. This is why his narrator confesses:

> Considerably more telling than any moral maxims appear to me to be simply the shards of an old mode of life: artificial flowers, vases, Christmas tree angels with lambs, worn-out coins, any kind of decadent jewelry, frayed garters, music boxes, birds' nests. Old aquariums interest me, petrified fish, soot-covered bathtubs and basins, of course, bottles excite me ... ⁴

Immersed in the past, such a character, it seems, has no limitations for his transgression. He is led by the romantic vision of the coming true of the future and the image of the happy West, which time and again come to his mind. "But I return to a picture that till this day appears to me. A small boy gazing over the river. Beyond the river the New World begins. America lies beyond the Danube, that is, the future, beyond the Danube lies everything that will come true (and not come true) in time."⁵

---

2   Ibid., 15.
3   Ibid., 67, 48.
4   Yuri Andrukhovych, *My Final Territory: Selected Essays*, trans. Mark Andryczyk and Michael M. Naydan (Toronto: University of Toronto Press, 2018), 10.
5   Ibid., 14.

He does not consider himself a post-Soviet person. For such post-Soviet people, "perplexed, angry, and tired," this beyond-the-Danube America appears as a "paradise lost." Not only does space separate; time separates even more. Sketching a utopian and apocalyptic vision of the progress of the New Europe, Andrukhovych admits that the past has become an abyss, and "the appropriation of time lies in this swallowing of it by the past."[6] However, the future, an anticipation of positive changes, is also an abyss, and the life of a Central European person runs between the two monsters, two chimeras: "between the unknown 'once' that no longer is, because it already happened, and yet one more unknown 'once', that still has yet to be, because it still has not happened."[7]

Unlike his Polish colleague, who sets for himself a limit of transgression, that is, chooses European identity, Andrukhovych remains a captive of time. He is fatally dependent on time and history, and, fleeing from them, he calls for dwelling "inside" the present. This "abiding 'inside' is always loftier, more grateful, and more noble, for it designates your connectedness, your being involved, your presence—in difference from a separated, torn away, and abandoned abiding 'between.'"[8] Family history only confirms that the intersection of eternity and time occurs in the present; precisely there "we reach that fullness at least by a half-note."[9]

Stasiuk's and Andrukhovych's visions of "my Europe" differ radically from each other, and not only because for Stasiuk it is an ideal geography and a space for travel, while for Andrukhovych it is the fatality of history, which is a delimitation that cannot be outflanked. Andrukhovych's existential present testifies to a non-belonging to Europe and the West; and the Central European Ukrainian person personified by him admits its subordination to history and fate.

Ukrainian postmodernism conducted a reevaluation of two myths: of modernism and of imperialism. Carnivalesque Bubabism can be compared with the communication theory of modernization, associated with an inversion of official and traditional forms of culture and an adoption of the structures of mass culture (kitsch, rock, and performance). From a psychoanalytical perspective, such a modernization of Bubabism appears as an overman-bohemian son's revenge against the "castrated" by totalitarianism maternal body of Ukrainian culture. Perverse images of Ukraine-the-mother, maternal culture with its unfilled and tabooed niches, and monstrous female characters gain popularity.

---

6 Ibid., 19.
7 Ibid., 25.
8 Ibid., 26.
9 Ibid., 50.

However, modernization in its peculiar Bubabist version is not the same as westernization. Irvanets's "Love Oklahoma," instead of "Love Ukraine," shifts the emphasis regarding westernization, jumping over the painful zone of the close relative (Europe) directly to the American relative. This reevaluation of the myth of Europe is the link that ties together Ukrainian postmodernists with the modernist tradition proper. The slogan "Away from Moscow!," formulated by Mykola Khvylovy in the 1920s, set a new perspective: "to the psychological Europe." Thus in the course of the entire twentieth century, the idea of Europe (or the myth of Europe) in Ukraine remained unchanged. And only in the 1990s, at the end of the century, does the Ukrainian cultural mindset turn to the demythologization of both the West and Europe.

This process rests in the realization that "Europe will not rescue us" and that modernization is not the only way of development. The fact that Europe, this desired "other," "is not waiting for us" became self-evident. It caused a complex of new longings and identifications among Ukrainian postmodernists. Andrukhovych expressed this complex in his observation of "Western people": "They were returning from the ball, stately, in pairs, older and younger, noble 'Western people'; they were flowing, without paying the least attention to us, and the only trick that could attract their attention was, throwing yourself on all four extremities, to desperately growl and painfully bite somebody's calf."[10]

Beside the carnivalesque, which was based on the idea of undermining and inverting official culture, Ukrainian literature of the 1980s also hosted a self-sufficient, neo-baroque, cultural-philosophic ideology—for example, LuHoSad, a group of Lviv poets professed a rearguard opposition to modernism and the whole modernization strategy ("we are but behind the whole development of literature; we are the rearguard of literary process").[11] The members of LuHoSad, in whose work all previous forms and traditions resonate, are related most of all to the premodern baroque (not without a futuristic tint) and its models of thinking and behavior. Their attitude toward national identity is best illustrated by Nazar Honchar's "The Apology of Indifference," in which an echo of demonstrations is offset by the Ukrainian immanent (in the spirit of Skovoroda) self-sufficiency—a safeguard against excessive emotions and protests:

> My Ukraine
> Gaping

---

10 Andrukhovych, *Dezorientatsiia na mistsevosti*, 60.
11 Taras Luchuk, "Literaturnyi ariergard. Poetychna kontseptsiia LUHOSADu," *Suchasnist'* 12 (1993): 16.

> You are full of joy
> Indifferent
> Scorning
> The fire extinguisher
> After all it is like all the rest
> A Ukrainian.[12]

Ultimately, self-sufficiency (rather than revolution or modernization) becomes the aesthetic ideal of LuHoSad. Ivan Luchuk's last novella, "Ulyssea," in principle rejects the idea of the imitation of the high modernist canon (that includes works such as Joyce's *Ulysses*), based on the synchronization of sacral and profane time. Instead of catching up with and emulating modern Europe, contingency and totality play at culture, as in a chess game. Changeable cultural contexts, adapted to the profiles on the map of culture or to the pieces on the chess board, are remote from the synchronization of culture that T. S. Eliot proposed as a modernist principle of filling in the gaps of literary history.

In the postmodern paradigm, Oksana Zabuzhko is the one who implements the criticism of the national colonial mindset, like feminist criticism. Without rejecting westernization as an intellectual discussion, she nevertheless is emphatically ambivalent in her perception of the West (or, more precisely, of America as a civilizational structure). She is interested precisely in the transformation of the *"culturally depraved,* 'dumb' ..., on the global historical scale exactly 'peripheral' reality—into the *full-fledged cultural one."*[13] It is from the perspective of this postcolonial inversion that Zabuzhko sees the European quality of Ukrainian literature. The paradigm of the West is changing: it is evolving from being an absolute to becoming a discourse of power. Zabuzhko notes,

> That is, when "the West is the West and the East is the East," the situation is not as bad; the colonization of spirit begins when "the West," with all its totally institutionalized power of its own ruling ... cultural egocentrism, falls on the "East" in order to prove to it that the latter is, in essence, a "sub-West," "a second-rate West," or, in the best case scenario ..., "a potential West."[14]

Perhaps the most dynamic form of cultural criticism is Bubabism, which through Yuri Andrukhovych proposed its own conception of Europeanization.

12　Ibid., 23.
13　Oksana Zabuzhko, *Khroniky vid Fortinbrasa* (Kyiv: Fakt, 2001), 294.
14　Ibid., 295.

At its most general, it is about a topology of identity, a specifically postmodern side of national cultural self-awareness. Andrukhovych thinks about Ukraine in terms of locality: he first of all is interested in Galicia, which, in his opinion, fulfills the function of a cushion between the West and the East.

The modernists wanted to appropriate Western culture by voluntaristically supplementing or inverting the national tradition in Western fashion (translating Western authors, looking for their other traditions or their Hellenic or Roman roots). Overall, this form of westernization was close to an ethnic conception, underpinned by an idea of the integrity and succession of tradition in which, according to Dmytro Chyzhevsky, all Ukrainian writers "felt like the members of one national-ideological family."[15]

Andrukhovych adopts from Polish postmodernist authors a nostalgia for "these spat upon margins" of Europe and supplements it with a nostalgia for institutionalized tradition in the form of the Austro-Hungarian Empire. This, as it were, is not the possible or desired "great" Western-ness, as in the modernists, but a historical-topographical one. "Just think about it—there was a time when my city belonged not to the same state entity that included Tambov and Tashkent, but to the one that included Venice and Vienna!"[16] Andrukhovych notes.

This postmodernist orientation toward the West is different. It is tantamount to establishing itself in Central-European space—in "entirely postmodernist territory."[17] This is not so much an adoption of the radically different as it is a nostalgic attempt to restore a mutual past (the Austro-Hungarian one), which is usually idealized through small narratives (a grandmother's memories, the appearance of a former map, the topography of memory, monuments). The context of such a Western-oriented topos of literary regionalism includes Rilke of the 1890s (but not of the later years—that is, the "Czech" Rilke), the Galician Joseph Roth, and the "Drohobychean" Bruno Schulz. Geographically, this topos is built around the history of Stanislav, a Galician city, deprived of "great history," which lies exactly between the glorious "kingly" past of Lviv and the "princely" glory of Chernivtsi (as judged in terms of the Habsburg Empire). From an ideological perspective, this topos contain a longing for empire—naturally, the Western and Austrian ones rather than the Moscow one.

Therefore, Andrukhovych's goal is "to assume the restorer's function and, aided by the most ephemeral material—language, to restore at least the fragments

---

15 Chyzhevs´kyi, *Istoriia ukraïns´koï literatury. Vid pochatkiv do doby realizmu*, 307.
16 Andrukhovych, *My Final Territory*, 58.
17 Andrukhovych, *Dezorientatsiia na mistsevosti*, 120.

of walls, towers, loves and dreams, which necessarily—this is how I wanted it—ought to have occurred here."[18]

Such Ukrainian postmodernism can be considered a post-European phenomenon (in the sense that modern Eurocentrism as cultural imperialism or egocentrism is reversed and looked at from a totally different perspective—from that of marginal and regional Europe). This is a crack on the map of Europe, projected onto the Mediterranean region, a hole containing the Central Europe of Milan Kundera, Czesław Miłosz, and Vaclav Havel, and the phenomenon of post-memory. This is also an opposition to a different myth—the Westernizational and ambivalent-modernizational one (those "endless and dozing, with a heavy afternoon dryness of the mouth, conversations about Europe, Europa, Eupropa, about Europeanness, the European meaning-vocation, European culture and cuisine, about the road to Europe, and that 'we also are in Europe.'")[19]

As Anthony Smith[20] notes, ambivalence, when westernization is combined with hostility toward the West and juxtaposing it with one's own philosophy and genealogy, is typical of modernization theory. Westernization is understood as a way to part with alterity—that is, with one's own national character. This character is romantic, like the Mystery of Man and Woman that is opposed to "Western" decadence in Andrukhovych's novel *Perverzion*. The vivacious carnival of the relocation of nations in Galicia, which opposes the "twilight" of classic Europe (Venice), is also related to romantic ethnocentrism. The subversive-ironic nature of the postcolonial character-Bubabist—the archetypal trickster-Aeneas—in the novel *Perverzion* in reality turns out not to be a new attempt to "correct" Europe by dint of Ukraine. Instead, it "spit[s] upon margins of Europe." As an alternative to an imitation of Western culture and the decadence of so-called Western phallogocentrism, Andrukhovych romantically proposes a Ukrainian orphism-eroticism and the idea of an eternal carnival, represented as the mystery of Man and Woman. However, ideally professing multicultural Europe and the conception of communicational modernization, which testifies to his rejection of the ethnic concept of national identity of the Bubabists as the carriers of the youth rebellion, Andrukhovych ultimately flirts with a Ukrainian romanticism sympathetic to ethnocentrism.

---

18   Ibid., 58.
19   Ibid., 119.
20   Anthony D. Smith, *Theories of Nationalism*, 2nd ed. (New York: Holmes-Meir Publishers, 1983), 106.

It is worth noting that Andrukhovych, irrespective of his sympathy toward the Third World, is emphatically aware of his own European roots; therefore, at the ritual of Asia's passage though the Great Germanic Gates, his character feels like a sacrifice rather than an acting person.

Ukrainian postmodernism features a topos of national identity that is different from a nostalgia for empire—it is called "A Farewell to the Empire."[21] While in the first case the strategy of reversal is closer to restoration, Andrukhovych's strategy toward the recent past and the Soviet Empire resembles a revenge, or, to be more precise, a post-imperial and necrophilous carnival, as in *The Moscoviad*. Primarily, it is about a deconstruction of the imperial discourse of "friendship among nations." On the one hand, this discourse stated the equality of nations and peoples in united, homogenized Soviet time-space. On the other hand, it reinforced the hierarchy of "the great nation" and of "national minorities," by which, obviously, the center dominates, alienating and turning other nations into an object of irony and criticism ("Chukchi," "Chuchmeky," "Khakhly").

Portraying an empire in agony, Andrukhovych demonstrates the ambivalence of the totalitarian mindset; and behind it, he exposes the illusion and myth of the Soviet idea of "friendship of nations." He uncovers a deep alienation between the "own" and the "other." The protagonist of *The Moscoviad* reflects,

> For what unites you with the Uzbek guy across the wall, except for the wall? With the Uzbek guy who always cooks pilaf and then lures with its rich smell many a hungry female student. ... And what unites you with Khudaidurdyev ... a Turkmeni who every evening, at nine o'clock, suddenly appears in his tracksuit pants in front of the TV, since he apparently cannot live without the evening news? And what unites you with Yezhevikin who enters all rooms without knocking, since he considers himself master of this entire land, from the Carpathians to the Pacific Ocean?[22]

The kitsch of "friendship of nations" is based here on the image of the total "ours" opposed to the total "foreign." The kitsch image of the "natsman" (a representative of national minorities, such as "Chechen," "Caucasian," "Chuchmek" ["darkie"], etc.) was created—as an emblematic image of the "not-entirely-Russian" type. Andrukhovych showcases such a collection of "natsmen" in *The Moscoviad*: "one of the Chechens (or most likely all of them together) yesterday gave a nice

---

21 For more detail about this see Olia Hnatiuk, *Proshchannia z imperiieiu: Ukraïnsʹki dyskusiï pro identychnistʹ* (Kyiv: Krytyka, 2005).
22 Andrukhovych, *The Moscoviad*, 46.

thrashing in the elevator to the phys. ed. guy, Yasha";[23] "The dorm superintendent turned out to be a darkie, although not a bad one, a Daghestani, and still rather young."[24] The image of the national (that is, non-Russian) poet is another component of totalitarian kitsch ("By some miracle these guys get admitted in large crowds, and move about also exclusively in large crowds. Nobody actually knows what they are up to, but only an endless laugh would be the appropriate response to the thought that they spend their time writing poetry. They buy boomboxes, leather jackets, girls, guns, grenades, gas masks, jeans, land, cognac").[25]

"Brothers" and "friends" (the "Vietnamese … dressed in Soviet children's clothing") and the "Mongols," who "snort like stallions, washing the dust of the Great Steppe off their muscular shoulders," as well as the post-imperial, post-Pushkin "friend of the steppes Kalmyk";[26] "[a]nd not a single Kalmyk will even greet you," only diversify with their local hue the great narrative of the "friendship of nations," foundational for totalitarian culture.[27] "Two little black boys," who "lead by the hands an old blind bandura-playing minstrel,"[28] also belong to this category. Depicting the fate of the mutants-Ukrainians, Andrukhovych talks about the price for the imperial ideal of the nation:

> For some reason new, different Ukrainians started being born: pig-eyed, with inexpressive round mugs, with colorless hair that exists only in order to fall out. Evidently, the natural desire of our ancestors to turn to Great Russians as quickly as possible led to certain adaptive mutations. Our glorious ancestors intensively tore off themselves the black eyebrows, brown eyes, lily-white feet, honey-sweet lips and other nationalist paraphernalia.[29]

Andrukhovych profanes the Soviet ideological kitsch of "universal human values" and "sense of one family," criticizing its homogenization of the multinational Soviet body:

> And what unites you with each of them and all the others?
> "Each of us breathes, drinks, loves, stinks the same way," announced Yura Golitsyn something secret.

---

23 Ibid., 7.
24 Ibid., 9.
25 Ibid., 23.
26 Ibid., 54, 21.
27 Ibid., 6.
28 Ibid., 12.
29 Ibid., 52.

"Hence we have been created in order to be together. The world unites in the name of universal human values," seconds him Roytman.[30]

The Moscoviad immerses the reader in the atmosphere of the apocalyptic city and the capital of the dead Empire. The wandering of the Ukrainian along the circles and levels of the ghost city is accompanied by irony and parody of the signs and symbols of the empire. At the same time, the colonial subject becomes reduced to a mere shadow, a victim of the KGB past; and the carnivalesque, with regard to the inexistent already empire, is boiled down to the virtually necrophilous bravado of the powerless colonial subject who wants to demonstrate his overman-erotic aggression, although the only sphere of his narcissistic-infantile freedom is language. Wordplay, parody of ideologemes-clichés, self-quoting, the whole array of literary codes, names, quotes—in short, masquerade becomes *the* mode of self-expression for the postcolonial Ukrainian subject on "foreign" territory.

Post-totalitarian and postcolonial revenge is not denied with regard to the "other," when even the name of the "foreigner" is annihilated and appropriated. The most expressive example of this overman conquering-revenge is the sexual taking of the brown-skinned woman in the shower room of the Literary Institute. The "low" place of this episode (the showers are located in the basement) hints at the subconscious location of the overman complex. It is worth noting, however, that it most explicitly reveals itself in the relationship of the postcolonial, post-totalitarian subject with women.

The cause of such an attitude, as Slavoj Žižek phrases it in Lacanian terms, is the jealousy toward the "other" who steals "my satisfaction." Thus all those listed and parodied "natsmen"—"others" in *The Moscoviad*—are the thieves of satisfaction, "my" satisfaction, "my" *jouissance*. This jouissance is what could be "mine" rather than "foreign" if I were free, independent, not split—by dint of the Soviet Empire—into a traitor and victim in one person. After all, the totalitarian subject, being a "castrated" victim of the empire, with his egocentric desires is a clandestine aggressor and conqueror. The situation of rape becomes archetypal. Overall, *The Moscoviad* reads as an almost necrophilous rape of Moscow, the empire, and its topoi.

Having passed through the numerous undergrounds of the Soviet Empire, *The Moscoviad*'s protagonist returns home (that is, to Ukraine) with a bullet in his temple, having committed a colonial subject's ritual suicide. Having

---

30 Ibid., 46.

experienced postmortem Western culture (represented by the classical beauty of Venice), his protagonist, Fish from the novel *Perverzion*, throws himself into the Grand Canal, thus also in a sacral-mysterious way coming back home. What does Andrukhovych's postmodern "home" mean? Is it the Europeanness, dreamed of by the Ukrainian author, who put Orpheus's mask on his hero and pushed him into the underground nonexistence of Ukraine in order to lead it out toward the light, as with Eurydice? Is it the orphic, baroque, quoting Ukraine, which the postmodern author wants to put together anew? Is it Ukraine, institutionalized in Central Europe, from which the bohemian postmodernist can "easily see everything, including New York or some Moscow."[31]

The return home after the trip to the West and to the East appears to be a new beginning for the lyrical subject in the fragments of Serhiy Zhadan's pseudo-novel *Big Mac*. His Europe is an image of a restless, mosaic land, populated by immigrants. The encounter with its past, namely with the tempestuous sixties, which heralded in a new culture of fashion and pop music, and which hitherto served as the ideal for the underground Soviet mindset, has now started to disappoint. Quasi John Lennon and old Balanescu with rotten gums cannot be idols. As Zhadan's protagonist confesses, "I worshipped him as God, and now what? … [W]hat good can come of a guy who has gum inflammation?"[32]

Old Europe is like a little house from the past, with a "smell of the strong wellbeing of the good old sixties, when a white man was still taken into consideration and one's own home was made comfortable very meticulously."[33] Now, however, it is stopped at by various casual and suspicious individuals—"outsiders of the united Europe," as Zhadan states. Conversely, for Izdryk and Andrukhovych, it remains a nostalgic and romanticized Central Europe—almost an intimate space, where one can escape from a modern reality whose essence is "a not-yet-fully formed, but already noticeable, post-totalitarianism."[34]

It is significant that Ukrainian authors more and more associate the image of a united Europe with a new totality—mass culture and its threats. For this reason, escapism—a typical feature of modern Ukrainian literature—signifies not only the situation between existence and nonexistence but also a resistance toward this new totality. Hence the escapes into childhood, to virtual Austria, alcoholic intoxication, virtual culture, and "ideal China." It is not a coincidence

---

31 Andrukhovych, *My Final Territory*, 86.
32 Serhii Zhadan, *Big Mak* (Kyiv: Krytyka, 2003), 103.
33 Ibid., 132.
34 Andrukhovych, *My Final Territory*, 85.

that Zhadan dreams about a place that is rid of mass culture, politics, civilization, and globalization:

> If I could, I would build some ideal Chinese People's Republic, something like China, only without the pederast Mao, with no boy bands, self-made men, middle-class, intellectuals and underground, instead with simple emotions, simple communication, sex without condoms, economy without globalism, parliament without the greens, church without Moscow patriarchate, and the most important—no cable television. ...[35]

---

35 Zhadan, *Big Mak*, 99.

CHAPTER 24

The Chornobyl Apocalypse of Yevhen Pashkovsky

Having reflected on a series of endless virtual transformations, bifurcations, and trifurcations, and thereby having realized the state of post-Chornobyl atomic apocalypse, Ukrainian postmodernism has put in doubt not only the existence of the sovereign subject and the integrity of Europe, but also the usefulness of the literary archive.

Yevhen Pashkovsky proposes his own injection to battle the virus of virtualization. In his novel-essay *Daily Warder* (1999), which is the author's answer to the crisis of authentic history experienced in the postmodern West, Pashkovsky, in essence, opposes Baudrillard. Pashkovsky proposes neither the cryonization of history, as in Baudrillard, nor the carnivalesque, as in Andrukhovych, nor virtualization, as in Izdryk. Instead, he suggests ethical, subjective experience.

Western philosophers also differ in their perception of Chornobyl. While Baudrillard, for instance, relates the Chornobyl explosion to the information explosion whose essence is the freezing of emotions and their simulation, Paul Virilio emphasizes first and foremost the new heroism demonstrated by the Chornobyl firefighters in their battle with the atomic fire, which threatened the very existence of the world, humanity, and civilization. He also talks about the time-world of virtual reality, which replaces geophysical reality, distorts time, and removes all that is human.[1] Yevhen Pashkovsky, with his catastrophic vision of the Chornobyl tragedy, as well as his opposition to the Western world, echoes Virilio, especially when it comes to anti-globalism, "the information bomb," and a distrust of Western democracy.

Pashkovsky powerfully experiences post-Chornobyl time—atomic time, as it were—and develops a unique nuclear discourse, whose prototype is

---

1   Paul Virilio, *The Information Bomb*, trans. Chris Turner (London-New York: Verso, 2005).

Chornobyl. "The day of the twenty-sixth of April will become the second date that has drastically altered the course of history," says Pashkovsky, echoing Baudrillard.[2] He then warns,

> all politicians and institutions will face the necessity of exerting every effort above the abyss of the unpredictable—a cauldron from which a wave of transformations, from new viruses to a new mindset, will spill on the space that is able, as half a century ago, to escape in short rushes to the nearest cemetery.[3]

For Baudrillard, Chornobyl has become a point of departure of global virtualization. The West itself, in this process, is undermined and threatened—it has nothing to hold out against the self-sufficient national cultures that flow into it from the East. Pashkovsky, on the contrary, criticizes the West for the colonial status of Ukraine and for the fact that the world, defending its own self-sufficiency in the face of the "Red Threat," left Ukraine to the mercy of the Soviets and—in recent times—neglected it after Chornobyl. The West also infects with the virus of consumption, with permanent spectacle, and with moral relativism. Hence Pashkovky threatens Europe, which is endangered by its lack of spirituality, with the Day of Judgement:

> [F]ate, which has seen so much and suffered mockery, will force their eyes wide open and make them discern the moment, from which they have hidden, extending military alliances, reinforcing borders, intercepting refugees, sympathizing with furry animals, luxuriating, and, as their Bismarck said, becoming lazy and obese from beer, when they kept watch, slept, and when someone scratched the board, threw over the side a chewing gum to the drowning men.[4]

Thus Pashkovsky creates one more image of Europe. Contradicting the rhetorical game of postmodernist authors, Pashkovsky—a hunter for definite meanings—wants to revive the authenticity of Ukrainian cultural space, opening it like a wound on the map of Europe. He combines modernist and postmodernist ideas. Instead of using baroque forms of stylization, which

---

2   Ievhen Pashkovs′kyi, *Shchodennyi zhezl* (Kyiv: Heneza, 1999), 71.
3   Ibid.
4   Ibid., 67.

would provide an opportunity for the author to hide himself, he leans toward a medieval sublimity of self-expression, which opens the liminal authorial subjectivity.

In his writing, Pashkovsky is first and foremost a witness. He is opposed to all the "great" stories ever told in the West, as well as to all Western ideals: progress, humanism, and emancipation. In so doing, he reinscribes Ukrainian reality, which was taken out of such narratives on purpose. He materializes the virtually unnoticed fluidity of time that merges with the narrator's tale and that transforms into an almost Proust-like body language. The condensation of language is achieved by dint of apocalyptic rhetoric: through lyrical invocations, appeals, threats, questions, and notes. This rhetoric serves to fortify the omnipresence of the subjective narrator. As such, in the novel *Daily Warder*, Pashkovsky is stylistically close to Ivan Vyshensky—a polemist of the sixteenth century, a denouncer of the West, and a critic of the anti-Christian "temptation" of Renaissance ideas: "one of the greatest Ukrainian writers of all time," according to Dmytro Chyzhevsky.[5]

Chornobyl "anti-time" becomes for the Ukrainian writer at the end of the twentieth century an opportunity to talk about "the time that is left with no answer and which remains like an unsolved crime and burdens the future" of Ukraine.[6] In search of lost ("abandoned") time, Pashkovsky adopts the logic of temporal nonlinearity and, returning to the past, fills it with his own childhood impressions, memories of his grandfather and father, reflections on the fate of his family, and musings on the fate of Ukraine in the twentieth century. He appeals to the "naked truth," without concealing his end-of-the-world metaphysical loneliness, which hunts you down like a beast with wolfish teeth. Here, as a leitmotif, reverberate the words of Paul the Apostle: "There is no one who seeks God! All have turned aside, together they have become worthless! There is no one who shows kindness, there is not even one."[7]

The unceasing monologue and jumping style and associations testify to the author's attempt to escape the siege of profane time, the one "in the television," received as a gift from mass culture, Western democracy, and the rule of sexual minorities—in one word, to escape the "feast with no end."

---

5 Chyzhevs′kyi, *Istoriia ukraïns′koï literatury. Vid pochatkiv do doby realizmu*, 225.
6 Pashkovs′kyi, *Shchodennyi zhezl*, 27.
7 Ibid., 260. The translation of "The Letter of Paul to the Romans" is taken from *The New Oxford Annotated Bible*, ed. Michael D. Coogan, 3rd ed. (Oxford: Oxford University Press, 2001), 247 [New Testament]); the translation is modified to reflect Pashkovky's punctuation (mainly the exclamation point [translator's note]).

The central theme of the novel is time—post-Chornobyl time, "the advent of the 'nuclear winter,'" "the Chornobyl calendar", "condensed time," "Chornobyl timelessness." Several times coexist alongside each other: childhood time, the lost time of the grandfather who wore himself out in the kolkhoz, as well as postmodern-profane time—the time of the "feast with no end."

The author-narrator—building huge sentences that run on for dozens of pages in a stream of consciousness, and skillfully bridging the gaps between the past and the present, between real details and abstract reflections—also holds the thread of time, hunts for it, and, like a sportsman, tracks it down. His expression gets automatized, crosses the border of the determined objective message, and grows into an absolute, emotional, monologue-accusation.

Pashkovsky's novel is anarchic, charged with an explosive negative force; therefore, in this sense, it is a purely Chornobyl-esque text. Words and suffering blaze with superhuman exertion, and the author-narrator perceives himself as "one of graphite rod, sweeping, a bit truncated, nailed into the reactor as though into a pencil sharpener, the pencil of the era."[8] Wrathful invective and angry accusations lash like a scourge, and thrash everybody with a word: the Westerners, the feminists, the homosexuals, the postmodernists, "the intellectual pederasts led by crypto-lesbians."[9]

In the essentially apocalyptic discourse that is created in the novel, there naturally arise biblical images and archetypes. Narrating about the last battle of chaos and truth, the storyteller identifies himself with the prophet, and the word-warden becomes his weapon: "as long as the word, despite its swoon, masters anxiety and yells, the evening is not turning into chaos, into something broken, dying, with an unbelieving face; the word only supports the warden, but this is also enough."[10] According to Derrida, this type of a discourse should be understood as an apocalyptic discourse, which combines testimonies against others and personal confessions, warnings and cautions, in order to stave off the Day of Judgment and save the world.

Such an ambitious role elevates the author above others, but at the same time brings him down to the very bottom. Chasing intensely the final word, the word of truth, he rejects the postmodern carnival and brings about its end—"the age of theatricalization." As the narrator states,

---

8   Ibid., 165.
9   Ibid., 450.
10 Ibid., 163.

the age, labelled as absurd and attractive for those eager to parade themselves and to show off, has come to its demise, leaving it to the next century to test the value of those parades and show-offs in the face of the ancient possibility to get serious in the eyes of the audience, because there is no escape from that scene into the audience, into the masses, to get lost before the finale.[11]

In his novel-essay, the narrator, who is very close to the autobiographical author, throws down the gauntlet before his colleagues, threatening them with the Last Day of Judgment: "my colleagues' texts and my own will be examined at the International Court of Justice; how much there is of juridical materials and educating dust in those manuscripts, how much, so to speak, of eternally active testimony, useful for those who like things to be systematized."[12]

He loads the word with such a hellish hatred that it becomes his weapon. Pain emanates through the words and becomes a pure mood. Negativity only sets off the affirmative dialectic of *Daily Warder*: in spite of all the encyclics, the main theme of the novel is the desire to save the world and culture from annihilation. Thus Chornobyl is associated with the last flash and explosion of culture:

> instead of the spectators there is senseless time sitting on the stone benches, gathered from the pure dust of the preatomic era, shaken out from the truly moral books with laugh-through-tears epilogues, crystallized for the show, like a valid benchmark, which everybody has to grow to, to nurture oneself in the dreamy culture, in luscious oblivion—if the two world wars and all the revolutions have diminished the value of a person to a zero, the last event has put it down below minus; the culture and prenuclear time are all around."[13]

The novel *Daily Warder* can be regarded as a tremendous metaphor for "the Chornobyl library." It is precisely in Chornobyl—the new Rome of the end of the twentieth century—that Pashkovsky gathers the remnants and debris of high culture, creating something akin to an archive or a library, in order to save culture itself. He also gathers quotations, authors, and the words of this literature, and very arbitrarily (like an equal) he disposes of them. His text preserves

---

11 Ibid., 74.
12 Ibid., 76.
13 Ibid., 102.

traces of the presence of the sacred Word, literature that resists entropic mass culture. Ultimately, in the face of the final apocalypse, he says that the whole Western world is going to flee there, to Chornobyl, which symbolizes simultaneously timelessness and a new time. Pashkovsky reminds us repeatedly that the last word of the literary message is always the finiteness of human life. No cryonization, congelation, virtualization, or simulation is able to annihilate this essential sense of human presence on Earth.

Daily Warder is an eschatological-ironical work. What can be more ironical than to look for salvation from the Last Judgment in Chornobyl? In spite of all the prophetic calls for truth, even the special mission of being a savior, which the narrator implicitly ordains for himself, becomes subject to irony: "Yesterday you greeted Noah, he informed that, ostensibly, for you and a few others, and someone else, with stock characters' IDs, places have been permanently booked, no worries, only from time to time, seldom, in the convenient, preferably holy Thursday, Monday you need to call."[14] Ultimately, the authors always talk about the Last Judgment, about the danger of culture's demise and the threat to the world library, but they always consider themselves as those in need of salvation rather than as those who save.

Pashkovsky, however, insists on his role as a savior who gathers the "Chornobyl library." Among those whom we encounter in the new Rome-Chornobyl, together with other renowned postmodernists, there could be Jean Baudrillard or Jacques Derrida—since this gathering features a selected society: there, Kundera "catapults with a slingshot at the apparitions of some tanks, delivering speeches and irritating everybody with such publicity," and the author-narrator confesses: "the day before yesterday we discussed a similar topic with Joseph Conrad, that is, Jerzy[15] Korzeniowski."[16] There, while shooting game he meets Turgenev and sees James Joyce—he is hurrying to the kindergarten ("later I learned that he teaches there").[17] Accidentally, in Georg's he runs onto Hermann Broch: "we drank a glass of wine and had a narrow escape from junkies."[18] "W. Faulkner stopped by, as usually neat, lean, in a new velvet jacket," he notes on occasion, "he shared your concern regarding the threatened freedom and in particular pointed out that 'we substituted patent for freedom—the patent on any

---

14 Ibid., 110.
15 Sic! (Translator's note.)
16 Ibid., 77, 90.
17 Ibid., 76.
18 Ibid.

action that is done within the law.'"[19] "We often correspond with A. Camus,"[20] notes the I-narrator and paraphrases *The Plague*, trying it on the Chornobyl situation. He receives a telegram from Marcel Proust and writes a letter to Salman Rushdie, offering him asylum from persecution in the Chornobyl zone. The whole text of *Daily Warder* is to the highest degree intertextual, filled with voices, quotations, and the names of men of letters, and his Chornobyl library is an important text in the novel.

The library is a refuge and time regained, which is so desired by the narrator. It is also the place of truth in a totally degraded and corrupted world, in which Chornobyl embodies for Pashkovsky the painful wound of truthfulness. In addition, the Chornobyl library metaphorically resonates with the time of childhood and innocence. Through it and in it there appears paradise lost and regained, which opens only in the moment of epiphany, when "inexpressible wonder, like a beehive that is lighted from the inside by the buckwheat honey, accumulates the stream of unending efforts, the habit of transformations into the immortal."[21]

The author calls his novel-essay a "novel-missive," "novel-encyclical," emphasizing the liminal subjectivity and partisan quality of his writing, and thereby juxtaposing himself to the "falsity" of mass culture and its simulacra. In his opinion, mass culture is also born together with Chornobyl, and in particular in the discourse about Chornobyl, which is boiled down to a false optimism ("everything is fine," as the official media maintained).

The desire to remake the world (the nation) with the word and to measure the power of the word in defiance of mass culture and in the era of advanced telecommunications, technologies, and cyborgs shows itself in the apologetic character of the book. *Daily Warder* is dominated by the rule of the word and an apology for reading. We witness "ship books" and book "archipelagoes." Ultimately, Pashkovsky-the-essayist would like to salvage high, literary culture in this "Chornobyl library."

To write after Chornobyl is to defend, says Pashkovsky,

> and, although from that April on, you have never slept again, you are waking up in the fog of all thinkable madnesses and in the disabled control room you are pressing the emergency button: it is not late, now twenty,

---

19 Ibid., 468.
20 Ibid., 76.
21 Ibid., 111.

now nineteen, fewer and fewer seconds are left, the earth is starting to shake as though someone is shaking the top of the globe, you are pressing so hard that blood is blackening under your fingernail (Ibid., 125).

The subjective truthfulness that goes as far as self-sacrifice (when the author's "I" is placed within the "unnamed graphitic rods"), passion and sincerity to the highest degree—this is what at the end of the twentieth century Pashkovky proposes as his version of nuclear discourse that is opposed to virtuality.

CHAPTER 25

The Postmodern Homelessness of Serhiy Zhadan

The development of youth subcultures, especially in the postwar period, is a phenomenon of mass culture. Punk culture, born in Britain at the end of the 1970s, adopted something from camp and Warhol's Pop Art, as well as from avant-garde performances and conceptual art forms. The shocking images created by punks aimed at the creation of an original costumed image—a persona. In addition, punk culture (with its differentiation between the aesthetized punk of the middle class and the coarse punk of the working class) was based on fantasy and the subversion of bourgeois values. At the same time, it grotesquely parodied popular culture and mass-produced style.

In Ukrainian literature and culture of the end of the twentieth century, the poetry of Serhiy Zhadan embodies both punk parody and youth defenselessness. Zhadan does not consistently follow punk style, but there are recognizable elements of punk alongside bohemian and avant-garde ones in his work. The bohemian quality somewhat upstages the middle-class aspect because it itself contains a radical, juvenile, punk mindset. This is where it differs from, for instance, performance, played out in the work of another author of postmodern games, Volodymyr Tsybulko, who represents the mindset of the middle class and popularizes it among the bohemia.

Zhadan's bohemian protest grows out of orphanhood and homelessness. A loss of trust in parents, an obsession with hallucinogenic present time, a broken belief in happiness, all of which would have been provided by home and family, a lack of desire to grow up, an infantile attitude toward reaching old age—all these attitudes were characteristic of the hippies and beatniks. It leaves the impression that the 1990s in Ukraine are a quotation of the Western 1960s (with its admiration for Pop Art, jazz, and sexual revolution). The mindset of 1960s radical youth reappears after some forty years in the new generation that grows up in Ukraine during the disintegration of the Soviet Empire. Like Benjamin's angel of history, time moves forward yet looks back to the 1960s.

An eloquent confirmation of this process is the book *The Day of Mrs. Day's Death: American Poetry of 1950–60s in Translations by Yuri Andrukhovych* (2006). Answering the question as to why exactly the 1950s and 1960s are so important, Andrukhovych explains that "this is the time when the new generation of artists enters the stage (the so-called *new avant-garde*)" and "the time when the American art ... abandons some *parochialism*, some shameful derivativeness with regard to the avant-garde Europe," as well as the time that becomes "a reaction to the *post-Eliot* situation"—that is, a reaction to an oversaturation with historical and cultural symbolism, as well as academism.[1] The appearance of these new, wild poets is the emergence of rebels coming out into the streets and to the public.

It is in the West in the 1960s that a new consciousness forms—one that creates, outside life as such, its own vision of a world in which one would like to live. Romantics, nonconformists, visionaries, and the homeless—all of these kind of people intertwine in the new hippie mindset. At the same time, one of the new consciousness's characteristics is a devaluation of the heroic model of the artist's life and creative work. Ultimately, it is hard to discuss the artistic "I" of artists such as Pollock or beatnik poets such as Allen Ginsberg by using the category of the heroic. Andrukhovych creates an eloquent portrait of a beatnik:

> The beat was first and foremost an existential practice, and only secondarily—a certain literary identity. ... The beatnik type often combined in itself asociality and marginality as a conscious and joyous life choice, anarchism, the vagrant mode of life (the paradigm that was reflected and reproduced anew in Kerouac's novel *On the Road*). Additionally it contains the fascination with jazz ..., eastern meditative schools, Zen Buddhism, experiments with the *broadening of consciousness*.[2]

For Andrukhovych, the beatniks are also "a radical experiment with self, sometimes bordering on death, and therefore it is predominantly an experience of

---

1   Iurii Andrukhovych, "Ameryka. Vidkryttia No 1001. Zamist´ peredmovy," in *Den´ smerti pani Den´* (Kharkiv: Folio, 2006), 11–12.
2   Ibid., 17.

physical self-destruction, which in a unique way is projected onto the social self-sacrifice."³

Pop Art becomes the sign of that time. Richard Hamilton, one of the participants of the exhibition *This Is Tomorrow* in the Whitechapel Art Gallery in London in 1956, called Pop Art "Popular (designed for a mass audience) / Transient (short-term solution) / Expendable (easily forgotten) / Low cost / Mass produced / Young (aimed at youth) / Witty / Sexy / Gimmicky / Glamorous / Big Business."⁴

Reflection on the Western 1960s is important against the background of the now open, now clandestine, reevaluation by the Ukrainian nineties of the experience of the Ukrainian sixties. It is worth repeating that the generation of the 1960s likewise becomes an object of reevaluation in the West; a testimony to this would be, for instance, Michel Houellebecq's scandalous novel *The Elementary Particles*. If it comes to the difference in worldviews, the metaphor for the sixties in Ukraine is "home," while for the beatniks and hippies it is "the road" that leads somewhere. The road that leads nowhere, homelessness, and orphanhood—these are the new coordinates of the world that the leader of the nineties writers in Ukraine, Serhiy Zhadan, chooses for himself, in spite of his infatuation with the Western 1960s.

In a sense, the Ukrainian sixties were rationalists-neophytes who believed in a new rationality. "It was a separation of the self from the amorphous mass of the mediocre Soviet citizen—as a thinking individual, as a Ukrainian and a citizen of the universe,"⁵ reflects Mykhailyna Kotsiubynska. The new rationalism was nonconformist, moral, and national. Thus the Ukrainian sixties phenomenon was first and foremost a version of intellectualism that, in contrast with bohemianism, is often political. Resistance against "the System," a cult of individual independence, and liberalism are the characteristics of this intellectualism.

Culturally and socially, the Ukrainian sixties are close to Western existentialism; however, there is an essential difference, which cultural anthropology labels as the difference between individualism and authenticity. The Ukrainian sixties were the authentic type, or, more precisely, populist. The worldview and the type of culture that became ideal for the sixties developed along the lines of

---

3   Ibid.
4   Thomas Crow, *The Rise of the Sixties: American and European Art in the Era of Dissent 1955–69* (London: Calmann and King Ltd., 1996), 45.
5   Mykhailyna Kotsiubyns´ka, *Moï obriï*, vol. 2 (Kyiv: Dukh i Litera, 2004), 6.

a synthesis of folk and modern culture. This is, as it were, a moderate modernism; a philosophy that is not determined by the creation of a new culture dictated by formal experiment, as we witness in the "high" modernism of the West. Rather, it is determined by the adjustment of modernist form to national-folk tradition. This tradition is based not so much on irrational mythology, as in the early ballads of Ivan Drach, as on the rationality of the Vasyl Symonenko type.

Thus the Ukrainian sixties phenomenon was not identical with Western existentialism. For the sixties, Being was not Nothingness. As opposition to the system, Being existed on the level of the ideal—the people, the nation, and the small person. The Western version of the sixties is first of all radical, bent on the destruction of social hierarchy and keen to introduce serial repetition and anonymous surfaces. This is the period of the decline of existentialism and the emergence of post-structuralist anonymity and the postmodern worldview. The Ukrainian 1960s, on the other hand, are a time of the formation of rational personalism.

Common for the both, however, was the problem of representation. Trust in aesthetic perfection of artistic phenomenon was undermined; totalitarian ideology had given birth to the so-called socialist realist monster of resemblance; and the gap between the ideal beauty of art and the brutality and evil of experienced reality in the postwar period had become huge and inhuman. In the West, in the hippie environment, a new genre was born—performance, which symbolized the principle of the incompleteness of the artistic act as well as being a sincere and direct involvement of the artist's "body" in the process of aesthetic experiment. The Ukrainian sixties, instead, wanted to revitalize aesthetics on the basis of the moral involvement of literature and art in the process of the nation creation.

The 1960s in the West were a time when once again reemerges the power of the unconscious and pop culture is formed. The social, political, rock, and sexual culture of those years are what Susan Sontag calls a new sensibility, or, more precisely, "a new (potentially unitary) kind of sensibility."[6] Sontag writes:

> This new sensibility is rooted, as it must be, in our experience, experiences which are new in the history of humanity—in extreme social and physical mobility; in the crowdedness of the human scene (both people and material commodities multiplying at a dizzying rate); in the availability of the new sensations such as speed (physical speed, as in airplane travel; speed

---

6   Susan Sontag, "One Culture and the New Sensibility," in *Against Interpretation*, 296.

of images, as in the cinema); and in the pan-cultural perspective on the arts that is possible through the mass reproduction of art objects.[7]

In 1957, Roland Barthes published *Mythologies*; Claude Levi-Strauss's *Structural Anthropology* appeared in 1958; in 1960 in England for the first time uncensored was published David Herbert Lawrence's *Lady Chatterley's Lover*. A book about how authority appropriates human madness and manipulates it in the course of centuries appeared in 1961 (Michel Foucault's *Madness and Civilization*). Rational Frankfurt criticism proclaimed—through the work of Herbert Marcuse—the birth of the "one-dimensional man." It is worth noting that all these publications, almost cult-like in their influence on the 1960s West, have been received by the Ukrainian reader only recently—in the last decade of the twentieth century.

Did this wave reach Ukraine in the 1960s, that is, forty years ago? The Ukrainian sixties also felt the addictively sugary and at the same time unbearable burden of being. It was entwined with resistance to the Soviet system, which in its essence was annihilating and deceitful, repressive and absolute. An unwillingness to turn into a "one-dimensional," total Soviet person underpinned the revolt of young Ukrainian intellectuals. The utopian worldview of dissidents developed both individually and in fraternal unity. Adopting the utopian belief that by way of reason one can change and "right" the world of disharmony, the sixties, as Ivan Svitlychny confesses, built their outlook on common sense, seeing the human in the a common ("small") person. They learned to tell people the truth, bringing back meaning to words. The difference between this "new rationality" and that which had been newly born in the West as postmodernity is obvious. However, sensibility was also one of the characteristics of the Ukrainian sixties. After all, the heroes of those days were young people who opened to the other not only their souls but also their bodies. Fraternal communities and Polissia colonies, flirting and chivalry, friendship and love also contributed to the unique atmosphere of the Ukrainian 1960s.

As long as in the West the 1960s were a time of the production of Pop Art, mass culture, and kitsch, the intellectual utopia of the Ukrainian sixties maintained faith in high culture. They thought of it not as much elitist as egalitarian. The merging of the cultures of the elites and the people is the foundation of the cultural utopia of the sixties. Shevchenko, as always, becomes a leading figure. The cultural, aesthetic, and individualist ideals of the sixties are crystallized in his work.

---

7   Ibid.

Shevchenko's poems, on the one hand, emphasize the conscious (rather than folk-mythological) foundation of the poet's language and style, its national character; on the other hand, they implicitly reject an elitist identification with their creator. Shevchenko's ballads are defined by their national character and, more importantly, by the absence of mysticism. Instead, there is an emphasis on the presence in the text of the poet himself, who feels sadness and joy together with his characters. The microanalysis of Shevchenko's poem "Don't Leave Your Mother, They Said" becomes a(n) (un)conscious illustration to Symonenko's "The Swans of Motherhood." What occurs here is the symbiosis of the "already said" and of what "is being said now." Literature moralizes and personalizes, and the focus here shifts toward the autobiographical intellectualism of Shevchenko's choice: the voice of the poet merges with the voice of the people.

Through Shevchenko's "people's" code sixtiers' ideals were revealed. Their essence was emphatic individuality. Thus the integrity, honesty, and morality of an individual are formed: from Sverstiuk's almost rigoristic moral directness to Stus's existential "filling oneself with oneself." Individuality becomes the core of intellectual self-therapy, which the sixtiers exercised over themselves and with which they enlightened society. Its emphasis is not as much a romantic conflict between the individual with society, as it is a moment of a potential synthesis of human individuality (immersed in identity) with tradition, folk language, and lore. "Home," "fullness," and self-sufficiency become a metaphor of such personalism.

It is exactly there where the differences are smoothed out (in the common home, that is), that the utopia of the sixties found its dwelling: at the crossroads of Ukraine and Europe, the folk and the avant-garde, the national and the universal, the honest and the human. This home took its roots in Being and was not a patriarchal idea of "Ukrainian home" ("Ukraïns′ka Khata"—the title of a journal that Ukrainian culturalists and philosopher-neoromantics published at the beginning of the twentieth century), but was something more in the spirit of Heidegger. Home—as revealed Being—was painted as "earthly" (Vasyl Symonenko) and "cosmic" (Mykola Vinhranovsky), and voiced by the expressive-female existentials of Lina Kostenko; it also loomed through the red and black embroidery of Dmytro Pavlychko. Ivan Drach stood near the Home, overcome and immobilized like an amazed sunflower, and within a loaf of homemade bread smelled fabulous. Across the road there stood "a cathedral without scaffolding," and a little farther Vasyl Stus was getting numb with pain in the baroque "merry cemetery." The metaphor of home became a refuge for the sixtiers.

In the 1990s, neither "homeliness" nor "high" culture in its patriotic-populist version are any longer ideal categories, especially for those who have just turned twenty. Even the thirty-year-old "bubs" at that time are totally engaged in buffoonery and eternal wandering. This is what their new superhero—a bohemian poet in a Chrysler Imperial—wants. Some "Bubabists," like, for example, Andrukhovych, tautologically would reproduce one and the same kind of adventurous protagonist, taking up translating American beatnik poets. Another Bubabist—Neborak—would write a book about Aeneas and the dangers of uprootedness, that is, of migration. Volodymyr Dibrova, in contrast, would write, emblematic for the beginning of the 1990s, a book of short fiction *The Songs of the Beatles* (1990)—a combination of the experience of freedom by an intellectual outsider at the time of perestroika and an expression of the unique spirit of the 1960s. Thus the return of the hippie on the post-totalitarian stage in the 1990s became particularly significant.

In Ukraine, this was a period of "war and restoration" for the generation that experienced a crisis of trust toward their parents and toward the heritage that their parents maintained and had held dear during the long years of Soviet rule. Dissatisfaction was felt not only with regard to faithful parent-Soviets, but also toward the oppositionist sixties. One of the first who expressed dissatisfaction was Oksana Zabuzhko in her *Field Work in Ukrainian Sex*. In fact, discussion about the fate of postmodernism in Ukraine unfolds around the heritage of the 1960s. Oxana Pachlovska would speak about faithfulness to the sixties and their ideals. Oles Doniy, one of the leaders in the nineties, would proclaim the bankruptcy of the sixties. Ultimately, this discussion did not receive a sufficient response from the public and faded away. Nevertheless, it is on the ground of attraction and repulsion between the 1990s and 1960s that the cultural philosophy of Ukrainian postmodernism is formed. At the basis of this philosophy is the condition of change, migration, and deterritorialization. In other words, Ukrainian postmodernism embodies a desire to be "on the road" rather than "at home."

The characteristic features of this ideology are infantilism and homelessness. Andrukhovych's characters travel around the world, Zabuzhko's heroine escapes to America, Izdryk's protagonist ends up at a mental facility, Dibrova's character is not able to find his home, and for Pashkovsky's protagonist hunting and traveling are the state of being. Even more expressively this homelessness is represented in the work of Serhiy Zhadan. The principal difference between the sixtiers and the nineters lies in the fact that the former did not have fathers because they had not returned from the Second World War. They lived with a desire to have a father The ninetiers had fathers, yet had lost trust in them and desired to return to the time of their parents' youth, to the hippie era, in order to alter, from there, their own lives. This union of unreturnability and the desire to return in order to wander in parallel worlds of virtual reality gives birth to a melancholy resentment of the lost generation that grows during perestroika. The metaphysical problem of the road, the return of the past, and fragmentary consciousness become the main attributes of this generation.

Having lost trust in their parents, the ninetiers, like escapists, no longer experience the Oedipus complex because the image of the father has been discredited without their assistance. Zhadan's characters generally flow against the current of life; they are deprived of history and of the past. As one of them admits, "this always irritates me, in the sense of seeing that someone already lived here before me and in contrast to me lived a real life—ate breakfasts, had sex, maybe even loved someone, visited the markets and stores."[8] For sure, Zhadan's homeless protagonist would like to be adopted by someone; however, it appears to be impossible: "why did no one adopt me, let's say, when I lived for several days at the bus station and slept on wooden chairs, or when I lived off boiled water for several days, come to think of it why doesn't anyone adopt me now, why doesn't this jerkoff adopt me?"[9]

In such a post-Soviet world the figure of the father is largely absent. For example, in Zhadan's novel *Depeche Mode*, the shadow of the father looms somewhere far away—under the guise of the stepfather, whose funeral one of the characters is supposed to attend. However, the father may exist as part of the system, for instance, Marusia's father-general, who has provided his daughter with "all that minimal collection of artificial limbs and false jaws that has been made to transport you more comfortably through your life."[10] Molotov's bust in this case easily replaces such a father, and "two unfortunate, wasted creatures, Marusia in her torn designer jeans and Rolling Stones T-shirt, and Molotov, central committee member, old hedonist, lover of cocktails."[11] Overall, Molotov's bust, stolen from a plant, easily replaces the image of father-stepfather in Zhadan's world of homelessness.

The mother is also absent in such a world. Probably, she was profaned due to Soviet society's cultivation of "love for the mother," which relegated this

---

8   Zhadan, *Depeche Mode*, 149.
9   Ibid., 149.
10 Ibid., 126.
11 Ibid., 141.

image to a popular cliché. This profanation is almost identical with cannibalism. "They ate the mother up," says one of Zhadan's characters, commenting on the lyrics to the song about a mother's salty eyes.

> "They ate the mother up," says Vasia with satisfaction, as though his suspicions about this world received confirmation.
> "Who ate up?"
> "Well, them—stepan haliabarda."
> "Come on. It's just a phrase."
> "Not just a phrase. Listen what they are saying—a little salty."[12]

This is the perception of the cliché about "a little salty" mother's eyes from the song "Mother's Eyes" by an anonymous stepan haliabarda.

Homelessness, however, is life in which there is a lack of somebody nearby. "I know that everything comes from him alone, because if I have felt alongside me someone's presence, it was precisely his, although personally I needed someone else's presence much more,"[13] confesses Zhadan's protagonist. Thereby the conflict between parents and children has not only a physical but also a metaphysical dimension. Without parental love, Marusia sits on the balcony, embracing her Molotov, who somewhat resembles her father general; Carburetor feels, instead, that "he's had to eat so much shit in his life that obviously he really doesn't need this extra load—what with his family, his one-legged stepfather, Uncle Robert."[14]

The state of infantilism in the 1990s was also fortified by the perception of the present as an apocalyptic time—the time of perestroika and the birth of "a new earth" and "new heaven." In the 1990s, virtually every author, regardless of age, in his or her own way responds to the situation of initiation: from Valery Shevchuk and Viacheslav Medvid to Serhiy Zhadan. At the threshold of the 1990s, literary critics even began to discriminate between the hermetic poets, that is, "adult poets" (Oleh Lysheha, Vasyl Holoborodko, and Vasyl Herasymyuk), and the poet-tricksters, or poet-adolescents. Among them there is first of all Serhiy Zhadan, recognizable by his persona of a punk and sad clown, a homeless teenager and the last Ukrainian futurist.

---

12 Zhadan, *Depesh Mod* (Kharkiv: Folio, 2004), 58. This fragment was not included in the Glagoslav Publications edition of the English translation of the novel (translator's note).
13 Zhadan, *Depeche Mode*, 191 (Glagoslav Publications edition).
14 Ibid., 195.

Serhiy Zhadan's poetry most expressively reflects the traumatic adolescent consciousness of post-totalitarian time. The author experiences this time as a period of orphanhood, when distrust and resentment toward the Soviet past, which has proved been utterly devalued, produces aggression and contempt toward their parents in teenagers. Perhaps homelessness can be regarded as the most productive metaphor that Zhadan has proposed. Homelessness means both endless wandering and distrust of the world of the adults who, concurrently with perestroika, acquired their own uprootedness of Being. This is the origin of punk—that culture of infantile adult men who play children's games.

Zhadan is a sad clown living at a railway station, a punk who does not want to grow up, and at the same time, a revolutionary of the "homeless" and "alcoholic" generation of the 1990s. To him, the nineties are when

> all these teenagers are helpless against the years
> and their hearts are hard as slate
> but at the same time as a slate pencil they are brittle
> so the only thing that is left is to listen to the winter
> that hovers from somewhere
> and to warm up the sky with a plastic lighter.[15]

Zhadan's poetic book *Tsytatnyk* (*A Collection of Quotes* [1995]) is the manifesto of the ninetiers' homelessness. The very principle of "a collection of quotes" recalls totalitarian and revolutionary times, perhaps those of Mao Zedong. Like the conceptualists, Zhadan demystifies the clichés, quotations, and slogans of the past era's culture—for example, "love for the fatherland," a formula from Kotliarevsky's *Eneida*. Love for the fatherland, or patriotism in its traditional understanding, appears to be fake, because a union can be both "the sun that is rising" and "a common washroom"; therefore, "love for the fatherland," as a principle of human life, is an artificial slogan because "as for me this method / successfully kills love / the very notion of 'sex.'"[16] Thus, sexuality, the base, undermines the rhetoric of pompous words.

However, Zhadan's conceptualism differs from the orthodox one. In the conceptualists' poetry, the subject itself disappears among the given objects-clichés; in Zhadan, it remains distant, and the poet's voice sounds childishly vulnerable. Quotes from Tychyna and Shevchenko, "implanted" into the alien

---

15 Zhadan, *Balady pro viinu i vidbudovu. Nova knyha virshiv*, 19.
16 Ibid., 64.

ideological body, become the formulas of a different consciousness—in particular, an infantile discourse. This rewriting of the classic occurs not without resentment toward the past. After all, that totalitarian past, corrupted by the false rhetoric of "The National Poet," "The People's Struggle," "Rotten Noblemen," the "The Suffering Mother," and "Mother-Ukraine," robbed the poet-teenager of his trust in language, fatherland, and the very process of growing up. Therefore,

> so much dregs in children's hearts especially at night
> school that has taught nothing
> and time that has not taught anything either
> and I guess—eternal waste because this shit persists[17]

For such people as Zhadan, perestroika coincided with "youth at the beginning of the nineties" ("this youth of the beginning of the nineties"—says the poet). Those "bad times" were "good times" because they were the times of youth, and the lyrical subject defends them:

> good times bad times—they gave a chance
> when you warmed yourself in the school backyard
> and puked the shabby Polish schnapps
> when all these winters and walls and pines and silence are around
> that you can hear birds' breathing in the sky[18]

Zhadan's collection *Ballads about War and Reconstruction* can be considered a biography of the generation that—regardless of its own desires—had to become mature and cross the river of time, even though the streets were a vortex of "dams washed away" and those "offended by memory" spoke at rallies. Thereby maturity came:

> and maturity like an adolescent girl who with no purpose
> is late disappears and does not let him know
> who knows that it is possible not to come at all
> but does not know yet why[19]

---

17 Ibid., 19.
18 Ibid., 18.
19 Ibid., 16.

Vladimir Sorokin prophesied that what comes after postmodernism is, probably, a new wave of literature of sincerity. Zhadan confirms that such literature existed as a possibility at the early stage of postmodernism in Ukrainian literature. The traumatic, existential, adolescent experience of the generation of the nineties is constructed by the realities of the common past: "old garages pipe of aquatic center destroyed by bombs Swiss military knives,"[20] the native city does not hold and does not defend any longer, "Kharkiv dances—the city of bastards."[21] The older generation, instead, is united by other "slogans," "inscriptions," and "cryptograms," dominated by the "slogans of struggle traces of communal depravity."[22] Along with a rejection of the socialist past, there is grown a distrust of parents and their heritage. The totalitarian past condemns people to an eternal struggle:

> from the very childhood this victory is like leprosy
> as though you are growing at an abandoned airdrome
> sewing together flying suits out of old parachutes
> attaching an empty machine-gun case to the keys
> thus the fatherland endows with the right
> to struggle[23]

Mother-Ukraine here is a country of "sausage and alcohol" ("love-for-humans and hatred, sausage and spirits— / This country is soiled by them to the lees";[24] it pushes away to the "labyrinth of steppes" and generates eternal escaping from the "sick provinces."[25] Painful and vulnerable existential experience undermines literariness as a game, and the homeless poet finds his refuge "where time shows like coal through the ground / where death begins and literature ends."[26]

Zhadan's "The End of Ukrainian Syllabo-Tonic Verse" sounds like defiance toward the linguistic devaluation perpetuated by totalitarianism. Avant-garde and surrealist primeval language, virgin and unsoiled, is tantamount to the birth of the prelapsarian semantic world that shows in the gaps of meanings and words:

---

20 Ibid., 20.
21 Serhii Zhadan, *Tsytatnyk (virshi dlia kokhanok i kokhantsiv)* (Kyiv: Smoloskyp, 1995), 18.
22 Zhadan, *Balady*, 43.
23 Ibid., 44.
24 Zhadan, *Tsytatnyk*, 19.
25 Ibid.
26 Zhadan, *Balady*, 24.

> when you will want to divide words
> by those you have used at least once and those you have never touched
> you must feel this silence tearing apart
> the heart of the night—the tormented circle—
> every time whenever you come back here[27]

The topography of interstitial time in which the nineties dwell consists of the stations of departure and arrival, where the travelers are a particular caste of homeless and uprooted people, all related by their homelessness. After all, with all the "love for the fatherland," which has always been in this country "a high," as Zhadan ironizes, things that really are able to unite people are different: "the rising sun / or common washroom," "and one can add, ultimately, folklore and alcohol."[28]

In this way, Zhadan's work asserts the immanence of being; nomadism and orphanhood as the common condition for all outsiders, homeless, and immigrants. This condition joins together Munich Turks and Irish students—"the angels of terrorism," "a kind negro pastor" and a blood-brother immigrant.[29] The lyrical subject associates himself precisely with those on the road, with those who undergo the rite of passage. The only thing that unites and protects in this homelessness is the sky, time, death, and God, "who within the limits of loneliness protects us"—meditates the lyrical subject of *A Collection of Quotes*.[30] Prayer about memory sounds like an incantation:

> I will pray for you and your route
> for your insurance and truck tires
> for water in cold rivers and foliage which is being burnt already
> for everything you have forgotten here
> and what you are forgetting now
> regardless where you end up
> for everything you will forget later
> and most important—
> for your memory for memory[31]

---

27 Ibid., 23.
28 Zhadan, *Tsytatnyk*, 50.
29 Ibid., 51.
30 Ibid.
31 Zhadan, *Balady*, 21.

Zhadan conveys the sense of the vulnerability of life in a surrealist way: it is "broken like a rose aviator's spine."[32] "Juvenile aviators," "juvenile transvestite," "vocational school," "little girl"—these are the homeless of the nineties about whom Zhadan often talks. Ultimately, and not surprisingly, there emerge a romance for the convict and a military nostalgia for the "occupation regime"; and there are stanzas about German fascist invaders and a romance for Latinos. Che Guevara becomes a blood brother for those who have not succumbed to "the regime and the dorm superintendent."

Zhadan's postmodernism has a neo-avant-garde complexion. The romance of avant-gardism with its revolutionary slogans combines with a juvenile distrust of anything "already said"—a typical object of postmodernism. Names that fix the traces of the past appear to be the most alive of anything: "blades badges films cartridges melted lipstick / trees hairpins steel tweezers" and "between sheath knives and reels somewhere at the very bottom / the picture you've been looking for."[33] Particular things attach us exactly to time because

> everything passes us
> only a few objects
> to preserve like heritage like fortune in order
> to show when asked[34]

For this reason, poems from the cycle "Pacific" sound like a transcendent pronouncement of the generation itself:

> time simply breaks everybody's necks
> with no question
> with no question whatsoever[35]

"Pacific" is pure postmodernism: it testifies that the passage from neo-avant-garde to postmodernism in Zhadan's work has already taken place. Echoes of quotations (Antonych's green gospel, Pavlychko's mystery of your face, melancholy decadence ["the dead hearts of vegetables"]), as well as the names Tychyna, Rylsky, Shevchenko, Sosiura, as well as a parody of Ukrainian liter-

---

32 Ibid., 13.
33 Zhadan, *Balady*, 41–42.
34 Ibid., 52.
35 Ibid., 15.

ature as a school subject, structure a new world and a new time for the post-Soviet teenager. The process of growing up resembles the river of life:

> In the same manner
> the year is a hard duration—
> a carefully filled
> niche—
> nourishing gratitude with years,
> becomes the deeper the quieter[36]

The traumatized generation "has survived the fatherland" "in which grown into men teenagers / were born."[37] Thus, as the poet-ninetier states, "a flock of words and a crowd of passersby / this is the dachau that is left of the whole generation."[38] Resentment toward the past, a distrust of the natural circle of life, a dislike of growing up, and a criticism of the Mother Ukraine—these are signs of the post-totalitarian era, borne by the homeless generation of the nineties. The generation does not want to grow old:

> why are you afraid of growing old
> after all we grew together
> there is an impress on our pajamas—
> made in Ukraine
>
> old age is always a failure
> this is not even a loss
> rather a possessing of something
> that quickly loses its value
> step by step the scaffolding of the soul is rotting and falling
> the linen is whitening the bridge bearings are bending
> the velvet on the carcass of the hot heart is being torn
>
> old age like deserter is in hiding
> in the underground[39]

---

36 Ibid., 31.
37 Ibid., 46.
38 Ibid., 11.
39 Ibid., 48.

Homeless teenagers, old tramps, and former hippies, get together in a single post-totalitarian space.

Ultimately, Zhadan's post-totalitarian homeless is aware of his homogeneity with the new European, the person in transit, and the immigrant. His collection of short stories *Big Mac* brings together former hippies and young poet-teenagers in the special zone called the zone of freedom. Such a zone exists everywhere: both here, in Ukraine, and abroad (for instance, in Germany). This zone gathers the homeless, that is, eternal migrants, those who treasure the space of freedom—for example, the Ukrainian poet, seventy-year-old immigrant from Czech Republic, former friend of Havel, and avant-garde artist Rudi. A creator of huge metal constructions, Rudi escapes, after Dubček's resignation, to Western Berlin, even though there too he has to struggle for freedom, this time not with the totalitarian regime but with the mass culture and commercials that invade every available meter of space. As Rudi writes, returning to the new Czech Republic is like a "new emigration" because

> although I am no cosmopolitan, I have understood here one thing: in reality, space is not divided into yours and somebody else's; space can be either free or under control, you see? I couldn't care less where I live; what's important is how I live. And here I live the way I want.[40]

Fatherland for Rudi, along with all immigrants, is the free space where they live. This is the rationale behind life both for "old Havel's friend in a military jacket" and for the "tens of thousands of Balkans and Turks who build this new Berlin, raising scaffolding everywhere and supporting with their help this cold Berlin sky that hovers low and mournful."[41] Creating a new image of the West, Zhadan records how the landscape of the "united Europe" is changing in the age of globalization under the influence of immigration.

Neither the meeting with the old Europe nor the meeting with the generation of Havel, John Lennon, and the hippies is as idyllic as one could imagine when looking from the 1990s to the 1960s. This becomes evident in Zhadan's "ten ways to kill John Lennon." The Vienna pacifist looks very much like Lennon, if you imagine "the late Lennon around eight in the evening, who has been drinking from as early as eight in the morning, his glasses misted over from his

---

40 Zhadan, *Big Mak*, 36.
41 Ibid., 38.

liquor breath, those round glasses, remember, real Lennon glasses."[42] Such a pseudo-Lennon is only one of those "freaks and losers"—"outsiders" who, as a character in Zhadan's *Big Mac* says, comprise the essence of "our torn to pieces by subsidies and defaults society."[43] Nevertheless, he is way more interesting than the so-called "successful people,"

> because what is the fun in communicating with, let's say, bankers or commercial organizations—they talk quotes from their own business plans; what can you talk about with young scholars, athletes, or dealers … another thing is such degenerates as John Lennon, the wistful stumps of the great European pseudo-revolution; they are all the same in essence—both here and in our country; both were painfully traumatized in their time with the stories of psychedelia and ecumene.[44]

Hence this is a tale about former homeless nonconformists who have preserved this quality to old age, unlike their teachers who "stealthily moved to new apartments."[45] Former hippies, "old schizoid European intellectuals," got mixed up with the new homeless—immigrants from Third World countries, and found their new relatives among them—Arabs, Hindus, Turks, Albanians, Ukrainians, whose fatherlands had thrown them into a new space—the "Berlin that we have lost."[46] The new migration of peoples, and Europe itself as migration, are the main themes of Zhadan's poetry. For this reason, the idols of the seventies appear quite naturally here: jazzmen, Buddy Rich's old friends, "Big Mac," "and everything falls into place, and everybody understands that life is wonderful and mysterious, and even approaching to its sheeting, almost touching its tight cool surface, you all the same are not a jot closer to the understanding of it."[47]

Such an "unbearable lightness of being" of the sixties is transposed onto the "whimsical life" of the perpetually homeless wanderers of the nineties, for whom life is not an appropriation of power and strength but a river in which you flow "without trying to drown somebody" with your deeds.

Zhadan's protest of the homeless is directed against the power of officialdom and mass culture, which dominated life in the 1990s: "if you are not

---

42 Ibid., 44.
43 Ibid., 50.
44 Ibid., 56.
45 Ibid.
46 Ibid., 11.
47 Ibid., 56.

interested in show business and macroeconomics, you are taken for … deserter who poses a direct and immediate threat to society."[48] It is an obligation of mass culture, democracy, and civil society to capture and restrain such an individual. For this reason, Zhadan creates a unique parallel reality: his characters all the time wander, get together, crash in someone's kitchen, or simply meet each other; and the lure of these spontaneous improvised acquaintances "lies in their sincerity, ingenuousness, and unobtrusiveness."[49] Zhadan's anti-world is not virtual, as in Izdryk, and not verbal, as in Andrukhovych. It rather subsists in the sincerity and fluidity of human existence, that is, in the life of "the last individual." The Eternal Teenager—this is the alternative that Zhadan offers to old Western culture. His protagonist, a perpetual migrant, recognizes the deceptive quality of the postmodern world; after all, both anti-globalism and the struggle with mass culture are elements of Western civilization.

Zhadan's accusations against the fruits of such civilization contain a terroristic and anarchic resentment, along with dissatisfaction and jealousy. His protagonist denounces a situation in which "heroic teenagers are trying to put up resistance, struggle, and organize underground work without ever noticing that the jungle from which they are trying to break through are made of a quality sham rubber covered in fireproof solution."[50]

He ironizes:

> you are falling short of your target, honestly; the barricades you built in the film studios, the struggle between respectability and rebelliousness that you stage, the generational, religious, and ethnic conflicts that you have paid for years ahead are just a part of the great soap saga, which you unsuccessfully try to endow with the qualities of epopee and tragedy.[51]

He defends himself from the total ideologization of democracy: "I don't like it when in everyday, let's say, alcoholism or everyday, let's say, hardship people discern a gesture and a sign."[52] He fantasizes about "simple emotions" and "simple communication," rejecting the globalism of the profane information age; and, at the same time, he himself—in accordance with the irony of sublime negation—succumbs to the magic of the slogans and empty words generated in this age.

---

48   Ibid., 77.
49   Ibid., 80.
50   Ibid., 96.
51   Ibid., 96–97.
52   Ibid., 98–99.

The homeless character of Zhadan from the 1990s ultimately knows that sham is not only the new European reality but also marks the brotherhood of the 1960s and 1990s. Like the hippies of the sixties, he is searching and waiting for an allusion, some transcendent prompt that has to be deciphered for him to go on living. For example, he believes that this sign is known to Balanescu, the jazz legend from the 1970s. Having discovered that the star drinks probiotic milk delivered from Britain, has rotten gums, and is physically an old ruin, Zhadan's protagonist, who considers himself an eternal teenager, wakes up from his daydreaming. "I regarded him as a god, and what now? Rotten gums, damn it, rotten gums. ... After all, blame yourself—*allusion, allusion*: what good can a guy with inflamed gums allude to?"[53]—this is the end of the hippie dream experienced by the hero of Zhadan's travels.

Zhadan's work confirms that the avant-garde (and the neo-avant-garde) has a social ferment. The avant-gardist reacts, opposes, and has utopian phantasies. Postmodernists, on the other hand, offer invitations to their virtual country—"an inner Mongolia," where they feel comfortable and from which they are reluctant to wake up. By combining the avant-garde and postmodern, Zhadan creates his own vision of the post-Chornobyl text. Zhadan's homeless teenagers, Andrukhovych's bohemian migrant, and Pashkovsky's "last man" get together at this post-Chornobyl picnic at the roadside.

---

53 Ibid., 103.

CHAPTER 26

Volodymyr Tsybulko's Pop-Postmodernism

> Or light
> Blue skite
> Mine kife
> Some shit
> —Volodymyr Tsybulko, "Mine Kife"[1]

The post-totalitarian mindset flourishes once official truth has been unmasked. It is post-totalitarian heterogenic discourse that undermines the official mindset and uncovers the deception of totalitarian ideology. This ideology penetrates all levels of the social hierarchy and is linguistically fixated, generating a new idiolect—"newspeak." The post-totalitarian mindset emerges out of the fundamental rejection of this ideology that is supported by the presupposition that there exists something higher than the everyday and visible sphere of ideas. This teleology of a totalitarian kind is opposed by post-totalitarian thinking, which understands the visible and tangible as self-sufficient reality and questions the existence of a reality that is common for all people. However, it would be an oversimplification to state that the post-totalitarian mindset is absolutely devoid of the characteristics of its opponent.

Postmodernism offers the clearest example of this. In fact, at the beginning of the 1990s, social art and conceptualism are post-totalitarian in their essence; they exploit socialist realist clichés only to ironize and deconstruct them. The conceptualists, using avant-garde techniques, actualize numerous stereotypes that circulate in mass consciousness, starting from the "beauty" of the native

---

[1] "Kaif" ("kife") means "high"; however, if translated as "high," the word loses its association with Adolf Hitler's *Mein Kampf* (translator's note).

landscape and socialist realist slogans and ending with, for instance, Dostoevsky's phrase that beauty will save the world. Such clichés, decontextualized, and hackneyed through mass use, become the target of the conceptualists' parody. The clichés are liberated from their sublimated ideal aura. In this way, the vulgar and phony essence of the phrases turned into the wrapper for popular ideas comes into the open. Borrowed from high literature, these phrases turn into kitsch, so mass consciousness easily reproduces them. Deprived of their authentic meaning, these empty and cynical stock phrases block access to real emotions and reflections. In his poetic cycle *Classics Lessons*, Oleksandr Irvanets plays with the popular slogans and clichés that constituted part of the dictionary of the cultured Soviet person: "Drop by drop to squeeze the slave out of oneself," "A person is born for happiness like a bird is born for flight," and "Everything in a person should be beautiful." He demonstrates the absurd meaning of the phrases:

> A person is born for happiness
> Like a bird is born for flight.
> Like a bird …
> Like a turkey,
> Like a rooster,
> Like a crake on the marsh?[2]

By bringing high phrases down to earth, that is, by profaning them and endowing them with bodily qualities, Irvanets demonstrates that beauty is not only a spiritual thing, as it was said to be in reference to the "classics." Hence

> Everything in a person should be beautiful:
> Thoughts and feelings,
> Overcoat and shirt,
> Socks and suspenders,
> Haircut, brows, eyelashes,
> Lips and teeth,
> Oral cavity, mucous membrane,
> Hair on the head,
> In the groin and armpits,
> Nails, skin, calluses on the heels,

---

2   Oleksandr Irvanets', "Uroky klasyky," in *"Bu-Ba-Bu" (Iurii Andrukhovych, Oleksandr Irvanets', Viktor Neborak): Vybrani tvory*, 223.

> Blood, lymph,
> Gastric juice,
> Genitals and feces.³

In the poetic opera *Chrysler Imperial*, Irvanets turns "classics lessons" into a monstrous triune character; Les Podervianskyi, in his own right, demonstrates how "the classics" can be combined with obscenities. Overall, the conceptualists defamiliarize slogans and clichés in order to reveal the dead zones of language—zones that are nonsensical and in a way aggressive. Such clichés, inscribed within an ideological image—say, a socialist realist one—are in fact the flip side of the empty ornate words produced by socialist culture.

Dmitri Prigov, one of the leaders of conceptualism, on purpose reduces his poems to the level of graphomania of the kind that Dostoevsky depicted in the character of captain Lebiadkin, and through it, as Mikhail Epstein points out, "emerges the tragedy of entire generations condemned to speechlessness, having swallowed their tongue and experienced the destruction of language down to its elementary signal system. Therefore, 'they speak us.'"⁴ Working with these "dead" zones of language, with styles and models of ready-made characters and plots, the conceptualists wreck the hierarchy of the official and unofficial and demonstrate that in such a culture the individual has been transformed into an objectified thing, an automaton, through which the social unconscious speaks. It speaks through Podervianskyi's argot, Zholdak's patois, and Zabuzhko's varnished phrases in Russian.

Vladimir Sorokin—once a leading conceptualist—says that

> working with texts that stylistically resemble socialist realism gives me an opportunity to get in contact with some literary language that is absolutely foreign to my mentality. Regarding an explosion that occurs later, it does not come as a shock to me. On the contrary, I try to find a certain harmony between the two styles and to combine the high and the low. The attempt to combine opposites for me is a kind of dialectical act that results in the symbiosis of textual layers.⁵

---

3   Ibid., 232.
4   Mikhail N. Epstein, *After the Future: The Paradoxes of Postmodernism and Contemporary Russian Culture* (Amherst: The University of Massachusetts Press, 1995), 34–35.
5   Vladimir Sorokin, "Literatura kak kladbishche stilisticheskikh nakhodok! (Interv´iu s Vladimirom Sorokinym)," in *Postmodernisty o postkul´ture. Interv´iu s sovremennymi pisateliami i kritikami*, ed. Serafima Roll (Moskva: Elinin, 1998), 115.

This confession is symptomatic in the sense that it testifies to the symbiosis of high and mass culture in literature of the postmodernist period. Satisfaction resulting from this symbiosis liberates writers, and readers, from outdated taboos and, at the same time, renders perpetual performance—the exchange of roles, masks, and clichés—familiar and acceptable. In the post-Soviet situation, this symbiosis is complemented by the eclecticism of traditional cultural and the adoption of Western pop culture with its hedonism, theatricality, and visuals; and in Ukraine all these factors combine with patriarchal and residual totalitarian structures of thinking—in particular, the cult of power and resentment toward the other. This symbiosis generates the explosive mixture of Ukrainian pop-postmodernism. This is the name for the special kind of postmodernism that arises from the combination of avant-garde pop culture and mass media popular culture. It is dominated by the unused surplus of satisfaction that accumulates in the phenomenon of body-sign satisfaction—a high which becomes almost perpetual.

Such satisfaction can be provoked, for example, by the so-called "pops" with its kitsch-like rhythms and nonsensical, automated refrains. All of this stands for a belligerent (and nihilistic) self-devouring narcissism and optimism. The whole of society seems to return to infancy. As Valeria Narbikova writes:

> Nowadays, people have quite aptly adapted to life, and some kind of an imitation of tragedy is emerging. This imitation of tragedy, as I see it, for the modern individual is even more horrible than tragedy. Literature now experiences exactly this imitation of tragedy. When Sorokin writes his terrifying texts, nobody is terrified; it is terrifying because nobody is terrified. My contemporary generation is spoiled. These are spoiled children who are allowed a lot and who receive satisfaction from various children's rogueries.[6]

The sense of all-penetrating satisfaction—the phenomenon of the "high"— is reflected in the title of Volodymyr Tsybulko's collection of poems *Mine High* (or *Mein Kife* ... 2000). The clear quote from Hitler's *Mein Kampf* has been turned into an apologetic sign of a new kind of literature ("newliterism," as the author calls it). The emotional and social foundation of this type of culture is the post-totalitarian mindset set up as high kitsch (let us call it that). Enjoyment of the

---

6 Valeria Narbikova, "Literatura kak utopicheskaia polemika s kul´turoi," in Roll, ed., *Postmodernisty o postkul´ture. Interv´iu s sovremennymi pisateliami i kritikami*, 125.

transgression of the conventional and canonical becomes the dominant condition. This liberation from the pressure of past totalitarian thinking naturally provokes a reevaluation of values and psycho-emotional tension; but the optimistic call for a happy socialist future is replaced here by the no less powerful anarchic high of destruction, aggression, and enjoyment. Thus the environment for the kitsch high is "prolific for creative work time of the rebirth of the closed society,"[7] which Tsybulko associates with newliterism.

The conception of newliterism is directed against postmodernism, and was called the "aesthetics of the owner's grandson" (73). Such a repartition of property—that is, of culture—becomes, as Tsybulko states, an attempt at a new generational appropriation (in other words, creation) of a lost inheritance.

> One more attempt at the appropriation of space and time by the generation that is only able to count the loss. But what if this generation also undertakes to create its way of appropriating something that someone else lost? In this case newliterism is the aesthetics of the lost; it is not the aesthetics of the owner yet but the aesthetics of the owner's grandson. (73)

In fact, *Mine Kife* demonstrates the emergence of a mass literature that is oriented not toward the aesthetic reception of the so-called "classics" but toward the consumption of someone else's images, emotions, ideas, and ideologies. This is a collection, or, roughly speaking, a pile of garbage, where aesthetic objects are recycled to produce "literature recycled."[8] Thus newliterism has one more association—"to litter."[9] Newliterism, which Tsybulko in his "Newliterism, Or Interpretation of 'New' Words" ("Novoliteryzm abo novykh sloves tolkovaniie") theoretically grounds as a form of high culture, in reality looks like an apology for the mass literature of the post-perestroika period.

Mine Kife allows for neo-avant-garde interpretations. However, the collection is dominated by mass culture (short narratives, jocular ditties, burlesque)—pseudo-literary "garbage," from which sometimes also grows a new high literature. However, *Mine Kife* offers texts for common readers, as the author himself notes. Not surprisingly, this literature is intended for "not the owner yet," that is, the one who dreams about the repartition of cultural

---

7   Volodymyr Tsybul′ko, *Main kaif. Knyha dla narodu* (Lviv: Kal′variia, 2000), 73. Subsequently, page numbers from *Mine Kife* will be given in brackets after the quotation.
8   English in the original (translator's note).
9   English in the original (translator's note).

heritage. As Tsybulko maintains, "the memory of appropriation rather than the memory oppression inhabits the new-literary text, which, however strange, *is an idea common for both the outsiders and the nouveau riche*" [emphasis added—T. H.] (74). It means that the appropriation and adoption of culture unites in this case the outsider, who is a man of letters, and the "new Ukrainian" (nouveau riche). The Bubabists, as is well known, also played at the carnivalesque and in their own way practiced the appropriation of the "classics"; but they associated themselves first and foremost with bohemia, outsiders, and performances that did not bear a raider-like character. Tsybulko, on the other hand, equates outsiders and the nouveau riche by using some market categories. This is where high kitsch grows from—it is a post-totalitarian pop, which is comprised of a symbiosis of high and mass culture.

Tsybulko views his literary experiment as the antithesis of postmodernism, which in his view amounts to heathenism. Postmodernism, in his opinion, features a "polytheism of aesthetic modes that repeat themselves" (75); his newliterism, on the other hand, has to become monotheism and post-heathenism. "Monotheism of language will perhaps end post-totalitarian paganism, which is a polytheism of aesthetic modes that repeat themselves," he declares (75). Monotheism here is a foil to neo-totalitarian conceptions. This is confirmed by the title, *Mine Kife*, which provides an opportunity to talk about the presence in Ukrainian millennial space of neo-totalitarian conceptions, such as Volodymyr Yeshkilev's demiurgism or Tsybulko's newliterism.

In "Newliterism, Or Interpretation of 'New' Words," Tsybulko quotes Johan Huizinga and emphasizes the "fairness" and "purity" of play as such: "In order to be healthy, culture-creative power, this element of play has to preserve purity" (71). He explains that the element of play "should not be sham, pretending, or masking political goals—the mere appearance of the true norms of play" (71). In this way, passionately in a Schiller-like manner, he identifies homo ludens with humanity[10] and reduces the sphere of play and laughter to that which is true—that is, to the canonical.

This interpretation of laughter is enlightenment-like rather than ambivalent—Bakhtinian, as adopted by the Bubabists. In this sense, Tsybulko's identification of postmodernism with its polymorphic paganism and newliterism, with its monological new Christianity, becomes understandable. After all, if the Bubabists' main character was a bohemian poet, Tsybulko's hero is a poet-bourgeois. "Prevailing these days postmodernism is nothing more than domination

---

10 In Ukrainian, "ludens" and "humanity" have homonymous roots (translator's note).

of the subjectivism of individuals, the rat race of the creative ego, polytheism, heathenism," states the author of newliterism (73). Admitting that this is salvation through culture and by culture, he specifies: "along with the experience of dehumanization and the experience of the paralysis of consciousness" (73).

The enlightenment pathos of newliterism is therefore directed toward the discovery of a fair, nonsubjective, and not paralyzed literature that is human and not dehumanized by culture. How is this ideal realized? Tsybulko's texts are immersed in the time-space of the post-totalitarian era, which brought "grim reality," the devaluation of values, and corruption, when it seemed that neither creativity nor spirituality were any longer in demand. A "paralysis of consciousness" —this is how Tsybulko calls this condition. He goes on to state that the "post-Soviet system has justified the executioner, aestheticized politics, idolized 'the people,' and crossed out the concept of the 'code of honor.'" As a result, the figure of the "new literature" is burdened with the function of a witness-mediator ("the state of the new literature lies precisely in this—in the search for resonance, in the recording of the condition of society in the current moment, in motivating a person's actions at the crossroads of ideology and morals" [73]).

However, the collection *Mine Kife* itself features a "play of powers" (73). For this reason, the newliteral claim that play guards against the threat of the "play of the swastika" is surprising. "In the case of the disappearance of newliteral play as the greatest threat in the eyes of its opponents," Tsybulko comments, "there will appear a play of other powers—not the play of intellect but the play of swastika" (79). When we take into account the duality of the image "mine kife—mein kampf" in the very title of the collection, not only does it erase Tsybulko but also recalls totalitarianism. The newliterary play of power is called upon to relegate the totalitarian-Nazi experience to the sidelines, to purge with truth and honesty contemporary society, in which is it still possible, Tsybulko writes, that "stripes made in Moscow would suit our local Schicklgrubers." At the same time, he proclaims that newliterism is directed toward the "absolute of the text": "Believing in the absolute of the text, we hope that in the Latin word *ludens* (from *homo ludens*) a Slav will sense the human. A naivereasoning of a naivecitizen in a pseudo-state" (79).

Such a "naivereasoning of a naivecitizen" is significant because it demonstrates a reversal of powers in the post-totalitarian space, resembling the negation of negation. Hence, not only play is opposed to power but also power subdues play in the newliterary text. Indeed, in spite of the ludic, playful element, in the aesthetic world of Tsybulko's collection *Mine Kife* we can see two forces: one stems from the Enlightenment monotheism that is imbued with aggression, resentment, and the voice of power, which controls the whole process; the other is first and foremost a chaos of alien voices, waste objects, and garbage, which by means of montage is transferred toward new figurative associations. There is nothing but copies; in other words, it is dominated by mass culture and kitsch.

But what, ultimately, does newliterism look like, and what exactly turns into the *"literature of waste"* in Tsybulko's texts? It is worth noting that the word "waste" is not used here negatively; it merely records the state of contemporary civilization, in which garbage, waste, and copies are of real value, like originals.

The watchword of newliterism is modernity—"the present" ("the present as aesthetic category"), and the form suitable for the presentation of "the present" is language. "The creation of language; ultimately, language is the protagonist of newliteral texts" (75), maintains Tsybulko, thereby repeating the common idea of all experimentalists-avant-gardists. What is new in Tsybulko, however, is the conscious use of kitsch in the process of the creation of this new language. "[E]ven the derailment of style is not a criminal offence anymore because the rails themselves are criminal—we didn't want it; we were taken by force" (78), he proclaims and thereby essentially legalizes the rights of nonstyle, copying, and kitsch in post-totalitarian literature. Kitsch is the repetition of the imposed—that is, "the rail." Repetition, however, as deconstruction proves, is at the same time dissemination, and in those shifts and interstitial spaces that open through repetition something anti-canonical is born. The anti-canonical appropriates performance, which turns into a mode of self-expression that accords with the new technological era.

In literature at the end of the twentieth century, performance becomes a peculiar form of rejection of kitsch. While kitsch is an essentially market art, performance strives to cross the limits of market art and avoid the zones of mass culture. Kitsch is visual, while performance is oriented toward process. Such contrasts can be further developed: kitsch is based on canon and convention—it is materialized (that is, it refers first of all to objects); performance, on the other hand, plays with the rules that delimit the canon and thereby destroys or suspends them. In other words, performance reinforces the creativity of art itself because it is not so much a search for new forms as it is a revelation of mechanisms of new communication through the act of art.

Kitsch and performance look like antipodes—regulated and reproduced aesthetic value versus aesthetic liberation. Juxtaposing different experiences and discourses, performance wants to free itself from the limitations imposed by the rules of the creation of discourses. As a kind of hypertext, it also transforms the audience's reaction, helping people escape the condition of conventionality. Not only as a genre but also as a form of representation, performance becomes an element of postmodern thinking and penetrates literary genres and conceptual thinking. It liberates people from their limited understanding of the difference between the copy and the original. It is said that from here—from the substitution of the original with a copy—mass culture begins.

This genealogy of mass culture is undermined by modern technology, digital in particular. What is the copy and what is the original in modern digital products? Theoreticians of new technological culture talk about the change in the very model of culture's functioning, in which the differentiation between the original and the counterfeit becomes problematic. Let us say, digital technologies and computers are linked together in a certain hypertext, and that this hypertext functions as a performance. Computers are, by the same token, connected to a computer network in order to carry out complex tasks supported by the hypertext. Ultimately, there emerges a type of writing that produces its own object rather than repeats structures that were previously outlined. In addition, there is no constant material mediator, and the material's role as a mediator between information and the mode of its transmission does not play the same part as it used to in modernist art. The nature of the aesthetic act also changes: communication occurs between the audience and the machine while textuality itself resembles a happening.

Certainly, the interactive hypertext does refer to Tsybulko directly—his textuality is constant, even though dislocated from its traditional place; it is collage-like. However, it plays on the surface of various styles and feeds from its shifts. Tsybulko does not imitate other styles—he deforms them, often with the help of burlesque.

Thus Tsybulko's hybrid newliterature is theoretically based on an attempt to combine authoritatively directed moral with play, that is, the ideology of text ("deed that is correlative with text") and play. Newliterism resembles a neo-totalitarian model of behavior, which opposes the ironic linguistic behavior materialized in Bubabism-carnivalism. After all, the deed (as a moral demand) is embodied, beside language, in other components of newliterism: "socium and power" (the medium level of newliterism) as well as "ethnos and spirit" ("the third, root level of the world tree of newliterism" [75]). This dangerous play with totalitarian symbols and attempt to play with them according to the rules of postmodern irony are highly explosive.

Authors construe newliterature as a testimony of the post-Chornobyl era, when "the sermon of reasoning" is opposed to "the simulacra of the atomic age." The representatives of "new literature" want to oppose tendency and deed to postmodernist virtuality and play with simulacra-counterfeits. To unmask the virtual refuge of the postmodernists means, for Tsybulko, to "side with the 'unsatisfied.'" Creating the image of the village of Absurdilovka, he projects the layout of the asylum where the "unsatisfied" could find refuge after the Chornobyl tragedy. However, the grotesque village of Absurdilovka is more like an antiworld, an "a-topia," that arises after Chornobyl than a shelter for the poor. Yevhen Pashkovsky suggests another place for human salvation—the "Chornobyl library."

Sweeping the post-Chornobyl reality with the train of his high, Tsybulko in reality proposes a sedative drug that in no component is different from the "simulacra of the atomic age." Sliding from the rails of high theory and ministry, Tsybulko represents the grimaces of the post-totalitarian mindset and ironizes the entire post-totalitarian situation, in which "the laughter of a sad natural" is reflected "in the eyes of a creative bumpkin" (93).

Tsybulko's poetry reaffirms the "*joy of being*" and emphasizes the "*imperishable*" ("netlenka"). The phrase that serves as an epigraph for the collection, anecdotal and of a mixed language, undermines the opposition of "being" and "consciousness" ("everything's shit—Father used to say—only bees are something, but if you think a little, bees are shit as well"). In reality, words in the post-totalitarian kitsch high mean little: they indicate that "sugary-yeasty / black-earth-independent / husbandry-cosmo-flying / wonderfulish ecstasy" (11) which defines postmodern Ukraine.

Tsybulko's *Mine Kife* is a play of improvisation: into one pool are thrown children's horrors-jokes, classical quotations, and verbal spells, because a lack of meaning sums up the condition of society. Therefore, Shevchenko's verses are paired there with such spells; after all, what matters in these improvisations is rhythmic echo:

> little beggars are walking and singing
> they found a Schmeisser in the field
> they are crying it's time to get it over with
> although it's not started yet
> the thing that it's time to get over with
> they themselves are not conceived yet
> and unconceived they have been sold
> for hire to the two-headed swine
>
> (52)

Shevchenko's burlesque becomes especially popular: it appears suitable not only for describing the travels of a Ukrainian poet-Bubabist to the center of the empire but also for describing the culinary drudgery of modern poets-newliterists:

> two Jews and I
> we wandered sorrowfully
> through Kyiv and darkness darkness
> let's go Vlodek to the deli
> to grab a coffee we came in
> grabbed each a glass of vodka
> and a tidy priest of Isis,
> black haired and gallant
> modestly reached with his hand,
> and the choir about the mania of a flunkey
> or of a pagan priest: "In Judea
> Was Saul the King." Then the choir
> Roared from Bortniansky: "O sorrow,
> My sorrow! O sorrow great!"
>
> (26)

The playful newlit quotes not only classics but also "a classic anecdote a discourse during hunting no hunting during game-shooting" (49):

> the quilted jacket's tossed upon the ground
> here's your classic anecdote
> a discourse during hunting no
> hunting during game shooting
> they scrutinize one another
> through the thin facets of the facers
> and also construe the senses of life
> if you drink then drink seriously
> what the hell they got a kick
> out of the caravaggio bone
> father hop enters Haifa
>
> (49)

This mixture of plots, images, and names is clearly designed as masscult imagery. In Tsybulko's texts, the mass-recognizable images refer to colors, sounds, and paintings:

> everything is blue and white
> green and a little yellow
> thus someone in October entered
> somewhere in an arhmored carh
>
> and some more of red
> very little teeny-tinily
> and little sickle with little hammers
> but in a distant corner
> (49)

The characters of socialist realist mass kitsch also figure here: for example, not only the one in the "arhmored carh" but also "hiron pheliks," "stierlitz, and "makhno":

> he sees everything and what he sees he forgives not
> yet it happens he forgets when it happens that he drinks
> in the company of stierlitz and makhno
> (50)

Naturally, this play yields satisfaction, because it allows for the mixing of hierarchies, the combination of the uncombined, the derision of the serious and the sacred. Burlesque becomes total style, patois is widely used, and colloquial language is employed. One can sense avant-gardist vitality along with the high of destruction and of violence toward the other.

Mine Kife is subtitled as *A Book for the People*. Mykhailo Moskal's collages give the project its expected stylistic effect: this is a play not only with ideological notions and clichés but also with aggression itself. Babak-Lymansky in his book *Devil's Children* offers a similar semiotic code. The illustrations to *Mine Kife* showcase its main images—the power lines of the poetic texts. Equated here are the female body and sexual machine ("hey you're a bodily machine my Gal"), geometry is combined with anatomy, and nature coexists with decomposition and dismemberment. Collage (an old picture overlaid with another card) affirms the presence of three generations—the future and past in the present. Generations who are destined not to meet in real life, perhaps.

The place where the post-totalitarian kitsch high unfolds is "the village of Absurdilovka." The protagonist is not a bohemian poet who reevaluates culture and inscribes himself into it but a nouveau riche, a former lumpenproletarian, hostile toward culture, who got rich. In "the village of Absurdilovka" there is a hybrid of Hitler-Voroshylov together with Brezhnev, and Soviet phallic symbols appear quite clearly ("with the name of Adolf-Voroshylov in the groin an idea stirred" [14]). Such kitsch renders mass satisfaction, but aggression is also one of its elements, along with invasion, transgression, defeat (and here Nazi "plan blitzkrieg" is combined with world revolution—"the collectivization in Texas / I had to start / from brothels." (14). A military ditty emphasizes the aggression of the "bottom" and lumpenproletarian resentment:

> o Adolf Illich o Adolf Illich
> issue a call issue a call to nations
> your fiddle sing-sling
> a red star on the wing
> (16)

The idea of the world revolution in the ditties becomes an idea of pure aggression:

> on the hill stands
> a shithead
> stares crookedly at me
> i will bring this shithead low
> so it stares at me not
> (15)

Ultimately, post-Soviet pop is rather eclectic: it has both para-religious and national-patriotic images, and military topoi are also combined with erotic symbols. Flying saucers ("three hundred irons flew toward Kyiv"), Gogol-and-Bulgakov-like angel-devils ("gnomes have caught an angel"), parapsychology, mass psychosis, formulas for weight loss and urine therapy (the idea of "diluting / orange gasoline with donkey / urine" (16)—all of these phenomena of the mass culture of post-totalitarian time are recorded in *Mine Kife*. There is also a technology for creating a new human (werewolves, robots, meccanos). This one

> needs to take fourteen thousand donkeys
> and feed them only conifer needles
> of the hard species of trees let's say

> of oaks
> once a week
>
> (17)

"Sexual machine"—Tsybulko's other collage—recalls the sexual revolution, while "hairy vampire bearded lout and other such yokeldom" observes the mass fascination with folklore kitsch. The underground, jokes, and ditties have moved from the periphery of culture and become fundamental for newliterature, as Tsybulko's collection establishes.

What is the main message of newliterism, then? Beside play, lumpen aggression, derision, and satisfaction, Tsybulko's texts reveal the drama of the post-Chornobyl nonexistence of Ukraine—a topic that is also significant for another newliteral author, Yevhen Pashkovsky. "O Lord God help Ukraine / help Ukraine that does not exist" (119). The lyrical subject cannot restrain himself and, like a shaman, invokes that this is "not that" Ukraine about which we dreamed and that we expected "would be fun" to live in

> if it were not so frightful
> the cherry trees shine out with phosphor
> the world is new but governed by the old
> the choir is singing but the thoughts are depressing
>
> (62)

It is natural, therefore, that, established by post-totalitarian kitsch, "not that" Ukraine is also a simulacrum, a counterfeit:

> This impasse is underbaked bread of causes
> death and decay honor and deed
> of all Ukraines this is not that Ukraine

The return of the prodigal son brings only disappointment:

> I am looking at the walls
> they are not white
> I am looking into the wells
> they are waterless
> I am shouting yet the wind
> tears my words apart

> these words like the burnt grass
> the ashes of fault and guilt
> this is not that Ukraine
> (63)

Thus Tsybulko's theory of newliterism, intended as serious, honest, and high literature, materializes in the form of mass culture. His lyrical subject so aptly has disguised himself in this world that it is hard to catch him at his word. The language of post-totalitarian society, enriched with a variegated and marginal vocabulary—argot, jargon, patois, and quotations and names from official, largely Soviet culture—is the main element of the masquerade and performance that are deployed in Tsybulko's works. Words can signify almost anything but in reality mean nothing. These features of totalitarian culture are demystified equally in Andrukhovych and in Tsybulko. Newliteral language, on the one hand, is a form of defamiliarization of official new language, demonstrating its absurdity, barrenness, and total meaninglessness; on the other hand, it reveals the individual freedom and playful linguistic behavior of the generation that in the post-totalitarian era takes over the role of the creators of culture. These are the parameters in which pop-postmodernism takes shape. It legalizes the underground, restores a taste for all kinds of hybrids (patois), and legitimizes argot and performance.

After all, it was already literary modernism that reaffirmed the doubt that there is a full correspondence between material objects and the grammar and syntax in which they are expressed. The Avant-garde even more consistently determines the obscurity of signification and reference. For the futurists, language is an expression of energy and power rather than reason; Dadaism associates it with spontaneity and intuitive imagery, while surrealism associates it with daydreaming and the subconscious. Postmodernism destroys linguistic hierarchy and restores the preference for tautological thinking, verbal lists, and the labyrinth of names. Tsybulko's newliterism, on the other hand, closes itself on collage.

One characteristic of even the earliest post-totalitarian literature is its attention to personal, live speech, which is perceived as an expression of real human existence rather than a medium of the official truth. For this reason, the works of Vaclav Havel, Milan Kundera, Aleksandr Solzhenitsyn, and Stanisław Barańczak were interpreted first and foremost as the testimony of a post-totalitarian culture: through their honesty, they aimed to undermine official culture, ideology, and politics. They preferred live speech to literariness and thereby testified that language became a realm of individual freedom. Later, the underground used language as a field of playful linguistic behavior.

Within the framework of totalitarian kitsch, as Kundera points out, "all answers are given in advance and preclude any questions. It follows, then, that the true opponent of totalitarian kitsch is the person who asks questions."[11] However, those who fight with totalitarianism also require "simple truths" that will provoke collective tears; the society of ostentation, in its turn, offers a mass culture devoid of questions. This is why post-totalitarian kitsch, which combines both undermining and affirmative functions, is directed simultaneously at power and pleasure. The post-totalitarian configuration, in short, is defined by the existence of an independent public sphere as well as the relationship between this sphere and that of officially recognized critical discourse.[12] Tsybulko's texts annihilate all these spheres: in the post-totalitarian "village of Absurdilovka" nobody asks anything anymore: it is dominated by gibes.

Pop, which adheres to postmodernism, does not pose questions either; it is based on the principles of burlesque, lowering the heightened by making the high play with the low and marginal. Satisfaction and the sense of the indefinite, absolute feast, ultimately, acquire a ritual subtext because they signify passage to a new condition of life. In this liminal, interstitial space, the taste for various hybrids gets a new life, and argot, fragmentariness, relativism, collage, and pastiche become lawful and legitimate. However, post-totalitarian space destroys the very hierarchy of the official/unofficial, while patois and argot penetrate speech on all levels: from the criminal world to the president's office.

Thus, within the post-totalitarian crisis, linguistic eclecticism and the formalization of the styles and discursive practices that can be used for a mask-game take shape. This creates a ground for mass literature, which grows out of the hybridization of genres, plots, and styles. A large portion of postmodern texts consists of vocabularies and collections of expressive forms. Gathering a collection of various discourses, initiated by Andrukhovych in *Perverzion* (the novel itself can be viewed as a practice of the formatting of discourses), has turned out to be an especially productive practice in the development of Ukrainian mass literature.

---

11 Milan Kundera, *The Unbearable Lightness of Being*, trans. Michael Henry Heim (New York: Harper Perennial, 1991), 253.
12 Jeffrey C. Goldfarb, *Beyond Glasnost: The Post-Totalitarian Mind* (Chicago: University of Chicago Press, 1989), 162.

CHAPTER 27

The (De)KONstructed Postmodernism of Yuriy Tarnawsky

Yuriy Tarnawsky is perhaps the most consistent avant-gardist in Ukrainian literature. He is focused first and foremost on form—so-called "pure" verbal form. In short, "art," "artificiality," is what he is attracted to in writing. Syntactic corporeality, along with the destruction of linguistic structures in general, which are characteristic of the "masculine" avant-garde, are especially significant in Tarnawsky's collection of dramas *6x0* (1998).

Against the background of Ukraine's somewhat infantile postmodern literature, preoccupied as it is with stylization and rhetorical morphology, the "strong" form of Yuriy Tarnawsky, his open epatage and metaphoricity, which has an open destructive-aggressive nature, look almost like the classics.

As is appropriate for developed modernism, this most hermetic and most aesthetic child of the continental avant-garde, Tarnawsky is very interested in the adequate interpretation of his experiment, which he regards as a type of elite theatre. The method he uses he calls metaphorical interpretation. Worth noticing is the open eclecticism of the theoretical references he cites in the afterword, as well as the fact that the author himself—almost in the spirit of postmodernity—is aware of this eclecticism.

What he proposes to his audience is a theatre-text. Intended for reading rather than staging, it is not a contingent collection of dramatic works but a coherent conceptual structure. One can succumb to the will of the willful Author, the protagonist in this theatre-text, and accept his theses that the texts follow (anti-drama?)—Greek drama with its exodos (exit) and parados (entrance), stasima and episodes, with various functional substitutes for the role of the choir (with the help of the Guide, the Telephone, the Author-Reader, and even the Viewer). The action as such occurs offstage, and the tetralogy encompasses two serious dramas and one grotesque (satirical) drama. Gradually, the actors become shallower, ultimately bearing a likeness to horses. And the main theme is the death of idealism (Love).

Offering extensive commentary in his afterword, the Author wants intellectually to guide his reader. At that, he appears more in accord with the modernist principle of the hermetic text, like Eliot in *The Waste Land*, than with postmodernist play with the reader. Tarnawsky's obtrusive commentaries bear a likeness to the overly detailed remarks of the Author (here and henceforth I mean character of the Author as it appears in the theatre-text), which direct the action in his dramas; in addition, they are overly serious, to the point of being supplemented by examples of "adequate" metaphorical interpretations of his own texts. It goes without saying that they do not even hint at the "inadequacy" of interpretation or commentary.

However, Tarnawsky unequivocally aspires to be regarded as a postmodernist, as they are still in vogue. Hence he wants to destroy traditional theatre—and so his plays turn into "texts of the stage." One can describe Tarnawsky's theatre-text as "de-KON-struction" (the Ukrainian "na KONu" means "on the stage"), or destruction undertaken on the stage, as well as construction that occurs DE(s´)—that is, somewhere else. Nevertheless, such postmodern deconstruction cannot overshadow (or hide) the avant-garde tendency that is present in it.

1. SOMETHING ABOUT THE "MALE" TEXT

Yuriy Tarnawsky's de(kon)struction replays ancient drama as well as the modern Theatre of the Absurd. As Tarnawsky points out, while "the theme of classical Greek tragedy is a hero, an individual elevated above his or her surroundings, and his or her tragic fate," the theme of *6x0* is "love, elevated above the everyday life of the modern individual, and his or her death, which seems to be the only event in the modern world that comes close to ancient tragedy."[1] Therefore, the ancient struggle between the individual and Fate is substituted here with the existential struggle between man and woman.

The tetralogy *Triangular Square*, comprised of the lyric-drama *Suka(skuka)-rozpuka* (literally: "Bitch(boredom)-dejection") and the plays

---

[1] Iurii Tarnavs´kyi, *6x0. Dramatychni tvory* (Kyiv: Rodovid, 1998), 334.

Female Anatomy, Not Medea, and *Dwarfs*, is an example of a text that can be called "male." In the broadest sense, it can be considered a mode of writing that explicitly (or implicitly) reveals the dominance of male meanings, intentions, and tastes. It is only natural that such domination takes shape against the background of the opposition between man and woman.

If we consider *Female Anatomy* the main thesis of the tetralogy—as "a description of some kind of female brutality, which, in its author's opinion, sharply differs from male brutality, whose existence by no means can be denied either,"[2] then the antithesis of the agon-deconstruction would apparently be a description of male brutality. After all, deconstruction (this is how exactly the genre of *Female Anatomy* is specified), as is well known, is not "neither nor" but "both and." The goal of deconstruction is the avoidance of antagonisms and oppositions characteristic of Western thinking. The most general emblem of such thinking is the subject identified with the "dead white European male," which bears witness to the domination in the Western cultural canon of the classics, the white race, Europe, and the male sex.

2. THE LOGIC OF (NON)IDENTITY

If we use the method of metaphorical interpretation, suggested by Tarnawsky-commentator, for the understanding of his text, the meaning of Love for Man in *Female Anatomy* is revealed as the (in)equality $4 \times 4 = 4$. This is the truth that, as Tarnawsky apparently ironizes, has an ability to change not only the life of man "but also the whole world, humanity!" This mathematical "equation-equality" signals the ideal of man and woman identity in that special mode of being called Love. This is an enlightening idea—equality, expressed in the equation $4 \times 4 = 4$. In this "male" game, the female fault lies in the fact that this equality disintegrates like ashes, and the cadaverous color of the metamorphoses of the table-body-skin and the "female dummy of a tailor" signal the apocalypse of lost male illusions.[3]

The absurdity of the identity of "male"—"female" is part of the title of Tarnawsky's book *6x0*. If the first case (the equation $4 \times 4 = 4$) was a clandestine tautology of male desire, the second equation (6×0) unequivocally hints at zero, or nothing, which becomes the object of idealizing male intentions. Zero is the

---

2 Ibid., 340.
3 Ibid., 93.

real meaning and place of woman in such a "male" text. Ultimately, the woman is demystified: when she is stripped of her hair (wig), clothes, and skin (a plastic mask beneath), she turns out to be "a doll of white plastic, empty inside."[4]

3. CUBISM

Beside the "equations of equalities," Tarnawsky's book contains a cube. 6×0 is the figure of a cube that, turning each of its sides in six positions (six plays), should ultimately yield a circle, that is, revive nature. It is, therefore, also female nature, which has been destructively spread out upon six squares. One is tempted to call Tarnawsky's collection as a whole an instance of literary cubism. This, by the way, is the name that the author provocatively suggests himself. "Speaking informally, the technique in which *Horses* is written can be defined as some kind of literary or dramatic cubism," he says.[5]

Picasso's experiment with cubism at the beginning of the twentieth century, which opened the abyss in the destruction of the Living, it seems, was not so frightful for Tarnawsky at the end of the century. The abyss that opened in Picasso's destructive experiments with the materiality of the world, as is well known, horrified the artist himself. However, Yuriy Tarnawsky, aspiring to be a postmodernist, wants to ironize over the darkness of Nothing, to slight it as Play. After all, in the post-Chornobyl text, as a kind of apocalyptic text, the end of the world has already begun and the abyss of Nothing should not be an object of horror: we live in it. Only in this context can the de(kon)structions Tarnawsky resorts to be perceived as playing at apocalypse.

This reference to cubism is not alone in Tarnawsky's collection. For example, the female body "acquires ... the stature of some abstract, objectless sculpture";[6] it is crying "like a moving cubist picture,"[7] crying "with a cubistic, as though artificial, cry, which amounts to a dance that is occurring in time yet not in space."[8] This emphatic breaking of nature and naturalness is, perhaps, the clearest characteristic of *Triangular Square*, in which the dismemberment and destruction of the organic world (body, matter, reality, human, and woman) takes place.

Female Anatomy destroys the corporeality (as traditionally idealized by men) of woman. Behind it looms the ugly cubist female images of Picasso's

---

4   Ibid., 79.
5   Ibid., 350.
6   Ibid., 44.
7   Ibid.
8   Ibid., 43.

paintings, which, by the way, are psychologically also marked by a struggle with nature and woman.

4. THE DE(KON)STRUCTION OF DECONSTRUCTION

The arsenal of Tarnawsky's mathematical forms is not exhausted by tetralogy, equality, the cube, and the triangle. There is also the Möbius strip, which is the one-sided surface that appears as a result of sticking together two opposite sides of a rectangle. The Möbius strip is a metaphor of Tarnawsky's male text, in which woman, projected on the one-sided surface of the text, is something that could be turned inside out to be viewed simultaneously from the outside and from the inside, while the focus of such visualization is, of course, man.

In the very first play *Female Anatomy*, Tarnawsky anatomizes woman by "turning her inside out." He also undertakes multiplication, in which the unit is male discourse, that is, linguistic-ontological structure and the orientation of speech toward a specific object of desire. The double-male projection is posed by Man, the play's protagonist, and Guide, the eternal witness and interlocutor, who fulfils the function of the Greek chorus. Woman here is an object that can be possessed, analyzed, and broken down into parts and details—in one word, deconstructed.

A few words about deconstruction. Although it is already in fashion to talk about the death of deconstruction, it has never unfolded in Ukraine. Perhaps, the only consistent attempt of such unfolding, revealing its own bankruptcy, is Tarnawsky's collection of plays. Let us discuss deconstruction, then. Jacques Derrida, the founder of the philosophy, wanted with his method to exceed Euro-phallocentrism, or, simply put, the "dead European white male." Ultimately, it becomes clear that deconstruction is also intended as a center-oriented theory, in which the center is occupied not by the subject of cognition but deconstruction itself.

Its mechanisms, as the founder testifies, are simultaneously destruction and reconstruction, erasing and reviving. Derrida constantly emphasized that deconstruction is not destruction; rather, it is a simultaneous process of de- and re-construction. He also associated deconstruction with the *chora*, the notion that, starting from Plato, denotes locality, the place of space appropriation, and refers to the pre-nominal realm of being, the first name, which carries the maternal-virginal nature of onto-logos.

Derrida, who wanted to overcome the opposition of male/female, also stated that the strategy of feminism and deconstruction is one and the same—

to reconsider Western (or male) thinking. Nevertheless, deconstruction is a male cognitive construct, which simulates female models of thinking. It seems that the reversibility of deconstruction starts to serve male whims and take the form of a purely male discourse. This happens in the case of Tarnawsky's play *Female Anatomy*, which is subtitled *Deconstruction*.

5. THE FEMALE BODY AS TEXT

What exactly is deconstructed in *Female Anatomy*? It is woman-object, viewed and broken down under the scalpel of the male gaze, armed with rational thinking. *Female Anatomy* can be interpreted as a collection of models for de-aestheticizing and de-idealizing woman. Just consider the highly visual scenes of the bathing of female breasts or of breastfeeding a man, or the archetypal image of a triumphant woman standing with her legs widespread while "a waterfall explodes from under her skirt and to the floor—this is her losing control over her bladder from ecstasy."[9]

Tarnawsky pitilessly strips the metaphorical veil from the erotic, unrolling the chain of female faults. He also discovers that male power is erotomania. For this reason, the "brutality" of Woman, according to Man's accusation (the play's protagonists are named in this way), lies in the fact that Woman escapes, diffuses, does not subject herself to male power, and most importantly, is unable to love.

At stake here is the grotesque quality of man's power, predicated upon sexual desire. Since woman is supposed to satisfy man's ideal, she has to be ideal, and therefore—immaterial. The manifestation of this metaphor leads to the grotesque transformations of eroticism, and woman becomes

> like a mist, in which he is dangling foolishly with his sex. Her whole body is like a cloud which he is lying upon and like an idiot is trying to embrace. ... He is looking at himself and sees the earth looming far away under him through this cloud, which he is struggling to embrace.[10]

This is the way Tarnawsky unfolds the de(kon)struction of male idealism, which in reality is based on the absolutization of woman's corporeality. It is worth noting that the idealism of love (and later of maternity in *Not Medea*) is consist-

---

9   Ibid., 105.
10 Ibid., 92.

ently presented as a male projection onto the female body (which cannot be hidden even by a series of identifications-inversions of man-woman when in *Not Medea* an actress plays the role of a male character).

6. "THERE IS NOTHING OUTSIDE THE BODY"

The difference between the so-called "male" and "female" texts becomes clearer. Tarnawsky plays with one of the deconstruction's principal theses—"there is nothing outside the text." According to him, it could be admitted about a woman: "your dear body exists outside the text"[11] (thus it is visible and accessible), but it is impossible for the woman's soul to be outside the body. Therefore, the man reiterates, "the soul also … [exists outside the text.—*T. H.*]. … But not outside the body."[12] (The man reaches out for the body of the woman, who is somewhere behind the canvas.) This response is also acceptable for Woman, who in Tarnawsky's text is utterly vulgarized: "There is nothing outside the body." Thus, the female body is precisely what the male text speaks about.

In such a deconstruction (in which the difference between male and female is supposed to be erased, and in particular their socio-cultural states), there is no capture or erasure of the sexual differences that Jean Baudrillard attributes to postmodern hyperreality. This makes it similar to the specifically female manner of appropriation—seduction. Through seduction there is created an illusion of power, a simulation of power-capturing-subjugation, in which the antagonism of male-female disappears.

Ultimately, Tarnawsky deliberately debates with deconstruction ("there is nothing outside the text"). As Man says, this thesis would presuppose that "the empirical author" is "a perfect connoisseur of language" and would encompass (predict) all possible understandings of the work. This is what the author is not able to be, states Tarnawsky logically. However, deconstruction hints at this, rejecting the structuralist idea that only a linguist is able to be an ideal recipient. In postmodernism, text and language play with the author, language reveals itself, and the "incorrect" interpretation of a text is at the same time the "correct" interpretation. The author in principle cannot control it. The Author is the phallic and logical center which deconstruction wants to get rid of. Tarnawsky's de(kon)structions, on the contrary, are preoccupied with the author's voice. Texts-remarks supplant the play (of the text) and become pure monologues.

---

11 Ibid. 90.
12 Ibid.

Is Tarnawsky's de(kon)struction deconstruction in its conventional sense then? Or is it not a coincidence that his texts reveal that the dominant avant-garde idea of authorship is male in essence?

7. TEXTS-DESTRUCTIONS

Female Anatomy presents the cruelest negative image of woman (to take it on trust, the metaphysical identity of woman in the images of girl—woman—broad). This destructive absolute gathers together the principal metaphors breaking down woman-as-sense-and-body—as a spiritual ideal, as an erotic object, and as an archetype of the original mother, which, ultimately, is associated with the realm of the infernal, supernatural, and thus with death.

Axes, swords, dismemberings—this series of images makes up the "anatomy" of the female role. Materially, or, more precisely, geometrically, woman is associated with a specific figure—for example, with the form of a triangle or a thing. Woman, therefore, is broken down both spiritually and physically—as a body that consists of "back, trunk, buttocks, thighs, calves, feet," as well as "the triangle of the flesh." This body is explored both from the outside and from the inside; it is broken down into objects and things, stripped naked and turned around—it is recognizable, revealed, and materialized in the "male" text.

Male visual "colonialism" penetrates woman's anatomy. At the same time, only for a second, as though demonstrating the consistency of all possible inversions in his theatre-text, Tarnawsky makes up his mind to show the nakedness of both the author and the actor. Thus the character Author appears on the stage naked, holding a rope and hinting at his traditional role of deus ex machina in classical theatre. However, from here it is a very long way to the "male anatomy."

8. TEXTS-KON-STRUCTIONS

Tarnawsky's man is a consistent victim: his connection with life existentially depends on his connection with woman, whom he is destined to struggle with. Derrida wrote that woman as the other is a distance, which man notices and by dint of which strives to realize or even elevate himself. Tarnawsky's woman is first and foremost a grotesque object. This once more brings him close to Picasso's cubism, which deforms the female body by giving it the shape of a guitar or fiddle and by comparing it with infernal things. In this way, the painter avenged himself on the female body, thereby overcoming the erotic power of a real woman.

By disposing of the female body, Tarnawsky identifies it with the "brutality" of woman, as he theoretically explains in the afterword. Hence the ontological différance, the distance between man and woman, remains unbridged, as a gap, as death. "The groups are placed separately, with a huge gap between them: men on the left side of the stage, and women on the right."[13] This is how this kon-struction on the stage ends.

De(kon)struction is also promised in the next play, *Not Medea*, in which an actress plays the role of man. This is about the relativity of not only the roles but also of the realm of the visible, that is, of the body. *Not Medea* is a male monologue in front of a mirror. The mirror-deconstruction of the androgynous body of the male-female in the play is multiplied by the mirrored repetition-reflection of the universal drama of man and woman in *Multi Medea* (text in text).

Such a reflection, however, brings into focus male narcissism. The de(kon) struction of the ideal desire for a woman on the stage turns out to be just a rationalization of self-cognition. Man, in his pursuit of the ideal, admits, "My life is total darkness, and I am moving with my hands stretched ahead, not knowing where. I cannot stand it anymore. I must understand you in order to understand myself."[14]

9. TEXT-MIRRORS

Beside performing woman as the mirror of man, texts-mirrors in Tarnawsky are also text-inversions (*Multi Medea, Hamlet, Prince of Damsels*, and *Hamliette*), which affirm the transformation of classical drama into kitsch. Further, mirror-characters such as Sibyl, Clytemnestra, Agamemnon, Medea, Nymph are the archetypes signifying the complexes of woman's fault, in particular, the murders of men and children.

The main idea is the axe with which Clytemnestra ostensibly killed her husband Agamemnon. A severed human arm or leg, a human trunk, knife and sword blades instead of the gaze—this is a metaphorical interpretation of the archetype of woman.

Let us recall Stefanyk. In Ukrainian literature there is no one, it seems, beside him whose writing is so existentially predicated upon the ontology of corporeality. "I would hone an axe with a whetstone and would cut them hands off up to the elbows. Just like one, two—and the hands whoops!"[15]—this is

---

13 Ibid., 106.
14 Ibid., 125.
15 Vasyl' Stefanyk, "U korchmi," in Vasyl' Stefanyk, *Vybrane* (Uzhhorod: Karpaty, 1979), 36.

how his character addresses the utter limit of dejection in the world in which a wife and children beat a husband-father-drunkard and in which the ties of patriarchal, natural corporeality, upon which family and the divine are predicated, break. This is the way corporeality is perceived in modernist texts. In the avant-garde postmodernist text of Tarnawsky, however, the body becomes a metaphor. It turns into an aesthetic form that entirely shuts down and supplants real flesh and blood. Only sometimes, as in *Dwarfs*, in which Bartholomeo picks up Homeless, and in which Dwarfs, who are getting ready to move to America, are still waiting for their Bot, the grotesque destructions of the body (since the dwarfs are the main characters of the play) echo sad lyricism.

10. THE "FEMALE" TEXT

Tarnawsky's plays are an anthology of anti-feminist concepts. As I mentioned, the "female" text (the text about a woman) is written by two (and even three) men: Man, Choir-Guide, and Author. This is the basis of Tarnawsky's literary cubism. Woman as an idea, an image, and a body is laid upon three squares and reproduced through double projections on/outside the stage. Guide plays the leading role here (because what he sees is not what exists independently from him, but that which the author puts in his view). Therefore, this is how he transmits to the audience (and to the character Man) all that (the character) Woman feels as she puts her hands on her breasts, thighs, and buttocks:

> She is in ecstasy! She in her entirety is a manifestation of her sex, femininity, and her power has totally possessed her. ... Now she is spreading and peeping between her legs. She sees a light tumor that grows toward the center. ... And it is the black, bottomless darkness of her mystery—a mystery that nobody will ever comprehend, even she herself. She almost literally diffuses in the ecstasy. She feels a limitless tenderness to herself, to her sex. She wants to reward it somehow. She comes to the sink and starts washing it.[16]

After Freud, who called woman a "castrated man," Tarnawsky states: woman is her sexual organ. This is the center of the female world and female textuality, as defined by Tarnawsky's so-called "male" de(kon)struction. Ultimately, sex—this Freudian "red room"—devours everything: "the doors throw open,

---

16 Ibid., 85

showing the unpleasant bright-red interior, like a dangerously inflamed internal bodily organ, and then close, as it seems, forever."[17]

Thus appearance, corporeality, is shown as illusory. Here, Woman is standing before a mirror and gradually is tearing off her hair, clothes, and skin. Ultimately, she "is standing in front of the mirror naked. This is a doll made of white plastic, empty inside."[18] There are dark slits in the body. Behind them there is void. Stuffed with colorful ribbons, she "is standing alone in the center of immense empty space."[19] It appears that the world in which Woman becomes the center is ugly and nonsensical. It recalls Olha Kobylianska's *Princess*, in which female emancipation, ultimately, cannot overcome its own limit. It is man who endows woman with value—be him a Nietzschean or the Princess's slave!

11. THE THEATRE-TEXT

Tarnawsky also wants to turn theatre inside out. The action is taken off the stage, and the audience acts instead of the players; the author forfeits his rights to the text (the classic works of Euripides, Shakespeare, and Beckett are being rewritten), and narrative supplants action. The stage also undergoes deconstruction—the action is removed, conversation and dialogue disappear, sounds are boiled down to breathing and to the rustle of paper, action becomes a reading of a different text or a retelling of the text over the phone. Instead of acting, the protagonist becomes a spectator. Dramatic perspective shifts and doubles.

Tarnawsky's metaphorical interpretation aims at transforming theatre into a play of adjacent series, of texts. This is the author's primary postmodern intention. It must be admitted that the grotesque intertextuality overfills Tarnawsky's plays. For example, the play *Dwarfs* is a series of ironical reinterpretations-lectures (in geography, history, mathematics, and English), and the implicit author does not even shun the grotesque of Ukrainian fables-absurdities or repetitions-replayings of one and the same scene with a dentist in Warsaw, Vienna, and Madrid, thereby mapping-aping the progress of the "Eastern" individual to the "West."

Dwarfs becomes a study, a course, in which rehearsal is inseparable from imitation, and irony and grotesque pass into kitsch. For example, in *Hamlet,*

---

17 Ibid., 278.
18 Ibid., 79.
19 Ibid., 80.

Prince of Damsels, "ham-let" is only a "little lout" ("kham" means "lout" in Ukrainian), and the drama of life is tantamount to "the happy triangle":

> Frau Trude and Hamlet are walking on both sides, while Ophelia (from the French "orpheline"—"orphan") in the middle. Now she has a huge pumpkin-like belly and she is pushing in front of herself an outmoded black children's stroller—who knows whether occupied or empty. Frau Trude is weaving, as usual, but her wool and braiding are not in the stroller. Hamlet is carrying in his left hand the skull of his friend Yorick. From time to time he raises the skull and looks at it, moving his lips. He has not forgotten his friend and has deep philosophical conversations with him.[20]

Indeed, "no need to talk about the relationship of this play with *Hamlet* because the similarities and differences between them are obvious and in abundance,"[21] says Author-Commentator.

However, the braveness of this grotesque rewriting of the classics notwithstanding, Author-Commentator's sympathies bear a footprint of modernist, avant-gardist ideals. They are expressively existential, since in modernistic texts, as Tarnawsky points out, "you appeal the death sentence by writing with every cell of your body, with every erg of energy that you have. There is no time or place for games."[22] When there is a game, it testifies to "society's illness," he says. Hence "modern theatre, in order to be of full value, has to be modernistic. Otherwise, it would be stylization, imitation of the theatre of the past."[23]

Author-Interpreter's argument against stylization and game-"illness" contradicts the postmodern theatre of Author-Artist in Tarnawsky's plays, which is based entirely on playing with references to, repetitions of, and grotesques of classic images. Allusions to Beckett's *Waiting for Godot*, Aristophanes's *The Frogs*, Aeschylus's *Agamemnon*, and Euripides's *Medea* often appear in Tarnawsky's texts, and the author does not conceal them—to the contrary, he defiantly emphasizes them.

Ultimately, Tarnawsky shows the passage of drama into prose, because the modern individual more and more becomes a consumer rather than a subject of action. This individual is surrounded by a plethora of signs and signals:

---

20 Ibid., 202.
21 Ibid., 349.
22 Ibid., 306.
23 Ibid., 307.

the "multi media" with "her many devices" powerfully dominates the stage of modernity.

Tarnawsky's avant-gardist experiment is in defiance of the not-yet-written (or maybe written?!) "female" text. Among the six "male" texts of Tarnawsky there is *Hamliette*—"another repressed theatrical figure," as Author-Commentator calls her. Based on the numerous replayings of the obtrusive theme of woman's "brutality," this play testifies to the exhaustion of Yuriy Tarnawsky's theatre-text. Hamliette—a woman with the skull of her not-yet-born child—symbolizes a female inversion of the humanistic code. However, this upper limit of the "male" destruction of the ideal of woman simultaneously becomes the end of the "male" discourse. When woman is absent, what is there to talk about? Thus, Actor, his mouth gagged with a bed sheet, "like an image of his silence that flows out of his mouth,"[24] turns into a metaphor of self-castration. Avant-gardist, postmodernist, de(kon)structions run out.

---

24 Ibid., 301.

Part Five
POSTSCRIPT

A Comment from the "End of Postmodernism"

Umberto Eco proposed—with regard to postmodernism—differentiating between "integrated" critics, who are immersed in the contemporary situation (characterized by the end of "Grand Narratives") and "apocalyptic" critics, who, on the contrary, wish to oppose postmodernism from a position outside it.[1] I can, it seems, regard myself as an "integrated" critic because I see myself inside the postmodern situation rather than outside. I am writing this commentary six years after the publication of *The Post-Chornobyl Library* in Ukraine. In this time, it has become a matter of course to talk about "the end of postmodernism." Therefore, this commentary is written from the point of view of "the end of postmodernism."

1. POST-SOVIET POSTMODERNISM

In 2010 Mikhail Epstein noted,

> In recent years the concept of postmodernism has often been deployed to explain the peculiarities of late Soviet and post-Soviet culture, such as the post-Utopian mentality, a critical attitude towards traditional notions of reality, and ironic playfulness with regard to the sign systems of various ideologies.[2]

---

1 As cited in Monica Jansen, "Has Postmodernism Ended? Dialectics Revisited (Luperini, Belpoliti, Tabucchi)," in *Postmodern Impegno: Ethics and Commitment in Contemporary Italian Culture*, ed. P. Antonello and R. Gordon (Oxford: Peter Lang, 2009), 49–50. See also: Monica Jansen, *Il dibattito sul post-moderno in Italia. In bilico tra dialettica e ambiguita* (Florence: Franco Cezari, 2002).

2 Mikhail Epstein, "The Philosophical Implications of Russian Conceptualism," *Journal of Eurasian Studies* 1 (2010): 65.

Roughly in those same years critics started to talk loudly about the end of post-Soviet postmodernism. In the preface to the special issue of *World Literature* (2011) dedicated to literature after the collapse of the USSR, it was concluded that postmodernism, which was "the dominant cultural movement" of the last decades after the collapse of communism, "seems, according to some observers, to be waning, and yet it is by no means clear what will replace it."[3]

The very term "postmodernism" in the post-Soviet space becomes very amorphous, and its meaning ranges from glamorous "coquettish postmodernism" (*Nezavisimaia gazeta* 69 [2008]) to negative "belligerent modernism" (*Izvestiia* 105 [2008]). At the same time, there are attempts to correct postmodernism in post-Soviet, in particular, Russian, space by emphasizing its real post-totalitarian content.

Mark Lipovetsky and Aleksander Etkind in 2010 reject the general term "postmodernism," which used to serve as a description of the specificity of post-Soviet literature, and offer a broad discussion of the newest Russian literature. They are looking for concepts, other than postmodernism, that can be of use when explaining "post-Soviet literary trauma." They come up with the notion of the literature of "Soviet catastrophe."[4] Lipovetsky, the author of well-known works on Russian postmodernism, states that he has never "reduced all of post-Soviet culture to postmodernism."[5] He writes that postmodernist discourse emerges in Russian culture in the 1960s and 1970s (predominantly in the underground and abroad) at the crossroads of two tendencies that seemingly complement each other: critiques of Soviet metanarratives and attempts to revive the disrupted (at least in official culture) discourses of the historical avant-garde.[6]

"[T]o me the postmodernist is both broader and narrower than the post-Soviet [culture],"[7] Lipovetsky writes more recently. Revision, with regard to the specificity of Russian postmodern situation, on the one hand, stems from "the trauma of cultural catastrophe" generated by modernism in the 1920s and 1930s; on the other hand, it arises from the fact that in the post-Soviet

---

3   Emily D. Johnson, "Twenty Years after the Collapse of the Soviet Union: Russian and East European Literature Today," *World Literature Today: Post-Soviet Literature* 86, no. 6 (2011): 133.
4   Mark Lipovetsky and Aleksander Etkind, "The Salamander's Return: The Soviet Catastrophe and the Post-Soviet Novel," *Russian Studies in Literature* 46, no. 4 (2010): 6–48.
5   Ibid., 9.
6   Mark Lipovetskii, "Transformatsii russkogo postmodernizma: Ot vzryvnoi aporii k vzryvnomu gibridu," http://do/gendocs. Ru/docs/index-41419.html.
7   Lipovetsky and Etkind, "The Salamander's Return," 9.

period postmodernist aesthetics constantly intersect with other, also active, discourses, such as realism, giving shape to hybrids such as "post-realism." One more tendency in post-Soviet literature is embodied, according to Lipovetsky, by postsots—as an offspring of the alliance between postmodernism and the socialist realist mythology.

At the beginning of the twenty-first century, researchers analyzing the phenomenon of post-Soviet literature resort to some new notions: "magic historicism," monstrosity, and melancholy. This discussion is notable for its active use of psychoanalysis as well as references to the theoretical categories of "the school of trauma"—in particular, to the work of Dominick LaCapra, which differentiates between two types of post-traumatic memory: "acting-out" (transmitting trauma into discourse, trauma's constant recirculation) and "working-through" (endowing trauma with a new meaning and thereby escaping it).

"The post-Soviet period is, undoubtedly, a time of melancholy,"[8] coauthors Lipovetsky and Etkind note, explaining that they view literature from the perspective of memory. In this regard, the sources of melancholy, in their opinion, can be found in the fact that memory has neither erased nor recycled the Soviet catastrophe: "The work of mourning is incomplete and unsuccessful; the loss has been incorporated into the subject, who cannot (meaning that he does not want to) free himself from it."[9] This accounts for the return in modern Russian literature of experience that was not reworked by melancholy in the form of monsters, vampires, sub- and super-human creatures—all of which are allegories of the catastrophic post-Soviet reality.

"Traumatic writing" becomes one of the central concepts of this discussion. Lipovetsky and Etkind, for example, propose considering one of the forms of such writing, which they call "magical historicism," as the common ground for different aesthetic languages, such as postmodernism, realism, modernism, postsots, and their hybrids.[10]

Indeed, Ulrich Schmid as early as in 2000 noted that, in the future, historical accounts of literary monstrosity will in all probability be regarded as one of the main traits of Russian literature at the end of the twentieth century.[11] Schmid maintains that this recourse to monstrosity and the use of aesthetic shock, which, in his opinion, is the basis of monstrosity, is something

---

8   Ibid., 9.
9   Ibid.
10 Ibid., 10.
11 Ulrich Schmid, "Flowers of Evil: The Poetics of Monstrosity in Contemporary Russian Literature (Erofeev, Mamleev, Sokolov, Sorokin)," *Russian Literature* XLVIII (2000): 206.

more than an infantile defiance with respect to the collective trauma of totalitarian Soviet culture. He sees an explanation of this phenomenon, following other researchers (Georgy Gachev, Mikhail Epstein), in the "semiotics of Stalinist culture"—a culture which was entirely based on the incorporation of monstrosity. One of the codes of post-Soviet monstrosity is the destruction of "reality," and its replacement by the absurdist-paradoxical fictional world of, for instance, Yuri Mamleev. "Here, nothing is hiding behind the world, 'things,'" the author repeats time and again, "because the real, even relatively real, cannot be a symbol of something that does not exist; on the contrary, the ineffable power of Trans-Darkness, Trans-Abyss casts its 'shadow' upon the embodied world, 'transforming' it into its own anti-analogue, anti-symbol, entering into an absurdist-paradoxical relationship with it."[12]

What is notable in the discussion of the known Russian researchers is that they avoid using terms that are connected to postmodernism and try to formulate a new theoretical paradigm based on the category of post-Soviet catastrophe, which is ostensibly the source of everything that characterizes contemporary literature. Post-Soviet literature is reduced to traumatic writing, and the nature of its imagery is explained with the help of the psychoanalytic notions of melancholy and "the work" of memory.

Such approaches are a characteristic feature of the post-postmodern revision of postmodernism, associated with the situation in various national literatures at the end of the twentieth century. This kind of revision in the case of Lipovetsky and Etkind was predicated upon the following:

> —a disavowal of general notions such as postmodernism;
> —an attempt to define the literary situation more specifically by appealing to a particular genre (novel) and particular stylistic (magical historicism, gothic) literary practices;
> —a replacement of postmodernist pan-aestheticism by the categories of cultural memory, trauma, and melancholy (categories that were also generalized to encompass all post-Soviet literature);
> —an appeal to the "Soviet catastrophe," which is a version of catastrophism (a particular model of cultural mindset, especially activated in the post-postmodern period).[13]

---

12 Iurii Mamleev, *Izbrannoe* (Moscow: Terra, 1993), 95.
13 I analyze the nature of this type of catastrophism using an example of the perception of the Chornobyl catastrophe in "Tranzytna cultura. Symptomy postkolonial'noi travmy" (Hrani-T, 2013, 338–457).

2. THE RETURN OF "GRAND NARRATIVES"

The revision of postmodernism occurs not only on post-Soviet ground. Western scholars also try to analyze the limitations reflected by the *end of postmodernism*. One of the central issues appears to be the issue of the death of "grand narratives," which Lyotard associates with the legitimization of postmodernism itself. The discussion around postmodernism clearly reveals the difference in ideological and methodological positions that are present today in humanities.

A representative of radical social criticism, Rosi Braidotti notices the return of neoconservative liberalism in the post-postmodern era. "At the end of postmodernism," she says, "in an era that experts fail to define in any meaningful manner because it swings between nostalgia and euphoria, in a political economy of fear and frenzy, new master narratives have taken over."[14]

On the one hand, there is a complete faith in the market economy; on the other, biological essentialism in the form of a new evolutionary biology (some experts mention in this regard "the genetic social imaginary") defines the character of new grand narratives whose main trait is a return of determinism. What happens here, as Braidotti maintains, is reification and an adaptation of one of the fundamentals of postmodern thinking. The category of difference, once a tool for thinking about representation, becomes a sign of the cultural and civilizational belonging of nations, classes, states, and races. Differences of identities, cultures, and religions are considered here a result of cultural-historical development and are formed hierarchically, using the method of exceptions, according to the principle "us and them."

As Braidotti notes, "the reassertion of differences introduces structural patterns of mutual exclusion at the national, regional, provincial, or even more local level."[15] Thus, borrowing from postmodernism the topos of différance, post-postmodern thinking introduces reification instead of the play of differences. The market and new technologies literally devour such a reified other—they package and sell "otherness" under the guise of new, hybrid, multicultural, and "other" identities. This consumption of "differences" has as its alternative the authentic otherness produced by feminist, postcolonial, youth, LGBT, and so forth, varieties of counter-subjectivity.

Appealing to the end of postmodernism, Rosi Braidotti talks about the potential of postmodernism, in particular, about various modes of producing

---

14 Rosi Braidotti, "A Critical Cartography of Feminist Post-Poststructuralism," *Australian Feminist Studies* 20, no. 47 (2005): 169.
15 Ibid.

differences, which are a source of cultural variability. In addition, she continues to develop the idea of nomadism as a form of affirmation of the different, because "the becoming-minoritarian, or becoming-nomadic," in her opinion, is a subversive matrix "open to both the empirical members of the majority (the 'same') and to those of the minority (the 'others')."[16]

Braidotti discusses the possibilities of postmodernism. Ewa Thompson, criticized by postmodernists for her essentialism, on the contrary stresses the limitations of postmodernism. She sees the cause of such limitations first of all in the fact that postmodernism appeals to constructedness, suspends essentialism, deconstructs the center, and expropriates moral judgement from theory. Working in the field of postcolonial (more precisely, anti-colonial) criticism, Thompson stresses the difference of approaches: "They [postmodernists— T. H.) wanted colonialism to be seen in the context of land, race, production, and consumption; I spoke of nations, of aggressive and defensive nationalisms, and of a desire to preserve identity."[17] In other words, otherness, like nationality, in Thompson's opinion, cannot be reduced to a constructive process: for her, they have both been grand narratives.

Mikhail Epstein, in his turn, rather than reverting to grand narratives, suggests looking for a solution to the internally contradictory aporia of the postmodern paradigm, in which two fundamental principles are mutually contradictory. One principle, in particular—deconstructionism—criticizes essentialism and the metaphysics of presence, demonstrating that there are neither physical nor historical origins, only signs; while another principle— multiculturalism—establishes a connection between cultural production and ethnic and physical origins, and thereby generates such cultural narratives as "'white male' culture, or 'black female' culture, or 'homosexual' culture."[18]

In order to overcome this contradiction and not cross the boundaries of the postmodern paradigm, Epstein introduces the notion of transculture as a particular mode of existence oriented toward establishing natural origins and at the same time overcoming their limits: "[t]ransculture presumes the enduring 'physicality' and 'essentiality' of existing cultures and the possibility

---

16 Ibid., 174.
17 Ewa M. Thompson, "Ways Out of the Postmodern Discourse," *Modern Age: A Quarterly Review* (Summer 2003): 196.
18 Mikhail Epstein, "Transculture in the Context of Contemporary Critical Theories," in *Transcultural Experiments: Russian and American Models of Creative Communication*, eds. Ellen E. Berry and Mikhail Epstein (New York: St. Martin's Press, 1999), 80–81.

of their further transcendence, in particular through interference with other cultures."[19]

He states that transculture in particular generates momentum for further transformations—a momentum that already dwells within postmodernism. The source of transculture, according to Epstein, is fluid and variable individual cultural practice—a children's experience, an experience of illness, and experience of love, loneliness, emigration, pilgrimage, and so forth. He maintains that due to the multiplicity of individual experiences various groups of people are able to communicate and feel otherness within themselves.

Thus, instead of *multi*-culture, which, in Epstein's opinion, emphasizes the multiplicity of collective identities, there is transculture, which is based on the ability of "multicultural" individuals to cross the boundaries of their collective cultural identities. Epstein formulated his theory of transculture back in 1999, and it testifies to the fact that already by then postmodernism was a complex and even contradictory discourse that could develop and ramify in different directions.

3. MODERNISM AND POSTMODERNISM

The problem of the relationship between postmodernism and modernism after the end of postmodernism remains particularly controversial. As is well known, one of the first interpretations of postmodernism—that of Fredric Jameson—was historicist: he called postmodernism the period that begins after the Second World War. The first manifestations of postmodernism are at beginning of the 1960s. These, he argued, coincide with the birth of postindustrial society, or, as it is also called, the consumer society, the society of the spectacle, the society of media, multinational capitalism, and so forth. Therefore, postmodernism was proclaimed to be not so much an aesthetic style as "the cultural logic of late capitalism."

> I believe that the emergence of postmodernism is closely related to the emergence of this new moment of late consumer or multinational capitalism. I believe also that its formal features in many ways express the deeper logic of this particular social system.[20]

---

19 Ibid., 84.
20 Fredric Jameson, *The Cultural Turn: Selected Writings on the Postmodern, 1983–1998* (London: Verso, 1998), 20.

Admitting that postmodernism is a restructuration of certain already present elements, Jameson states that the content with which postmodernism deals is already present in modernism; what matters is the specific way in which this content shifts from the margin to the center. Postmodernism, for Jameson, then, emerges *after* modernism in times and uses elements that modernism developed.

This conception of postmodernism from time to time became (and often is now) the basis for defining certain literatures as unprepared for the emergence of postmodernism. Special emphasis is laid here on the absence of (or underdevelopment) of postindustrial society and on the incompleteness (or absence) of a modernist canon. The rationale behind this way of thinking is that postmodernism is not possible if there was no full modernist cycle (as there was in the West). This logic is maintained by Ukrainian critics of postmodernism. For example, Oleh Ilnytzkyj, initially accepting Jameson's thesis that "postmodernism is a reaction to the seventy or eighty years of modernism,"[21] draws the conclusion that "unlike the Americans or the French, the Ukrainians have not experienced modernism."[22] Therefore, the emergence of postmodernism in Ukraine is not possible. Further, he considers postmodernism first and foremost "a consequence of the particular historical circumstances ... cultural, intellectual, artistic, and social" of countries like France and the USA. Consequently, according to his logic, postmodernism developed out of an attempt to make sense of the "specific problems these countries encounter—technological ... and informationally oversaturated societies in which art acquires new forms and functions."[23] This is the origin of Ilnytzkyj's reservations, which are quite understandable:

> My observations compel me to the unfortunate conclusion that postmodernism ... in no way can be brought into accord with Ukrainian cultural experience of the last seventy years. ... I doubt that now Ukraine is coming to or approaching those cultural conditions that are dominant in the West."[24]

Viktor Neborak, in his turn, notes that the "one thing that can be stated based on what is in the term itself is that postmodernism appears after modernism."[25]

---

21 Oleh Il′nyts′kyi, "Transplanatatsiia postmodernizmu: Sumnivy odnoho chytacha," *Suchasnist′* 10 (1995): 112.
22 Ibid.
23 Ibid.
24 Ibid., 113.
25 "Bubabists′kyi khronopys Viktora Neboraka: shche odna intryga z pryvodu suchasnoï ukraïns′koï literatury," *Literatura plius* 9–10 (2001), http://maysterni.com/ukrart/poetskr414.html.

However, the historical conception of postmodernism has not stood up to the task, and now an idea that Fernando Burgos expressed with respect to Latin American literature has become prevalent:

> postmodernism is not an artistic occurrence that is positioned after modernism, but on the contrary, cohabits with it. Their coexistence should not be alarming if one takes into account the foundational moments of modernity. Clearly, it seems to us that it is not necessary to wait for postmodernism until the final stage of capitalism or the manifestations of the most daring technological innovations, as Jameson claims.[26]

Postmodernism to a great extent fulfils the functions with which modernism was burdened: renouncing the pursuit of the new, which used to be the basis of the modernist sensibility, it simultaneously expedites the modernist impulses of innovation, thus in its own way rediscovering the "transformative aesthetics" of modernism.

4. POST-POSTMODERNISM

Critics' attention is very much drawn to the limits of postmodernism itself and to what comes after it. Among various versions of post-postmodernism are: post-humanistic literature; the literature of trauma; and literature of "new sentimentality." Posthumanism, for instance, tries to overcome postmodernism by resorting, in the works of late postmodernists, to a sensual-sentimental experience, "new sincerity," and autobiographism. The fundamental principles of posthumanism are based on Donna Haraway's "Cyborg Manifesto" (1985), which states that at the end of the twentieth century we all are chimeras, theoretical and fabricated hybrids of machines and organisms. Equally important is Katherine Hayles's thesis about the posthuman stage of contemporary culture: material objects are penetrated with information patterns, and computerization, rather than a biological organism, individualism, or social community, is the basis of being.

Nicoline Timmer explains the character of this *new post-postmodern sensitivity* in her analyses of David Foster Wallace's *Infinite Jest* (1996), Dave Eggers's *A Heartbreaking Work of Staggering Genius* (2000), and Mark Danielewski's *House of Leaves* (2000). Post-postmodern sensitivity here is a "symptom," or a series of feelings, needs, and problems that do not fit into the standard postmodernist

---

26 Fernando Burgos, "Postmodernism and Its Sighs in 'Juegos Nocturnes' by Augusto Roa Bastos," in *Postmodernism's Role in Latin American Literature: The Life and Work of Augusto Roa Bastos*, ed. Helene Carol Weldt-Basson (New York: Palgrave Macmillan, 2010), 56.

paradigm. Among these symptoms are: an inability to choose because of lack of tools to do so and because we now live in the world oversaturated with choice; a self that is separated from feelings and is not able to appropriate them because feelings are not tied to individuality and are "vaporous"; and a structural need to be a "we"—that is, to be connected with someone, to be social, to cross the limits of solipsism, but not in the ironic way postmodernism proposed.

Hal Foster, in his turn, discusses various types of sensitivity in postmodernism itself. In his opinion, postmodern art at the end of the twentieth century is dominated by a melancholy structure of feeling; before that, postmodernism had an ecstatic structure.[27] It is worth noting that as early as in the 1980s, Foster tried to define various cultural-ideological tendencies within postmodernism—in particular, "critical" (or resistant) postmodernism—that are present in both cultural forms and social institutions.[28]

5. THE LITERATURE OF TRAUMA AND THE RETURN OF THE "REAL"

When it comes to cultural postmodernism, beside the issues of "sentimentality," "constructedness," "grand narratives," and "ethical reading," the problem of the "return of the real" is of special importance. For example, Robert McLaughlin associates post-postmodernist dissatisfaction with the fact that postmodernism has brought the self-reflectivity of art to its utmost level. This alienation from the social world and immersion in self-referential language has proven catastrophic for literature itself.

Analyzing John Barth's essays about the "literature of exhaustion" and the "literature of replenishment,"[29] which have become fundamental for theories of literary postmodernism, McLaughlin states that, as a matter of fact, Barth also stood in opposition to high modernism, with its orientation toward complexity and codes that only experts can understand, and to "the lobotomized mass-media illiterates."[30] Barth was inclined to see the origins of new novel "in middle-class popular culture" and in writing referring to simple human feelings and lives. His point was the rehabilitation of reality and its liberation from

---

27 Hal Foster, *The Return of the Real: The Avant-Garde at the End of Century* (Cambridge, Mass.: MIT Press, 1996), 165.
28 Hal Foster, *The Anti-Aesthetic: Essays on Postmodern Culture* (Port Townsend: Bay Press, 1983), 3.
29 John Barth, "The Literature of Replenishment," cited in Robert L. McLaughlin, "Post-Postmodern Discontent: Contemporary Fiction and the Social World," *Symploke* 12 (2004): 58.
30 Ibid.

the demands of intertextuality and self-referentiality. In McLaughlin's opinion, modern television has already adopted the postmodernist ruse of "erasing" (annihilating) reality; for this reason, postmodernist authors and consumers need to find other modes of constructing reality—ones that differ from self-referential irony.

In fact, the real has already returned in late-stage postmodernism. Admittedly, it is the real not in its regular understanding but as in the repressed content that arises with feelings of loss—through trauma. As consequence, there emerge notions and conceptions that, while based on postmodernist approaches, are interpreted as transgressions of the boundaries of postmodernism—in particular, theories of post-traumatic writing and post-memory. Jay Winter, analyzing the "boom" in work on memory in the last decade, uncovers a multiplicity of contributing factors: the rediscovery of the Holocaust as a theme; attention to the victims of various traumatic twentieth-century events; the increased role of memorials in national narratives; the expansion of "generational memory" due to demographic growth after the 1960s; the development of cultural consumption and the technologies of its production; and the transmission of psychoanalytic discourse into public sphere.

Already in postmodernism, then, we witness a growing dissatisfaction with a total textuality and desire which relegate the real to the realm of consumption and the commodity. There is also a certain dissatisfaction with the fact that the social contract among people is broken and the individual is locked in a world of repetitions, imitations, and labyrinthine meanings. These circumstances explain the preoccupation of postmodernism with trauma. It can be argued that the narrative of trauma, which has recently attracted the attention of both scholars and authors alike, is one of the consequences of the return to reality which in later postmodernism receives a peculiar understanding.

A correction of the postmodernist paradigm starts in the 1990s. Postmodernist irony is so widespread in culture that people argue that any new experimental art ought to develop in a different direction. The opinion takes hold that although spiritual and emotional self-reflection may seem hackneyed, naive, and melodramatic, such art offers readers tangible feelings instead of empty, ironic self-repetition. Thus, sentimental and sensual postmodernist writing becomes of interest—writing typified by the neo-Victorian prose of John Fowles or the posthumanist prose of David Foster Wallace.

Worth noting is also a peculiar cult for post-postmodernist authors associated with early, Borgesian postmodernism—for example, the cult of Danilo Kiš among the new post-Yugoslavian prose fiction writers (Drago Jančar from

Slovenia, Svetislav Basara from Serbia, and Aleksandar Hemon and Muharem Bazdulj from Bosnia). Readers are first of all attracted to Kiš's stylistics—in particular his positioning of the narrator vis-à-vis his or her characters. Readers also respond to the author's emphatic assertions of the historical truth of documented facts, pictures, and books, in spite of the fact that all of these items are the narrator's invention.[31]

6. POSTMODERNITY AND POSTMODERNISM

During the period of the supposed end of postmodernism, discrimination between postmodernity as a period and postmodernism as an aesthetic practice becomes vital again. In this regard, the work of Ihab Hassan is central. In the 1970s, he initiated a literary, rather than philosophical, theory of postmodernism. His article "From Postmodernism to Postmodernity: The Local/Global Context" (2001), asks: *"What was postmodernism? What was postmodernism and what is it still?"*[32] Rhetorically emphasizing—by way of the question mark—the finality and at the same time incompleteness of postmodernism, he asserts two things: on the one hand, it was not an immutable concept; and on the other hand, postmodernism itself is a metahistorical phenomenon that keeps coming back, haunting like a ghost. Even noting the so-called end of postmodernism, we nevertheless again and again formulate its principal characteristics. Even when we realize that a whole era whose main task was to discover the truth of postmodernity is now becoming history, we understand that this does not mean that all the concepts, narrative strategies, and types of sensitivity generated by postmodernity will disappear. What happens is, rather, a reformulation, transformation, borrowing, incorporation, transgression, and inversion of the codes and mechanisms of postmodern thinking. In short, postmodernism becomes a system of ideas even though it opposes systematicity as such.

Thus Hassan takes note of the theoretical elusiveness of postmodernism and admits that he knows less about it now than thirty years ago when he started to write about it. Ultimately, we can believe Hassan's assumption that if we lock in one room all those who have discussed postmodernism—namely

---

31 Andrew Wachtel, "The Legacy of Danilo Kiš in Post-Yugoslav Literature," *The Slavic and East European Journal* 50, no. 1 (2006): 137.
32 Ihab Hassan, "From Postmodernism to Postmodernity: The Local/Global Context," *Philosophy and Literature* 25, no. 1 (2001): 1.

Leslie Fiedler, Charles Jencks, Jean-François Lyotard, Bernard Smith, Rosalind Krauss, Fredric Jameson, Marjorie Perloff, Linda Hutcheon, and Hassan himself—and throw the key away, no consensus would ever be reached.

Hans Bertens's statement that there exists more than one postmodernism is useful for understanding its fate.[33] Postmodernism cannot be reduced to one version because each model mirrors the others, and thereby creates a field of changing meanings and references. It consists of different versions of early and late postmodernism, as well as different variants with regard to locality, nation, gender, ethnicity, the world culture, and pop culture.

Hassan introduces one more point that is important for understanding mature, late postmodernism. Certainly, postmodernism diversifies, acquires a formal character, and even encroaches upon the realm of popular culture; and certainly it offers more complex hybrid forms that cooperate with both traditional and new techno-informational forms. Yet, Hassan emphasizes, it becomes increasingly obvious that postmodernism relates *equally* to modes of life, thinking, and artistic representation. That said, it is appropriate to differentiate between postmodernism proper and its peripheral offshoots, as well as to discuss the adoption of its devices and their transitioning into other spheres, for example, into informational technology (cf. the phenomenon of Facebook).

Some experts regard postmodernism as an umbrella that covers multiple realms—culture, philosophy, scholarship, politics, ideology, fashion, culinary art, etc. Such an umbrella, in Hassan's view, unites postmodernism in art, poststructuralism in theory, feminism in social discourse, and postcolonial and cultural studies in the sphere of academic scholarship. It also encompasses multinational capitalism, cybertechnology, international terrorism, and ethnic, national, and religious movements.[34]

Hassan considers the fact that postmodernism from the very beginning is not limited to the realm of aesthetics and that it is also by no means a homogenous phenomenon. He proposes discussing postmodernism as a cultural sphere, which includes literature, philosophy, architecture, and the fine arts; and further, suggests conceiving of postmodernity as a geopolitical scheme. As he admits, he did not make this distinction in this early work. The rationale behind this differentiation is that postmodernism has turned into an interpretive category and hermeneutical instrument: "More

---

33 Bertens, "The Postmodern Weltanschauung," 26.
34 Hassan, "From Postmodernism to Postmodernity," 3.

than a period, more even than a constellation of artistic trends and styles, postmodernism has become, even after its partial demise, a way we view the world."[35] Postmodernity, after all, evolved into local-global conflicts (the Balkans, Chechnya, Tibet, Sudan), while "cultural postmodernism itself has metastasized into sterile, campy, kitschy, jokey, dead-end games or sheer media stunts."[36]

Arthur Danto, in his turn, suggests one more perspective on the heterogeneous stylistic nature of postmodernism, which he considers a separate period in the development of art. Observing the history of Western art, Danto discerns—up to the mid 1960s—two basic phases: the realist (Vasarian, post-Renaissance) and the Greenbergian (so named after the well-known twentieth century critic Clement Greenberg). In the first phase, art actively appropriates external visual experience; in the second, it is preoccupied primarily with the circumstances of its own production and becomes self-referential. In the 1960s, together with Andy Warhol, art enters the stage of "after the end of art,"[37] as Danto calls it—that is, after the completion of the linear, progressive stage of art's development. The art of this third phase is characterized by a lack of stylistic uniformity that would serve as a criterion for differentiating art as such. Now, artists can do anything and it will still be construed as art. Postmodernism introduces this third phase, but it occurs only in the 1970s–80s. Thus, postmodernism is a "post-life," or an embodiment of what an artistic product can be in the absence of any conception of what it should be.

7. "THE END OF POSTMODERNISM"

However, discussions unfold also around the very thesis of "the end of postmodernism." They now have a local rather than general character, and are no longer based on the universal logic of late capitalism described by Jameson. Statements about the end of postmodernism emerge at the beginning of the 1990s. To a great extent, they stem from the fact that postmodernism is understood differently in different disciplines and contexts, and that even its chronology is determined only approximately.

---

35 Ibid., 9.
36 Ibid., 5.
37 Arthur Danto. *After the End of Art. Contemporary Art and the Pale of History* (Princeton University Press, 1996).

At the same time, criticisms of postmodernism in the post-postmodern period are not free from what Josh Toth and Neil Brooks call "mourning." Looking at the origins of postmodernism, they note that "this emergent epoch seems to 'mourn' the apparent loss of the very idealistic alternatives that postmodernism strove to efface."[38] The terrorist attack on September 11 is pronounced to be the most expressive marker of that period, yet the cultural shift to "mourning" had already occurred in the 1990s in reference to a new form of realism. A radical break with postmodernism is witnessed, however, after September 11, 2001, when, as Maurizio Ascari notes,

> a certain kind of postmodernism—with its jocular manner, its ostentatious irresponsibility, its deconstructive frenzy—suddenly appeared frivolous against the enormity and terrible novelty of this tragedy, which was immediately replicated on every television screen throughout the world.[39]

The form of postmodernism that appeared frivolous was popular rather than philosophical, and was associated with the apolitical and asocial. It seemed that it was opposed by not only realistic but also ethical and political literature. However, it was precisely within the framework of the postmodernist paradigm that both postcolonial criticism and postcolonial literature emerged and unfolded—their political component (just take one example, Salman Rushdie's *Midnight's Children*) was very important. By the same token, the postmodernism that emerged in post-totalitarian spaces—from Russian conceptualism to Yuri Andrukhovych's and Dubravka Ugrešić's novels—is ideologically tinted.

And yet, admittedly, an ideological component is much more expressive in the works of opponents of postmodernism—for example, in those of the Italian Marxist Romano Luperini. In his *La fine del postmoderno* (2005), Luperini states that the fall of the Berlin Wall and the reinforcement of ethical values after September 11 testify to the fact that capitalism dominates absolutely and needs to be resisted, and that postmodernism is unequal to the task. Consequently, as Luperini writes, the importance of neo-modernist themes rises. In his opinion, this does not mean regression back to modernism or a

---

38 Josh Toth and Neil Brooks, eds., *The Mourning After: Attending the Wake of Postmodernism* (Amsterdam: Rodopi, 2007), 3.
39 Maurizio Ascari, *Literature of the Global Age: A Critical Study of Transcultural Narrative* (Jefferson: McFarland & Company Publishers, 2011), 25.

restylization of postmodernism. Instead, it marks the end of the age of "general anesthesia."[40]

At the same time, Luperini suggest forgetting about postmodernism and replacing it with postcolonialism, which, ostensibly, offers an open dialogue with the other, contrary to postmodernist isolation. Here, one element of postmodernism (postcolonial criticism) is simultaneously borrowed and aimed at the rest of the postmodernist paradigm. However, Douwe Fokkema had already brought attention to the fact that the first stage of postmodernism ends somewhere in the middle of the 1990s, with the publication of Umberto Eco's *The Name of the Rose*. New, or "late postmodernism," in his opinion, begins when extra-textual reality is newly accepted and there emerge five new types of writing, which "signify the end of postmodernism or, at least, have drastically modified the concept."[41] Among those types (feminist literature, historical fiction, autobiography, and the fiction of cultural identity), Fokkema also notes the emergence of "postcolonial writing."[42]

In discussions of the end of postmodernism, apocalyptic criticism becomes especially popular. However, the very understanding of apocalypse differs. Luperini sees apocalypse as full and complete destruction while his compatriot Marko Belpoliti sees in it a new beginning. Belpoliti appeals to the idea of catastrophe as a dynamic paradigm and the core of postmodern philosophy. Consequently, the apocalyptic time of postmodernism, according to Belpoliti, is a time "of the end that does not end," a time that is able potentially and creatively to unfold even after destruction. Examples of such a new post-postmodern art are, for him, "posthuman" and "informal" artforms, as well as reflections on the "double" connection between ethics and aesthetics. It is worth noting that, zeroing in on the theory of catastrophe, the Italian critic locates the double connection of ethics and aesthetics in Susan Sontag and Milan Kundera—in their separated assessments of kitsch. Neither of these intellectuals repudiates kitsch in favor of higher values; instead, they both understand it as an ambivalent designation of fear, which in a catastrophic situation is quite emotionally experienced. Thus, as Italian criticism testifies, the resolution of the end of postmodernism yields the discovery of its alternative

---

40 See Monica Jansen, "Has Postmodernism Ended? Dialectics Revisited (Luperini, Belpoliti, Tabucchi)," in *Postmodern Impegno: Ethics and Commitment in Contemporary Italian Culture*, ed. P. Antonello and R. Gordon (Peter Lang, Oxford, 2009), 49.
41 Douwe Fokkema, "The Semiotics of Literary Postmodernism," in *International Postmodernism*, ed. Hans Bertens and Douwe Fokkema (Amsterdam-Philadelphia, 1997), 30.
42 Ibid.

forms—postcolonialism and catastrophism. They are, in essence, components and variations of the postmodernist paradigm that separate and direct themselves against postmodernism itself.

This emphasis on the ethical moment in post-postmodernist thinking is worth noticing. Quite a few critics state that ethics returned via studies of the Holocaust, which generated theories of post-memory, trauma, and responsibility. At the same time, ideas of nostalgia (in Jameson's understanding) and irony (Hutcheon) collide. The first is associated with escapism and aestheticism, while the second is associated with critical disposition and innovation. Overall, the literature of trauma restores attention to the ethics of both literature and reading.

8. TYPES OF POSTMODERNISM

Discriminating between early and late modernism has become a fundamental for theories of postmodernism. When looking at the history of the development of postmodernism itself, Hans Bertens's reflections are helpful. He notes the moment of the passage from numerous postmoder*nisms* to a single postmoder*nism*. He states that in the course of the 1950s and 1960s, critics were looking for new tendencies, which they marked as postmodern. In the 1970s, however, with the publication of Ihab Hassan's *The Dismemberment of Orpheus: Toward a Postmodern Literature* (1971), postmodernism becomes more and more an inclusive term that encompasses all literary and cultural phenomena that are not called realist or modernist. In this exact same period, William Spanos identifies various postmodern cultural modes, such as pop art (in Leslie Fiedler's understanding), the *nouveau roman* (in Richard Wasson's understanding), and counterculture (in Susan Sontag's understanding). Spanos unites them into one notion of postmodernism. For Spanos, iconic modernism transforms existential time into artistic, eternal spontaneity; while postmodernist writing, for example, involves turning literature into an ontological dialogue with the world, which thereby reveals the authentic historical essence of the modern individual.[43]

However, to understand the evolution of the postmodernist paradigm in its entirety, it is important to include the idea of the core and the periphery of postmodernism, along with its influence on popular culture. In the course

---

43 William Spanos, "The Detective and the Boundary: Some Notes on the Postmodern Literary Imagination," *Boundary* 2 (1972): 165–66.

of the 1990s in the West, postmodernist aesthetics and techniques become a component of the mass media, thereby generating a phenomenon that Douglas Kellner calls "a popular form of postmodernism."[44] Kellner takes as his example of pop-postmodernism the TV series *The X-Files*, in which forms and figures of media culture are used to comment on dangerous aspects of contemporary society, such as the loss of government power, unregulated sciences and technologies, threats to the body and integrity of the individual, and the emergence of technologically and genetically constructed creatures.

The appropriation of textualized images of history, culture, and everyday experience, and the mythology of modern fashion, associated with the emergence of poststructuralism in the 1980s, led to the appearance of a new type of postmodernism. "Interliterary postmodernism," despite "the death of the author," maintained the position of the author as the primary creator of the text. The characteristic feature of interliterary postmodernism was an intertextual play that involved mostly cultural and literary codes. The development of the tech and computer industries in the late 1990s facilitated the further unrolling of postmodernist artistic models and led to the emergence of one more type of postmodernism: inter-media. Its constructive principle lies in a play of virtual realities, media spheres, and morphological forms; simultaneously, the role of the recipient as a participant of the creative process rises, even though the author's part remains important. Currently, post-postmodernism is discussed as an entirely interactive artistic model in which it is consumers who define the existence of a text because they are endowed with an exceptional capacity to combine someone else's texts, to write their own text, and to rewrite the texts that are already written.

Ultimately, Thomas Oord differentiates among various trends inside the postmodern paradigm—in particular, deconstructive postmodernism, liberationist postmodernism, narrative postmodernism, and revisionary postmodernism. In Oord's opinion, the most popular form of postmodernism is deconstructive. Not only does it form a separate tradition but is also the basis for postmodernism as a whole. It is founded on Derrida's notion of différance, whose meaning is predicated upon the polysemantic French word "différer," which means "to differentiate," "to differ," and "to "defer." If this characteristic is projected onto a literary text, according to Derrida, the text acquires meanings that are different from those that the author intended; and these meanings,

---

44 Douglas Kellner, "The *X-Files* and the Aesthetics and Politics of Postmodern Pop," *The Journal of Aesthetics and Art Criticism* 57, no. 2 (1999): 161.

although deferred and postponed, undermine the meaning of what is said in the text. Différance, in short, gives an opportunity "to think a writing without presence and without absence, without history, without cause, without *archia*, without *telos*."[45]

Thus "text" does not have a definitive meaning; it does not refer to reality and it is not reality's progeny. Instead, it contains words and phrases that undermine what it says—defers it. Text does not have only a singular meaning: it can be interpreted in various ways. It lacks a center and splits an utterance into multiple meanings, which resist rational cognition. Language becomes a field of indefinite play—of signifiers rather than signifieds. There is no truth in the words that are said. One can only talk about an unending multiplicity of voices. There is no certainty that language is linked with the objects it speaks about; and if there are ties, it is doubtful that they are truthful.

Another tradition within postmodernism itself, despite the versatility of the term, is characterized by an emancipative desire—a desire to break free from the limitations imposed by modernism. Oord calls this tradition "liberationist postmodernism" and includes here feminist, ethnic, and ecological trends.[46] Gender, nation, race, culture, and the environment are regarded in terms of diversity, an absence of hierarchy, accord with nature and the environment. Attention shifts from the dominant center to the polyphony of the margins.

The next type of postmodernist tradition that Oord lists is "narrative postmodernism."[47] This recalls those linguistic games which reveal the life of individuals and communities, as well as the failure of grand narratives. The narrativization of history, of the everyday, and of traumatic situations is an important feature of the postmodernist mindset. Language itself creates its own rules, and the mode of speaking defines the mode of life; in addition, narrated tales, or narratives, can exist without the reality that they narrate.

Another type of postmodernism is revisionary. Its representatives want to overcome the limitations of postmodernism itself:

> Revisionary postmodernism overcomes the modern worldview by offering what it considers the most viable worldview for our time. This worldview accounts for a variety of sensibilities, including religious, scientific,

---

45 Jacques Derrida, *Margins of Philosophy* (Chicago: University of Chicago Press, 1982), 67.
46 Thomas Jay Oord, "Breaking Free: Liberationist Postmodernism," http://thomasjayoord.com/index.php/blog/archives/breaking_free_liberationist_postmodernism.
47 Thomas Jay Oord, "Postmodernism—What Is It?," http://repository.mnu.edu/sites/default/files/Didache%201-2.pdf.

ecological, liberationist, economic, and aesthetic. By contrast, deconstructive postmodernism overcomes the modern worldview through an antiworldview.[48]

The emphasis here on sensibility, along with pan-experimentalism, holism, synergetic organicism and bio-zoo-humanism, is a prominent feature of revisionary postmodernism.

A special issue of *Twentieth Century Literature* (2007) defines the situation after postmodernism in various ways. Paul Giles's notion of "sentimental posthumanism" is particularly useful.[49] In his introduction, editor Andrew Hoberek notes that the positions of anti-postmodernists and those who proclaim the end of postmodernism are "fundamentally postmodern gestures."[50] First of all, postmodern techniques do not disappear: they are now part of contemporary fiction. Secondly, the diversity of contemporary fiction is thoroughly postmodern:

> the middle-class realism of Susan Choi and Jhumpa Lahiri, the books of John Updike; ... the comic-book magical realism of Jonathan Lethem and Michael Chabon, the more traditional version practiced by Toni Morrison; ... the picaresques of Han Ong, Jonathan Safran Foer, and Benjamin Kunkel, those of Saul Bellow.[51]

In addition, there is no such notion as the purity of postmodernist style; on the contrary, postmodernist authors combine experimental, realistic, and autobiographical trends.

Many characteristics that Fredric Jameson, say, regarded as key to postmodernism (the excessive integration of aesthetic products into commodity production, the cannibalization of styles from the past, the absence of a position for potential resistance against the dominant culture, for example) are today no less topical than at the time they were formulated. Yet, despite the fact that postmodernism legitimized popular literature and actively employed its structures, the works of leading postmodernist authors have never become phenomena

---

48 Ibid.
49 Paul Giles, "Sentimental Posthumanism: David Foster Wallace," *Twentieth-Century Literature* 53, no. 3 (Fall 2007): 327.
50 Andrew Hoberek, "Introduction: After Postmodernism," *Twentieth-Century Literature* 53, no. 3 (2007): 236.
51 Ibid.

of popular culture. Instead, as Hoberek notes, authors usually associated with high modernism have recently resorted to the forms of popular culture: Cormac McCarthy's *No Country for Old Men* (2005) and John Updike's *Terrorist* (2006).

Contemporary criticism, focusing on the decline of postmodernism, encounters a few problems as well. Firstly, the aesthetic component proper is ignored, and instead of noting the exhaustion of artistic forms, critics discuss the circumstances and situations that shaped those forms and that exhaust them; secondly, criticism is reduced to the reproduction of modernist experience—in particular, reproduction of the formal, experimentalist experience that is associated with "real" literature; thirdly, the revival of realism and forms that have an expressive emotional content does not mean that postmodernist experience fades away; and fourthly, the experience of postmodernism widens the very understanding of modernism, and vice versa.

A Commentary on the "End of Ukrainian Postmodernism"

Lipovetsky and Etkind's reflections on the trauma and melancholic sublimation, which form the basis of post-Soviet literature, make sense to me because I have attempted to apply this approach in my own studies for quite a while. In *Femina Melancholica* (2002), I develop a psychoanalytic interpretation of melancholy in the work of Olha Kobylianska. Melancholy here is born out of an invincible longing for the ideal of the "complete individual." This leads to a repression of the maternal instinct, the appearance of androgynous characters, and an appeal to the classic image of the "strong man," as well as variations of perverted sexuality (homoerotic, masochistic). Therefore, the whole conception of *The Post-Chornobyl Library* (2005) is based on the idea of a Chornobyl trauma, which facilitates the manifestation and deployment of a new cultural phenomenon—a postmodern event, which literally explodes on the post-totalitarian margins of Europe, even though certain elements of this type of mindset and its cultural models had already existed within totalitarian culture long before Chornobyl.

However, the point of the matter is not just Chornobyl trauma, which is compensated for by the post-Chornobyl library. I understand the whole concept of Ukrainian postmodernism as a recycling (in the sense of working through) of both post-Soviet and postcolonial traumas. When I was writing that book, it was clear to me that the era of postmodernism was on the decline, that the artistic practices discovered by postmodernism were passing into popular culture, and that the emerging literature did not quite fit the categories of postmodernism. Then, in 2005, it was also important to note that postmodernism, despite its constant confrontations with official culture, brought a new mentality and new imagery. The whole cultural field in post-Soviet Ukrainian literature was changed and—most importantly—genuinely *Ukrainian* literature was created.[1]

---

1   It was manifested both theoretically and artistically in *Pleroma. Mala ukraïns'ka entsyklopediia aktual'noï literatury*.

Personally, for me, it was an attempt to sum up more than a decade's interest in postmodernism. First, I applied postmodernism as a methodology for my doctoral dissertation, which later became the basis for *The Emerging Word: Discourse of the Early Ukrainian Modernism: A Postmodern Interpretation* (1997). At the time, postmodernism became the interpretive basis and optic that enabled me to discuss early twentieth-century Ukrainian modernism without an inferiority complex. As a result, I discovered the incredible diversity of artistic practices of the period. In *The Emerging Word* I tried to draw attention to a new postmodernist theory and devoted a whole subchapter to Western concepts of modernity and postmodernity. I also addressed the perspectives that were opening due revisionary work on the enlightenment project in Ukrainian literature. I proposed replacing the formula "European modern*ism*" with "European modern*isms*." It was also important for me to examine the collision of high and low culture in terms of Nietzsche's idea of high culture. I discussed Jürgen Habermas's communicative reason, and the radical poststructuralism and deconstructionism of the French philosophers. In short, postmodernism in that book was a methodological frame, which did not escape John Fizer's eye when he reviewed the monograph.[2]

In 1996, I went to a conference at Urbana-Champaign to deliver a presentation on Yuri Andrukhovych's *Perverzion*, and since then I have been paying close attention to postmodernist practice, which has profoundly changed Ukrainian literature. Observing these processes, I witnessed how postmodernism altered, provoked resistance, weakened, degenerated, and so forth. At the threshold of the new century, Ukrainian postmodernism was over. For me, it was the end of a whole period that was now becoming history in our presence.

This is the spirit from which in *The Post-Chornobyl Library* was born; its thesis and rough conception ripened in 1998. In 2001–2002, I spent eight months at Harvard on a scholarship, and my project was called *The Post-Chornobyl Text: Ukrainian Literary Postmodernism*. However, it happened I dedicated most of my time there to writing a monograph about Olha Kobylianska, which was published in 2002. Nevertheless, while still at Harvard, I managed to prepare some notes for my post-Chornobyl text. A couple of years later, I developed them into *The Post-Chornobyl Library*. I include these autobiographical notes here to show that postmodernism was not merely an academic interest but part of my personal story, a motif in my life.

---

2   John Fizer, review of *The Emerging Word: Discourse of the Early Ukrainian Modernism: A Postmodern Interpretation*, by Tamara Hundorova, *Slavic and East European Journal* 42, no. 3 (1998): 571–73.

In the first edition of *The Post-Chornobyl Library*, I emphasized the process of the unfolding of postmodernist Ukrainian literature—its carnivalesque (Bu-Ba-Bu) and apocalyptic (Izdryk, Taras Prokhasko) stages. There is a chronological shift between them, since the second comes a little later. But what is most important is that the carnivalesque version is more closely tied with the inversion of totalitarian signs and channels vital energy toward play, aesthetics, and masquerade. In other words, it is not free from bohemian escapism. The second version, on the other hand, is directed toward the micronarratives of a split consciousness and disrupted history, and on the recombination of the narrative traces of human presence. The focus is on the morphology of presence of the individual in the postapocalyptic world, in other words.

I tried to make clear that not only groups and vectors—the Stanislavian and the Zhytomerian—define the ideological field of Ukrainian postmodernism, but also that it has various stages and phases; and that Ukrainian postmodernism defines the whole literary situation of the 1990s and influences authors who are seemingly remote from it and deem themselves its opponents (like Yevhen Pashkovsky and Viacheslav Medvid, for instance). It was also important for me to demonstrate that postmodernism is by no means a homogenous phenomenon—that there are various postmodernist practices: national practices; generic practices; and the inventions of individual writers (such as Umberto Eco and Julian Barnes, John Fowles and Milorad Pavić, Viktor Pelevin and Yuri Andrukhovych or Izdryk, and many others).

I also tried to show that there are specific stages in the development of Ukrainian postmodernism: first the authors of the Kyiv grotesque, who can be considered pre-postmodernists; then the Bubabists, who represent the emergence of the postmodernist era in Ukrainian literature; and then the arrival of postmodernist metafiction writers (Izdryk and Prokhasko). By the end of the 1990s, postmodernist fiction proper appears. *Pleroma* emerges as an example of a self-referential encyclopedia, and a new, clearly postmodernist wave in Ukrainian literature—intertextual and polymorphous—comes to life. I call this type of postmodernism "apocalyptic." At the same time, one more version of postmodernism takes shape—the one represented by Serhiy Zhadan, whose work is a hybrid of postmodernism, the avant-garde, and pop culture. The work of Yuriy Tarnawsky is a sample of the Western type of postmodernism, which develops on the basis of avant-gardism. Popular postmodernism experiences the transmission of postmodernist techniques of intertextuality and fictitiousness into the genres of popular culture. Representative of this is the work of Yuri Vynnychuk's novel *Malva Landa*—a Joycean work, radically rewritten with the help of decadence, pastiche (from *The Odyssey* to Andrukhovych), and trash. This type of popular postmodernism includes also various forms of fantasy, which is, to use Andrukhovych's description, "a kitschish-masscultish branch within the postmodernist situation" ("The Return of Literature").

Thus, Ukrainian postmodernism is represented by the following types: the grotesque (Volodymyr Dibrova, Les Podervianskyi, and Bohdan Zholdak); the carnivalesque (Bu-Ba-Bu); the feminist (Oksana Zabuzhko); the apocalyptic (Izdryk, Taras Prokhasko); the pop-cultural (Serhiy Zhadan) and the popular (Yuri Vynnychuk).

Later, after *The Post-Chornobyl Library* had been published, while teaching a course on postmodernism at Taras Shevchenko National University in Kyiv, I determined for myself the national origins of Ukrainian postmodernism: the avant-garde and modernism of the 1920s, from Mykhail Semenko to Maik Yohansen and V. Domontovych; Ukrainian whimsical fiction from Oleksandr Ilchenko's *Cossack Mamai* to Volodymyr Drozd's and Valery Shevchuk's fiction; the literature of the diaspora—such as the existential, metafictional prose of Emma Andijewska, the historical, adventure metafiction of Yurii Kosach's *Regina Pontica*, and Andrukhovych's "novel-forerunner"; and the radical avant-gardism of the New York School, represented, in particular, by Yuriy Tarnawsky.

Today, talk about various types of postmodernism is common: first and second waves; early, late, classic (deconstructive), and neoclassic (constructive) stages. Bertens and Fokkema call the second wave of postmodernism "political."[3] Tomo Virk even points out a reverse process of the development of postmodernism in Slavic literatures as compared with Western literatures. His hypothesis is predicated upon the fact that Western postmodernism first emerged in metafictional and deconstructive versions, and only later in a political form. By contrast, Slavic literatures followed a reverse order.

Initially, works of early postmodernism such as Milorad Pavić's *Dictionary of the Khazars*, Danilo Kiš's *A Tomb for Boris Davidovich* in Serbian literature, Drago Jančar's Slovenian fiction, and Andrei Bitov's Russian *The Pushkin House* had an explicitly political tendency. Only later were they superseded by metafictional and intertextual writing. Tomo Virk concludes that "political involvement is a characteristic feature of early postmodernism in a number of Slavic literatures."[4]

---

3   Bertens and Fokkema, eds., *International Postmodernism*, 34.
4   Tomo Virk, "'Politicheskii' postmodernizm v proze nekotorykh slavianskikh literatur," *Slavianovedenie* 6 (2006): 60.

This hypothesis seems quite convincing. In any event, we can hardly deny the fact that the first authors who use postmodernist techniques in the late Soviet period do not simply borrow from the West: they create forms that allow them to slip by official socialist realism and properly explore how to liberate the individual, society, and history from totalitarian domination (in other words, from totalitarian grand narratives). It can be said that the works of Volodymyr Dibrova, Bohdan Zholdak, and Les Podervianskyi have a clearly political tendency, which lies first and foremost in their reflection of the absurdity and monstrosity of totalitarianism. The following wave of postmodernists—the Bubabists—recorded the liberation of the individual from the ties of the totalitarian mindset with the help of jocular carnivalesque culture. They too were not devoid of political involvement, and anti-colonial and anti-totalitarian motifs were often an essential part of their work. Only later did the political component fade and the role of metafictionality grow. Hence, it is no coincidence that probably the apotheosis of Bubabist postmodernism is Andrukhovych's novel *Perverzion* (1996), the last work in his Bubabist trilogy. The feminist novel, represented by Oksana Zabuzhko, is also a type of political postmodernism.

Beside political involvement, emotionalism is another characteristic of Slavic versions of postmodernism. More precisely, there is a certain sensibility that does not fully succumb to deconstructionist reduction. This suggests that in Slavic postmodernism the elimination of the subject does not occur—that is, the decentering of the subject is absent. The emotions of a character, a sentimental narrative, melancholy loss, sensual and corporeal travel, narcissism, and so forth—these morphemes are typical in Ukrainian postmodernist authors. Ultimately, reflecting on postmodernism in "The Return of Literature," which serves as a coda to *Pleroma* and argues against Volodymyr Yeshkilev (one of the leading "Kulturträgers" of Ukrainian postmodernism), Andrukhovych makes three points that undermine classic postmodernism itself: about the return of literature (as a grand narrative); about the presence of the author as an individual (rather than a textual function); and about literature as ideal "spiritual and emotional" communication ("the necessity of understanding and intimacy," "empathy and mutual penetration").[5]

Considering the unique development of postmodernism in Ukraine, it is important to take into account that it was perceived there predominantly (1) as a product of Western culture, (2) as theoretical-philosophical rather than aesthetic-literary (and the theory itself was limited to Eco and Pavić), and (3)

---

5   Iurii Andrukhovych, "Povernennia literatury," in Ieshkiliev and Andrukhovych, eds., *Pleroma. Mala ukraïns′ka entsyklopediia aktual′noï literatury*, 19.

as a universal, essentially homogenous style. Nevertheless, in post-totalitarian Europe, postmodernism showed its multileveled and multifunctional quality. This revealed itself also in (4) a hybridity of forms, and (5) a saturation with ideological categories.

Epstein states that conceptualism—the Russian version of Western postmodernism—"might be called a meta-ideological approach to art, or, meta-aesthetic approach to ideology."[6] This second approach reveals itself in the quite broad functioning of sots-art and in the use of conceptualism as a tool for undermining trust in "reality," since the specificity of conceptualism lies in showcasing the emptiness of signs. Debunking ideologically constructed reality through empty signs, ironic repetitions, and pastiche in post-Soviet space plays an exceptionally important role and begets the phenomenon of Russian conceptualism. Some elements of conceptualism are also present in the works of the authors of the Kyiv Ironic School (Volodymyr Dibrova, Bohdan Zholdak, and Les Podervianskyi). In general, empty signs and the intensified intrusion of the fantastic into mimetic-realistic narratives perform the function of undermining trust in "reality," particularly in Slavic literatures, as is best demonstrated in Serbian postmodernism (Milorad Pavić).[7]

It may seem interesting that discussions about Ukrainian postmodernism had a predominantly ideological character. For nationalist critics, postmodernism embodied "rotten demo-liberalism" and a borrowed phenomenon ("Ostensibly a certain group of authors came to an agreement: 'We shall be postmodernists, because they write like this in the West, and since we love everything Western, we are going to imitate this too,'" ironized Andrukhovych). More sophisticated critics thought that postmodernism in Ukrainian literature was impossible for several reasons: it requires the context of postindustrial capitalism; postmodernism is mainly (or even entirely) an ideological phenomenon—a reaction to the Soviet version of modernization; literature should be regarded, first of all, from the point of view of post-genocide; and, finally, there is no modernist canon from out of which postmodernism can develop.

Nevertheless, postmodernism effected a revolutionary shift in the Ukrainian cultural mindset. It brought new models of cultural behavior, new modes of constructing identity, new worldviews, and new kinds of cultural consumption. Carnival was a way to win back individual freedom, and performativity was a

---

[6] Mikhail Epstein, "The Philosophical Implications of Russian Conceptualism," *The Journal of Eurasian Studies* 1, no. 1 (January, 2010): 65.

[7] See Dagmar Burkhard, "Culture as Memory: On the Poetics of Milorad Pavic," *The Review of Contemporary Fiction* 18, no. 2 (1998): 164.

form of involvement with reality itself; linguistic play was a tool for the construction of self-identity; masks and fragmentation (also) allowed the individual to recover their freedom; the classics could be rewritten; new images of Europe appeared; gendered writing emerged; and monadic existence was revealed—postmodern homelessness.

Postmodernism was born in the national literatures of the Second and Third worlds as an encounter between the familiar and the strange, and it was postmodern techniques that provided the opportunity to modernize national literary traditions. Therefore, when it comes to the origins of postmodernism in various national literatures, it is worthwhile heeding the words of Nobel Prize winner Orhan Pamuk who, answering a question about the roots of his own postmodernism, offered the following:

> Borges and Calvino liberated me. The connotation of traditional Islamic literature was so reactionary, so political, and used by conservatives in such old-fashioned and foolish ways, that I never thought I could do anything with that material. But once I was in the United States [1985—T. H.], I realized I could go back to that material with a Calvinoesque or Borgesian mind frame.[8]

Ukrainian literature survived postmodernism and became richer for it; and the Ukrainian authors integrated in postmodernism became "ill" with it in order to grow stronger. "Postmodernism can truly be considered an illness," Andrukhovych notes. "But what can we regard as 'healthy' in literature (except for socialist realism?)"[9]

However, one cannot but notice that contemporary authors prefer not to be associated with postmodernism and have a rather abstract understanding of it. For instance, Petro Midianka writes that he does not feel like a postmodernist and that in Ukrainian literature "it is pretty hard to find 'postmodernism' of the pure kind." In fact, he associates postmodernism itself with "the glass bead game," buffoonery or parody, and comic or absurdist effects.[10]

Marianna Kiyanovska, admitting that postmodernist discourse has emerged in Ukraine, hides behind terminological vagueness, which seems to get in the way of serious talk about postmodernism of the Ukrainian kind. She

---

8   Orhan Pamuk, "The Paris Review Interview," in Orhan Pamuk, *Other Colors: Essays and a Story*, trans. Maureen Freely (New York: Alfred A. Knopf, 2007), 367.
9   Andrukhovych, "Povernennia literatury," 18.
10 Petro Midianka, "'Ia ne pyshu postmodernizm'. Ia pyshu te, shcho vidchuvaiu …'" Interviu z Petrom Midiankoiu z nahody rishennia pro prysudzhennia iomu Shevchenkivs'koï premiï, http://zakarpattya.net.ua.

rhetorically asks: " Herasymyuk, Rymaruk, Bilotserkivets, Malkovych—not postmodernists? What about Ivan Andrusiak? Halyna Kruk? Slyvynsky? And is the postmodernist Andriy Bondar—postmodernist?"[11]

For many Ukrainian authors, postmodernism has turned into a name for poor taste, relativity, fashion, and a lack of spirituality. For example, Ihor Pavliuk, in accord with the tastes and recipes of culinary art a là Poplavsky, proposes the following components to identify a postmodernist work: "[a] sex shop, built from the construction debris of an old ruined temple and dog kennels with the inclusion of, let's say, car doors and windows (hatches) from an atomic submarine. … Or—rich proper Ukrainian borsch, with an addition of exquisite French perfumes Chanel. …"[12]

Contrary to postmodernism, the idea of "a deep national style in all of its artistic and socio-political manifestations"[13] for the representatives of nativism acquires a sacred meaning and appeals to images of the national spirit-temple and "immanent archetypes of God, Ukraine, and Freedom."[14] Ultimately, it is easy to agree with the idea that "in Ukrainian media we find more publications that evaluate Ukrainian postmodernism as a negative phenomenon than as a positive one."[15]

In the West, postmodern culture was clearly outlined already in the 1980s in the works of Jean-François Lyotard, Fredric Jameson, and Jean Baudrillard. They identified the defining features of postmodernism: a distrust of those grand narratives that symbolize the ideals of social and individual liberation; an overproduction of cultural signs; and the emergence of simulacra, which precede reality. Turning toward aesthetics, Lyotard pointed to one of postmodernism's general features when he observed that "the disappearance of the Idea of rationality and freedom" begets a special style, which is based on a peculiar "bricolage," with an excess of quotations borrowed from the previous styles and periods, classic and modern.[16]

Postmodernism, born in the West and moving to the East, acquires new characteristics by superimposing itself onto various national cultures. It is an

---

11 Marianna Kiianovs'ka, " Postmodernizm i natsional'no-dukhovna identyfikatsiia," *Dzerkalo tyzhnia* 2 (2007), http://gazeta.dt.ua.
12 Ihor Pavliuk, "Kryza postmodernizmu: shcho dali?," http://bukvoid.com.ua.
13 Ibid.
14 Petro Ivanyshyn, "Postmodernizm i natsional'no-dukhovna identyfikatsiia," *Ukraïns'ki problemy* 1–2 (1999): 123–30.
15 Daryna Popil', " Ukraïns'kyi postmodernizm u dzerkali mediia," *Visnyk L'vivs'koho universytetu. Seriia zhurnalistyky* 34 (2011): 54–56.
16 Jean-Francois Lyotard, "Note on the Meaning of 'Post,'" in *Postmodernism: A Reader*, ed. Thomas Docherty (Routledge: London and New York, 1993), 47.

undeniable fact that Ukraine happened to be in the forcefield of postmodernism and that the period of postmodernism is already in the past. What intersecting trends define the postmodernist constellation in Ukrainian culture? This is the question that, to my mind, Ukrainian intellectuals have not yet answered. There is a lack of ideas, notions, and metaphors that would describe the Ukrainian version of postmodernity. However, the main tone of, and the main tensions within, Ukrainian discussions about postmodernism have clearly taken shape: national, essentialist criticism has stood in active and complete opposition to postmodernist criticism. The positive (as it were) aspects of postmodernism, and positive defenses of it, with rare exceptions have not been discussed. Moreover, a distrust of postmodernism can be seen on all levels: from determined antimodernists to the creators of postmodernism themselves. There is a widespread opinion that postmodernism is fundamentally alien to the Ukrainian mentality; that it destroys national naturalness with cosmopolitanism and globalism; that its significance may only lie in its confrontation with socialist realism; that it means the "surrender of intellectuals before the totality of mass media and mass culture" (Larysa Briukhovetska); that de-canonization does not stand up to the task of the nation building; and that postmodernism is impossible in Ukrainian culture because the national canon has not taken its final shape there.

The most expressive trend criticizing postmodernism was, ultimately, catastrophism—Oxana Pachlovska formulated this position back in 2001—which maintained that "in the greater part of Eastern Europe, in particular in Ukraine, Catastrophe is able to stop Culture if Culture does not try to stop Catastrophe."[17] As a consistent critic of postmodernism, Pachlovska found a few more shortcomings in the Ukrainian reception of postmodernism. First of all, it is a consequence of "an uncritical appropriation of Western culture" and "an inorganic adoption of its historical dynamics" (in other words, "so-called postmodernism is one of the outcomes of the non-meeting between Ukraine and Europe").[18] Secondly, "for the Orthodox East of Europe the problem of 'postindustrial society' is not relevant" because "the orthodox version of Christianity, with its mysticism and hierarchical stagnation, for many

---

17 Oksana Pakhliovs'ka, *Sytuatsiia postmodernizmu v Ukraïni. Kruhlyi stil*, http://www.ukma.kiev.ua/2001/6/postmodern.html.
18 Ibid.

centuries has blocked the development of technical and natural sciences."[19] In other words, informational technologies ought to have omitted Ukraine entirely. Thirdly, for Pachlovska, "Slavic tradition," in accord with old essentialist views, is the ground for the idea that Ukraine should not have "a postmodern civilization." Postmodernism can only be a "stylistic practice," alienated from the "socio-historical content of its era."[20] And, finally, she juxtaposes postmodernism with evolutionary and positivistic versions of the national culture because, she is convinced, postmodernism in Ukraine can only lead to "a persistent struggle against native culture and political history," while "Ukrainian culture and history only begin to take shape on the theoretical level as evolutionary systems."[21] We can conclude that anti-postmodernist criticism is based on the principle of the "naturalness" of the national culture, an emphatic idea of its evolutionary development, Slavic essentialism, and an ideology that supports the dissident intellectual, social responsibility of an individual, and appeals to catastrophism as a new national idea.

At their most general, these arguments, if devoid of politicization and total negativity, are rooted in the internal tensions of postmodernism itself. The discussions that occur, so to speak, in the post-postmodern period, testify to an important shift from globalization theories, which grounded the assumptions of the theoreticians of postmodernism, to issues of a local and national character. Almost all (including Western) philosophical thought is now returning to questions that were dealt with by Second and Third World cultural critics, who perceived postmodernism as both a challenge and an opportunity to look anew at the social and cultural conditions of their existence.

In 2006, Fredric Jameson described the post-postmodern situation as "a struggle of universality against singularity," in which "the most interesting consequence of globalization is a revision of the boundaries of national cultures and the phenomenon of their interaction."[22] This very question about the role of the nation and growing national self-awareness, as well as how to combine national culture with the postmodernist ideology of globalization of consumerism, adds a special pungency to the discussions of Ukrainian intellectuals about postmodernism. Quite a few references appeal to the post-totalitarian and postcolonial

---

19 Ibid.
20 Ibid.
21 Ibid.
22 Fredric Jameson cited in Vladimir Mironov, *Istoriia kak tsep´ katastrof*, http://yarcenter.ru/articles/society/istoriya-kak-tsep-katastrof-622/.

character of postmodernism. Roksana Kharchuk, giving an outline of various discussions about postmodernism in Ukraine, which she identifies with the phenomenon of "the total reevaluation of values," sees the essence of postmodernism in the Eastern European territories in the fact that it "found food in those societies that were passing from the totalitarian model to democracy, from the singular socialist realist canon to the freedom of artistic creation and cultural polyphony."[23] Oxana Pachlovska, on the contrary, insists on the "conservatism and inertia of our cultural situation," which "localized" the phenomenon of postmodernism and adapted the global phenomenon for national needs, having begotten "the 'local' perspective itself, cultivating and encoding it as a mere 'neocolonial' space."[24]

Ukrainian postmodernism "explodes" on the margins of Europe—explodes so powerfully and originally that there is a shortage of Western theoretical signifiers to give it a full description. Ukrainian postmodernism develops out of an experience of epistemological drama, in which the individual of the end of the twentieth century encountered the excess of the archive containing the records of the whole world, as well as the national history and its culture. This individual attempted ironically and somewhat playfully to free itself from this excessiveness by creating its own cultural intertext. This intertext was created selectively and subjectively and had an existential character. It was built on principles other than those of existing models of cultural creation: it was not a romantic ideal of the creation of the world-organism; it did not presuppose a full separation from the canonized past, like the avant-garde; and it did not, like modernism, juxtapose the sublimated and aestheticized image of the world in the text with reality. Postmodernism led to a unique recreation of past experience and was based on ironic self-awareness; and in this process both components were equally important—recreation as well as creation. In addition, the postmodernist mindset was not an aesthetic utopia, which symbolized dreams of liberation from the total and rational ideologies formulated in the period of the Enlightenment, but developed as a socio-psychological and cultural practice that not only promised, but actually discovered, new possibilities of experience and intellectual transformation, and begot other forms of creative activity.

At the same time, it is hard to disagree that postmodernism was a fashion, an experimental artistic field, testing and unfolding new modes of cultural production and reception. The development of postmodernism also depended

---

23 Kharchuk, *Suchasna ukraïns′ka proza. Postmodernyi period*, 8.
24 Pakhliovs′ka, *Sytuatsiia postmodernizmu v Ukraïni*, Kruhlyi stil, http://www.ktm.ukma.kiev.ua/2001/6/postmodern.html.

upon new technological discoveries, which sharply changed the idea of the author and changed how one creates and reads texts. Along with the passing of the postmodern situation, the very techniques of postmodernism, which were actively developing in the course of a few decades, became mainstream. Advertising, the movies, PR technologies, pop culture, and mass literature all show the influence of postmodernism.

The culture and art that have superseded postmodernism do not have a unanimous name: there are discussions about performativity, postmillenarianism, transmodernism, post-realism, new sentimentalism, email-modernism, and IMmodernism. The main question is, is it possible to overcome postmodern irony, and what can it be replaced with: faith, dialogue, interaction, performance, new sincerity? This question remains without an answer or, to be more precise, has multiple answers. What is obvious is that today's aesthetics, after postmodernism, aim at gathering—from broken fragments, discarded debris, and discredited forms—a new art, reviving the authority of high art yet simultaneously preserving the high status of popular art. In other words, new literature wants to create order from chaos or, simply, find form in the middle of chaos.

In the course of some five or six years, the world has drastically changed. It is not postmodern anymore—that is, it is not a world governed by ironic self-awareness and by the playful individual. It has become a world in which wars have no end, fear of terrorism nests down, civilizations clash, there is resentment and protest, and in which global and ecological catastrophes are ever more threatening. The postmodern mindset overcame resistance by suggesting in its place models of decentralization, deterritorialization, and the rhizomatic play of differences. It thus tried to supplant antagonism, aggression, dissatisfaction, and alienation from the other and from "the Real" (understood in Lacanian terms as the inaccessible thing-in-itself). Globalization not only causes a homogeneity of cultural codes, as did postmodernity, it also aggravates the opposition of "self" and "other."

Contemporary literature is not postmodern anymore. It is entirely different. However, postmodernism's dream has come to fruition. In Ukraine too—and not only in literature and culture. The phenomenon of the Orange Revolution is the last act of Ukrainian postmodernism, and symbolized its completion and conclusion. Maidan is a carnival, which turned from a cultural into a political action. It was only because postmodernism had taught people playful linguistic behavior and accustomed them to the carnivalesque—which for the post-Soviet individual had served as a form of liberation—that Ukrainians came out to join Maidan rather than taking up arms.

Bibliography

Abrams, M. H. "Apocalypse: Theme and Variations." In *The Apocalypse in English Renaissance Thought and Literature: Patterns, Antecedents, and Repercussions*, edited by C. A. Patrides and J. Wittereich. Manchester: Manchester University Press, 1984.

Adamovich, Ales´ (Ales). *Imia sei zvezde Chernobyl´*. Minsk: Kovcheg, 2006.

Adorno, Theodor. *Negative Dialectics*. Translated by E. B. Ashton. London and New York: Routledge, 1973.

Aheieva, Vira. *Poetesa zlamu stolit´. Tvorchist´ Lesi Ukraïnky v postmodernii interpretatsiï*. Kyiv: Lybid´, 1999.

Alexievich, Svetlana. *Voices from Chernobyl: The Oral History of Nuclear Disaster*. Translated and with a preface by Keith Gessen. New York: Picador, 2006.

Anderson, Benedict. *Imagined Communities: Reflections of the Origin and Spread of Nationalism*. London: Verso, 1983.

Andrukhovych, Iurii, Andrii Bondar, and Serhii Zhadan. *Maskul´t. Eseï ta poeziï z novykh knyzhok*. Kyiv: Krytyka, 2003.

Andrukhovych, Iurii (Yuri). "Ave, 'Kraisler'! Poiasnennia ochevydnoho." *Suchasnist´* 5 (1994): 8–18.

———. *Dezoriientatsiia na mistsevosti. Sproby*. Ivano-Frankivsk: Lileia-NV, 1999.

———. "Autobiohrafiia." In Iurii Andrukhovych, *Rekreatsiï, Romany*, 31. Kyiv: Chas, 1997.

———. *Rekreatsiï. Romany*, 33–112. Kyiv: Chas, 1997.

———. "Povernennia literatury." In *Pleroma. Mala ukraïns´ka entsyklopediia aktual´noï literatury*, edited by Volodymyr Ieshkiliev and Iurii Andrukhovych, vol. 3, 14–21. Ivano-Frankivsk: Lileia-NV, 1998.

———. *Recreations*. Translated and with an introduction by Marko Pavlyshyn. Edmonton and Toronto: CIUS, 1998.

———. "Vona robyt´ mynule zhyvym i nezavershenym." *Komentar* 2 (2003): 4–5.

———. *Perverzion*. Translated and with an introduction by Michael M. Naydan. Evanston: Northwestern University Press, 2005.

———. *Twelve Circles*. Translated by Vitaly Chernetsky. New York: Spuyten Duyvil, 2005.

———. "Ameryka. Vidkryttia No 1001. Zamist´ peredmovy." In Iurii Andrukhovych, *Den´ smerti pani Den´*, 11–12. Kharkiv: Folio, 2006.

———. "Apologiia blazenady (Dvanadtsiat´ tez do sebe samoho)." In *"Bu-Ba-Bu" (Iurii Andrukhovych, Oleksandr Irvanets´, Viktor Neborak): Vybrani tvory: Poeziia, proza, eseïstyka*, 23–24. Lviv: Piramida, 2007.

———. "Zahybel kotliarevshchyny, abo zh Bezkonechna podorozh u bezsmertia." In *"Bu-Ba-Bu" (Iurii Andrukhovych, Oleksandr Irvanets´, Viktor Neborak): Vybrani tvory: Poeziia, proza, eseïstyka*, 101. Lviv: Piramida, 2007.

———. *The Moscoviad*. Translated by Vitaly Chernetsky. New York: Spuyten Duyvil, 2008.

———. *My Final Territory: Selected Essays*. Translated by Mark Andryczyk and Michael M. Naydan. Toronto: University of Toronto Press, 2018.

Arndt, Melanie. "Memories, Commemorations, and Representations of Chernobyl: Introduction." In *Anthropology of East Europe Review* 30, no. 1 (2012): 1–12. http://scholarworks.iu.edu/journals/intex.php/aeer/article/view/2009. Accessed February 20, 2019.

Ascari, Maurizio. *Literature of the Global Age: A Critical Study of Transcultural Narrative*. Jefferson: McFarland & Company Publishers, 2011.

Bakhtin, Mikhail. *Rabelais and His World*. Translated by Helene Iswolsky. Bloomington: Indiana University Press, 1984.

Baran, Ievhen. "Literaturni devianosti: pidsumky i perspektyvy." *Kurier Kryvbasu* 116 (1999): 3.

Barth, John. *The Floating Opera*. Garden City: Doubleday, 1967.

———. "THE END: An Introduction." In *On with the Story*. New York: Little, Brown, and Company, 1996.

Baudrillard, Jean. *Simulacra and Simulation*. Translated by Sheila Faria Glaser. Michigan: Michigan University Press, 1994.

———. *The Illusion of the End*. Translated by Chris Turner. Stanford: Stanford University Press, 1994.

———. "Fatal Strategies." In *Selected Writings*, edited and with an introduction by Mark Poster. Stanford: Stanford University Press, 2001.

Ber, Viktor. "Zasady poetyky (Vid 'Ars poetica' I. Malaniuka do 'Ars poetica' doby rozkladenoho atoma)." *MUR* 1 (1946). 7–23.

Berger, James. *After the End: Representations of Post-Apocalypse*. Minneapolis: University of Minnesota Press, 1999.

———. "Twentieth-Century Apocalypse: Forecasts and Aftermaths." *Twentieth Century Literature* 46, no. 4 (2000): 387–95.

Bertens, Hans. "The Postmodern Weltanschauung and Its Relation to Modernism: An Introductory Survey." In *A Postmodern Reader*, edited by Joseph Ntoli and Linda Hutcheon, 25–70. New York: State University of New York Press, 1993.

Bertens, Hans, and Douwe Fokkema. *International Postmodernism: Theory and Literary Practice*. Amsterdam and Philadelphia: J. Benjamins, 1997.

Biden, Joseph R., Jr. "A Comprehensive Nuclear Arms Strategy." http://www.voltairenet.org/article164856.html. Accessed February 20, 2019.

Bilotserkivets´, Natalka (Bilotserkivets, Natalka). "Bu-Ba-Bu ta inshi. Ukraïns´kyi neoavangard: portret odnoho roku." *Slovo i chas* 1 (1991): 42–52.

Bloom, Harold. *The Western Canon: The Books and School of the Ages*. New York: Riverhead Books, 1994.

Bondar, Andrii (Andriy). "virsh iakyi nikoly ne perekladut' inshymy movamy." In Iurii Andrukhovych, Andrii Bondar, and Serhii Zhadan, *MASKUL'T: Eseï ta poeziï z novykh knyzhok*, 30–32. Kyiv: Krytyka, 2003.

Braidotti, Rosi. "A Critical Cartography of Feminist Post-Poststructuralism." *Australian Feminist Studies* 20 (2005): 169–80.

"Bu-Ba-Bu." In *Pleroma. Mala ukraïns'ka entsyklopediia aktual'noï literatury*, edited by Volodymyr Ieshkiliev and Iurii Andrukhovych, vol. 3, 35. Ivano-Frankivsk: Lileia-NV, 1998.

"Bubabists'kyi khronopys Viktora Neboraka: shche odna intryha z pryvodu suchasnoï ukraïns'koï literatury." *Literatura plius* 9–10 (2001). www.uap.iatp.org.ua/litplus/lit34-35.php. Accessed February 19, 2019.

Burgos, Fernando. "Postmodernism and Its Sighs in 'Juegos Nocturnes' by Augusto Roa Bastos." In *Postmodernism's Role in Latin American Literature: The Life and Work of Augusto Roa Bastos*, edited by Helene Carol Weldt-Basson, 51–80. New York: Palgrave Macmillan, 2010.

Burkhard, Dagmar. "Culture as Memory: On the Poetics of Milorad Pavic." *The Review of Contemporary Fiction* 18, no. 2 (1998): 164–71.

Chyzhevs'kyi, Dmytro (Chyzhevsky, Dmytro). *Istoriia ukraïns'koï literatury. Vid pochatkiv do doby realizmu*. Ternopil: Femina, 1994.

Coogan, Michael D., ed. *The New Oxford Annotated Bible*, 3rd ed. Oxford: Oxford University Press.

Crow, Thomas. *The Rise of the Sixties: American and European Art in the Era of Dissent 1955–69*. London: Calmann and King Ltd., 1996.

Danylenko, Volodymyr. "Zolota zhyla ukraïns'koï prozy." In *Vecheria na dvanadtsiat' person*, edited by Volodymyr Danylenko, 5–12. Kyiv: Heneza, 1997.

———. *Misto Tirovyvan*. Lviv: Kal'variia, 2001.

———. *Lisoryb u pustyni. Pys'mennyk i literaturnyi protses*. Kyiv: Akademknyha, 2008.

Deleuze, Gilles. *The Logic of Sense*. Translated by Mark Lester with Charles Stivale, edited by Constantin V. Boundas. London: The Athlon Press, 1990.

Derrida, Jacques. *Margins of Philosophy*. Chicago: University of Chicago Press, 1982.

———. "No Apocalypse, Not Now (full speech ahead, seven missiles, seven missives)." *Diacritics* 14, no. 2 (1984): 20–31.

———. "Of an Apocalyptic Tone Recently Adopted in Philosophy." *Oxford Literary Review* 4, no. 2 (1984): 3–37.

———. *The Post Card: From Socrates to Freud and Beyond*. Translated and with an introduction and additional notes by Alan Bass. Chicago: The University of Chicago Press, 1987.

Dibrova, Volodymyr. *Peltse and Pentameron*. Translated by Halyna Hryn, foreword by Askold Melnyczuk. Evanston: Northwestern University Press, 1994.

———. "Prynts Hamlet khams'koho povitu." *Krytyka* 5 (2001): 27.

———. *Vybhane*. Kyiv: Krytyka, 2002.

Drach, Ivan. "Chornobyl's'ka madonna. Poema." *Vitchyzna* 1 (1988): 46–62.

Drahomanov, Mykhailo. "Literatura rosiis'ka, velykorosiis'ka, ukraïns'ka i halyts'ka." Lviv: T-vo im. Shevchenka, 1874.

Dziuba, Ivan. "Metod—tse nasampered rozuminnia." *Literaturna Ukraïna* 3 (January 25, 2001): 3.

Eaglestone, Robert. "From Behind the Bars of Quotation Marks: Emmanuel Levinas's (Non) Representation of the Holocaust." In *The Holocaust and the Text: Speaking the Unspeakable*, edited by A. Leak and G. Paizis, 97–108. London: Macmillan Press Ltd., 2000.

Eagleton, Terry. "Estrangement and Irony in the Fiction of Milan Kundera." In *The Eagleton Reader*, edited by Stephen Regan, 90–96. Oxford: Blackwell, 1998.

Epstein, Mikhail (Epshtein, Mikhail). *After the Future: The Paradoxes of Postmodernism and Contemporary Russian Culture*. Amherst: The University of Massachusetts Press, 1995.

———. "Posle karnavala, ili Vechnyi Venichka." In Venedikt Erofeev, *Ostav´te moiu dushu v pokoie: (Pochti vse)*, 18–19. Moscow: Izd-vo HGS, 1995.

———. "Informatsyonnyi vzryv i travma postmoderna." http://old.russ.ru/journal/travmp/98-10-08/epsht.htm. Accessed February 19, 2019.

———. "Transculture in the Context of Contemporary Critical Theories." In Ellen E. Berry and Mikhail Epstein, *Transcultural Experiments: Russian and American Models of Creative Communication*, 80–81. New York: St. Martin's Press, 1999.

———. "The Philosophical Implications of Russian Conceptualism." *Journal of Eurasian Studies* 1, no. 1 (2010): 64–71.

Ferguson, Frances. "The Nuclear Sublime." *Diacritics* 14, no. 2 (1984): 4–10.

Fialkova, Larisa. "Chornobyl's Folklore: Vernacular Commentary on Nuclear Disaster." *Journal of Folklore Research* 38, no. 3 (2001): 181–204.

Fizer, John. Review of *The Emerging Word: Discourse of the Early Ukrainian Modernism: A Postmodern Interpretation*, by Tamara Hundorova. *Slavic and East European Journal* 42, no. 3 (1998): 571–73.

Fokkema, Douwe. "Metamorfoza postmodernizmu. Europejska recepcja amerykańskiego pojęcia." In *Postmodernizm w literaturze i kulturze krajów Europy Środkowo-Wschodniej*, edited by H. Janaszek-Ivaničková and Douwe Fokkema. Katowice: Śląsk, 1995.

———. "The Semiotics of Literary Postmodernism." In *International Postmodernism: Theory and Literary Practice*, edited by Hans Bertens and Douwe Fokkema, 15–42. Amsterdam and Philadelphia: J. Benjamins, 1997.

Foster, Hal. *The Anti-Aesthetic: Essays on Postmodern Culture*. Port Townsend: Bay Press, 1983.

———. *The Return of the Real: The Avant-Garde at the End of Century*. Cambridge, Mass.: MIT Press, 1996.

Freud, Sigmund. "The Uncanny." http://web.mit.edu/allanmc/www/freud1.pdf. Accessed February 20, 2019.

Giles, Paul. "Sentimental Posthumanism: David Foster Wallace." *Twentieth-Century Literature* 53, no. 3 (Fall 2007): 327–44.

Goldfarb, Jeffrey C. *Beyond Glasnost: The Post-Totalitarian Mind*. Chicago: The University of Chicago Press, 1989.

Grois, Boris (Groys, Boris). "Polutornyi stil´: mezhdu modernizmom i postmodernizmom." *Novoe literaturnoe obozrenie* 15 (1995): 44–53.

———. "A Style and a Half: Socialist Realism between Modernism and Postmodernism." In *Socialist Realism without Shores*, edited by Thomas Lahusen and Evgeny Dobrenko, 76–90. Durham and London: Duke University Press, 1997.
Hassan, Ihab. "From Postmodernism to Postmodernity: The Local/Global Context." *Philosophy and Literature* 25, no. 1 (2001): 1–13.
Hnatiuk, Olia (Ola). "Avantiurnyi roman i povalennia idoliv." In Iurii Andrukhovych, *Rekreatsiï. Romany*, 9–26. Kyiv: Chas, 1997.
———. "Wirtualna perwersja. Z Jurijem Andruchowyczem rozmawia Ola Hnatiuk." *Dekada Literacka* 19 (1998): 6–7.
———. *Proshchannia z imperiieiu: Ukraïns′ki dyskusiï pro identychnist′*. Kyiv: Krytyka, 2005.
Hoberek, Andrew. "Introduction: After Postmodernism." *Twentieth Century Literature* 53, no. 3 (2007): 233–47.
Horbachov, Mykhailo (Gorbachev, Mikhail). "Chornobyl′s′kyi perelom." *Den′* 65 (2287), April 18, 2006, 1.
Hrabovych, Hryhorii (Grabowicz, George G.). "Semantyka kotliarevshchyny." In Hryhorii Hrabovych, *Do istoriï ukraïns′koï literatury*, 316–32. Kyiv: Krytyka, 1997.
Hrinchenko, Borys. *Narodnyie spektakli*. Chernihiv: Tipografiia Gubernskogo Zemstva, 1900.
Hrycak, Alexandra. "The Coming of 'Chrysler Imperial': Ukrainian Youth and Rituals of Resistance." *Harvard Ukrainian Studies* 21, no. 1–2 (June 1997): 63–91.
Huizinga, Johan. *Homo Ludens: A Study of the Play-Element in Culture*. London: Routledge, 2002.
Hundorova, Tamara. "Postmodernists′ka fiktsiia Andrukhovycha z postkolonial′nym znakom pytannia." *Suchasnist′* 9 (1993): 78–83.
———. *ProIavlennia slova: Dyskursiia rannioho ukraïns′koho modernizmu. Postmoderna interpretatsiia*. Lviv: Litopys, 1997.
———. "'Bu-Ba-Bu,' Karnaval i Kich." *Krytyka* 7–8 (2000): 13–18.
———. *Femina melancholica. Stat′ i kul′tura v gendernii utopiï Ol′hy Kobylians′koï*. Kyiv: Krytyka, 2002.
———. "Sotsrealizm iak masova kul′tura." *Suchasnist′* 6 (2004): 52–66.
———. "Post-Chornobyl′: Katastrofizm iak nova natsional′na ideia." *Ukraïns′ka pravda*, April 26, 2012. Accessed February 20, 2019. life.pravda.com.ua/columns/2012/04/26/101231.
Husserl, Edmund. *Logicheskie issledovaniia. Prolegomeny k chistoi logike*. Kyiv: Venturi, 1995.
Iarovyi, Oleksandr (Iarovy, Oleksandr). "Skazhu, iak ie." *Literaturna Ukraïna* 8 (March 1, 2001): 3.
———. "Lyst samomu sobi. Kviten′ 19, 2001." *Literaturna Ukraïna* 15 (April 20, 2001): 3.
Iavorivs′kyi, Volodymyr (Yavorivsky, Volodymyr). "Mariia z polynom pry kintsi stolittia." *Vitchyzna* 7 (1987): 16–139.
Ieshkiliev, Volodymyr, and Iurii Andrukhovych, eds. *Pleroma. Mala ukraïns′ka entsyklopediia aktual′noï literatury*, vol. 3. Ivano-Frankivsk: Lileia-NV, 1998.
Ieshkiliev, Volodymyr (Yeshkilev, Volodymyr). "MP-dyskurs u suchasnii ukraïns′kii literaturi." In *Pleroma. Mala ukraïns′ka entsyklopediia aktual′noï literatury*, edited by Volodymyr Ieshkiliev and Iurii Andrukhovych, vol. 3, 35. Ivano-Frankivsk: Lileia-NV, 1998.

———. *Votstsekurhiia bet. Komentari do "vnutrishn′oï entsyklopediï" romanu Izdryka "Votstsek."* Ivano-Frankivsk: Vydavnytstvo Unikomus, 1998.

———. "Tin′ stanislavs′koho fenomenu." *Literatura plius* 9–10 (1999): 4–5.

Il′nyts′kyi, Oleh (Ilnytzkyj, Oleh). "Transplantatsiia postmodernizmu: Sumnivy odnoho chytacha." *Suchasnist′* 19 (1995): 111–15.

Iovenko, Svitlana. "Zhinka u zoni." In Svitlana Iovenko, *Liubov pid inshym misiatsem*. Kyiv: Ukraïns′kyi pys′mennyk, 1999. 121–223

Irvanets′, Oleksandr (Irvanets, Oleksandr). "Iak vono bulo …" In *"Bu-Ba-Bu" (Iurii Andrukhovych, Oleksandr Irvanets′, Viktor Neborak): Vybrani tvory: Poeziia, proza, eseïstyka*, 17–19. Lviv: Piramida, 2007.

———. "Keske bubabu?" In *"Bu-Ba-Bu" (Iurii Andrukhovych, Oleksandr Irvanets′, Viktor Neborak): Vybrani tvory: Poeziia, proza, eseïstyka*, 27–29. Lviv: Piramida, 2007.

———. "Uroky klasyky." In *"Bu-Ba-Bu" (Iurii Andrukhovych, Oleksandr Irvanets′, Viktor Neborak): Vybrani tvory: Poeziia, proza, eseïstyka*, 223–32. Lviv: Piramida, 2007.

Ivanyshyn, Petro. "Postmodernizm i natsional′no-dukhovna identyfikatsiia." *Ukraïns′ki problemy* 1–2 (1999): 123–30.

Izdryk (Izdryk, Yuri). *Podviinyi Leon. Istoriia khvoroby*. Ivano-Frankivsk: Lileia-NV, 2000.

———. *Wozzeck*. Translated and with introduction by Marko Pavlyshyn. Edmonton: CIUS, 2006.

———. "Biohrafiia artysty." In Izdryk, *Fleshka-2GB*, 223–32. Kyiv: Hrani-T, 2009.

———. "Teoriia vidmovy." In Izdryk, *Take*, 267. Kharkiv: Vydavnytstvo "Klub simeinoho dozvillia," 2009.

Jameson, Fredric. *Postmodernism, or, The Cultural Logic of Late Capitalism*. Durham: Duke University Press, 1991.

———. *The Cultural Turn: Selected Writings on the Postmodern, 1983–1998*. London: Verso, 1998.

Janaszek-Ivaničková, Halina, and Douwe Fokkema, eds. *Postmodernism in Literature and Culture of Central and Eastern Europe*. Katowice: Śląsk, 1996.

Jansen, Monica. *Il dibattito sul post-moderno in Italia. In bilico tra dialettica e ambiguita*. Florence, Franco Cezari, 2002.

———. "Has Postmodernism Ended? Dialectics Revisited (Luperini, Belpoliti, Tabucchi)." In *Postmodern Impegno: Ethics and Commitment in Contemporary Italian Culture*, edited by P. Antonello and R. Gordon, 49–50. Oxford: Peter Lang, 2009.

Johnson, Emily D. "Twenty Years after the Collapse of the Soviet Union: Russian and East European Literature Today." *World Literature Today: Post-Soviet Literature* 86, no. 6 (2011): 32–33.

Keats, Jonathon. "Apocalypse Made Easy." https://www.salon.com/2002/02/07/doomsday. Accessed February 2, 2019.

Kellner, Douglas. *Jean Baudrillard: From Marxism to Postmodernism and Beyond*. Stanford: Stanford University Press, 1989.

———. "The *X-Files* and the Aesthetics and Politics of Postmodern Pop." *The Journal of Aesthetics and Art Criticism* 57, no. 2 (1999): 161–75.

Kermode, Frank. *The Sense of an Ending*. Oxford: Oxford University Press, 1968.

---. "Ending, Continued." In *Languages of the Unsayable: The Play of Negativity in Literature and Literary Theory*, edited by Sanford Budick and Wolfgang Iser, 71–94. New York: Columbia University Press, 1989.

---. *The Sense of an Ending*. Oxford: Oxford University Press, 2009.

Kharchuk, R. B. (Kharchuk, Roksana). *Suchasna ukraïns'ka proza. Postmodernyi period. Navchal'nyi posibnyk*. Kyiv: Vydavnychyi tsentr "Akademiia," 2008.

Kiianovs'ka, Marianna (Kiyanovska, Marianna). "Meni vystachaie intryh u literaturi." *Dzerkalo tyzhnia* 2 (2007). http://gazeta.dt.ua. Accessed February 19, 2019.

Kosovych, Leonid. "Postskript." In Izdryk, *Podviinyi Leon. Istoriia khvoroby*, 176–78. Ivano-Frankivsk: Lileia-NV, 2000.

Kotsiubyns'ka, Mykhailyna (Kotsiubynska, Mykhailyna). *Moï obriï*, vol. 2. Kyiv: Dukh i Litera, 2004.

Kozhelianko, Vasyl' (Kozhelanko, Vasyl). *Defiliada v Moskvi*. Lviv: Kal'variia, 2000.

Kravchenko, Ihor. "Chas diiaty (Literatura iak haluz' ukraïns'koï kul'tury)." *Dnipro* 7–8 (2001): 82–99.

---. "Priorytet Krytyky." *Literaturna Ukraïna* 2 (January 18, 2001): 3.

Kristeva, Julia. "Word, Dialogue, and Novel." In *The Kristeva Reader*, edited by Toril Moi, 34–61. New York: Columbia University Press, 1986.

Kulish, Panteleimon. "Kharakter i zadachi ukrainskoi kritiki." In Panteleimon Kulish, *Tvory v dvokh tomakh*, vol. 2, 517–18. Kyiv: Dnipro, 1989.

---. "Prostonarodnost' v ukrainskoi slovesnosti." In Panteleimon Kulish, *Tvory v dvokh tomakh*, vol. 2, 522–32. Kyiv: Dnipro, 1989.

Kundera, Milan. *The Unbearable Lightness of Being*. Translated by Michael Henry Heim. New York: Harper Perennial, 1991.

Kuprina, N. A. *Iazykovoe soprotivlenie v kontekste totalitarnoi kul'tury*. Ekaterinburg: Izdatel'stvo Ural'skogo universiteta, 1999.

Kvit, Serhii (Serhiy). "U mezhakh, poza mezhamy i na mezhi." *Slovo i chas* 3 (1999): 63–64.

Lemarshand, Frederik. "Topos Chornobylia." *Dukh i Litera* 7–8 (2001): 372–82.

Levinas, Emmanuel. *Entre Nous: Thinking-of-the Other*, translation by Michael Smith and Barbara Harshav. New York: Columbia University Press, 1998.

Lipovetskii, Mark (Lipovetsky, Mark). "Transformatsii russkogo postmodernizma: Ot vzryvnoi aporii k vzryvnomu gibridu." http://gendocs.ru/docs/index-41419.html. Accessed February 19, 2019.

Lipovetsky, Mark, and Aleksander Etkind, "The Salamander's Return: The Soviet Catastrophe and the Post-Soviet Novel." *Russian Studies in Literature* 46, no. 4 (2010): 6–48.

Lohvynenko, Olena. "Shcho za fasadom 'uspishnoho pys'mennyka': spozhyvats'kyi ehoïzm chy osobysta vidpovidal'nist'." *Literaturna Ukraïna* 8 (February 28, 2002).

Luchuk, Taras. "Literaturnyi ariergard. Poetychna kontseptsiia LUHOSADu." *Suchasnist'* 12 (1993): 16.

Lueckel, Wolfgang. *Atomic Apocalypse: "Nuclear Fiction" in German Literature and Culture*. Dissertation, Graduate School of the University of Cincinnati, 2010.

Lyotard, Jean-François. "Answering the Question: What Is the Postmodern?" In *The Postmodern Explained to Children: Correspondence 1982–1985*. Translated by Julian Pefanis and Morgan Thomas. Sydney: Power Publications, 1992.

———. "Note on the Meaning of 'Post.'" In *Postmodernism: A Reader*, edited by Thomas Docherty. Routledge: London and New York, 1993.

Mamleev, Iurii (Yuri). *Izbrannoe*. Moscow: Terra, 1993.

McLaughlin, Robert L. "Post-Postmodern Discontent: Contemporary Fiction and the Social World." *Symploke* 12 (2004): 58.

Medvid´, Viacheslav (Medvid, Viacheslav). "Selo iak metafora." In *Desiat´ ukraïns´kykh prozaïkiv*, edited by Viacheslav Medvid´, 63–83. Kyiv: Rok kard, 1995.

———. "Imperiia ludens." *Kurier Kryvbasu* 119–21 (1999): 16.

Midianka, Petro. "'Ia ne pyshu postmodernizm. Ia pyshu te, shcho vidchuvaiu … '" *Interv´iu z Petrom Midiankoiu z nahody rishennia pro prysudzhennia iomu Shevchenkivs´koï premiï*. https://zakarpattya.net.ua/News/93155-IA-ne-pyshu-postmodernizm.-IA-pyshu-te-shcho-vidchuvaiu. … Accessed February 19, 2019.

Mikhalchuk, Larisa. *Bikfordov shnur*. http://www.br.minsk.by/index.php?article=30134. Accessed February 20, 2019.

Mironov, Valdimir. *Istoriia kak tsep´ katastrof*. http://yarcenter.ru/articles/society/istoriya-kak-tsep-katastrof-622/. Accessed February 19, 2019.

Morenets´, Volodymyr (Morenets, Volodymyr). "Poetychna epika Chornobylia." *Radians´ke literaturoznavstvo* 12 (1987): 3–15.

Motyl, Alexander J. *Dilemmas of Independence: Ukraine after Totalitarianism*. New York: Council on Foreign Relations, 1993.

Narbikova, Valeriia. "Literatura kak utopicheskaia polemika s kul´turoi." In *Postmodernisty o postkul´ture. Interv´iu s sovremennymi pisateliami i kritikami*, edited by Serafima Roll, 125. Moscow: Elinin, 1998.

Neborak, Viktor. "(Pliashkosmoktach)—(Voda). Rozdil iz romanu *Pan Bazio ta reshta*." *Suchasnist´* 5 (1994): 55.

———. "Z vysoty Litaiuchoï Holovy, abo Zniaty masku. Rozmova z V. Neborakom," *Suchasnist´* 5 (1994): 57.

———. "Literaturna orhanizatsiia i literatura (Prynahidni mirkuvannia kolyshn´oho 'zvil´nenoho' sekretaria z ideino-vykhovnoï roboty komitetu komsomolu L´vivs´koho medychnoho instytutu)." *Literatura plius* 5–6 (1999): 16.

———. *Litostroton. Knyha zibranoho*. Lviv: LP-Vydavnytstvo, 2001.

———. "Dekil´ka utochnen´ z pryvodu napysannia zvukospoluchennia [bubabu]." In *"Bu-Ba-Bu" (Iurii Andrukhovych, Oleksandr Irvanets´, Viktor Neborak): Vybrani tvory: Poeziia, proza, eseïstyka*, 23–24. Lviv: Piramida, 2007.

———. "Iurko i Sashko, Sashko i Iurko." In *"Bu-Ba-Bu" (Iurii Andrukhovych, Oleksandr Irvanets´, Viktor Neborak): Vybrani tvory: Poeziia, proza, eseïstyka*, 23–24. Lviv: Piramida, 2007.

———. *Perechytana "Eneïda": sproba sensovoho prochytannia "Eneïdy" Ivana Kotliarevs´koho na tli zistavlennia ïï z "Eneïdoiu" Vergiliia*.

Norris, Christopher. "Versions of Apocalypse: Kant, Derrida, Foucault." In *Apocalypse Theory and the Ends of the World*, edited by Malcolm Bull, 227–49. Oxford and Cambridge: Blackwell, 1995.

Oliinyk, Borys (Oliynyk, Borys). "Sim." *Literaturna Ukraïna* 38 (1988). 4

O'Neil, Patrick. *The Comedy of Entropy: Humour, Narrative, Reading*. Toronto: University of Toronto Press, 1990.

Onyshkevych, Larysa, "Echoes of Glasnost: Chornobyl in Soviet Ukrainian Literature." In *Echoes of Glasnost in Soviet Ukraine Ukrainian*, edited by Romana Bahry, 151–70. York: Captus University Publishers, 1989.

Oord, Thomas Jay. "Postmodernism—What Is It?" http://repository.mnu.edu/sites/default/files/Didache%201-2.pdf. Accessed February 19, 2019.

Pahutiak, Halyna. *Zakhid sontsia v Urozhi*. Lviv: Piramida, 2003.

Pakhliovs'ka, Oksana (Pachlovska, Oxana). *Sytuatsiia postmodernizmu v Ukraïni. Kruhlyi stil*. http://www.ukma.kiev.ua/2001/6/postmodern.html. Accessed February 19, 2019.

———. "Ukraïns'ki shistdesiatnyky: filosofiia buntu." *Suchasnist'* 4 (2002): 65–84.

Pamuk, Orhan. "The Paris Review Interview." In Orhan Pamuk, *Other Colors: Essays and a Story*. Translated by Maureen Freely, 353–78. New York: Alfred A. Knopf, 2007.

Pashkovs'kyi, Ievhen (Pashkovsky, Yevhen). *Shchodennyi zhezl*. Kyiv: Heneza, 1999.

Pavliuk, Ihor. "Kryza postmodernizmu: shcho dali?" http://bukvoid.com.ua/library/igor_pavlyuk/kriza_postmodernu_shcho_dali/. Accessed Febraury 19, 2019.

Pavlychko, Solomiia. *Dyskurs modernizmu v ukraïns'kii literaturi*. Kyiv: Lybid', 1997.

Pavlyshyn, Marko. "Ukraïns'ka kul'tura z pohliadu postmodernizmu." *Suchasnist'* 9 (1993): 117–21.

———. "Shcho pere-tvoriuiet'sia v *Rekreatsiiakh* Iuriia Andrukhovycha?" *Suchasnist'* 12 (1993): 115–27.

———. "Zasterezhennia iak zhanr." *Suchasnist'* 10 (1995): 116–19.

———."Chornobyl's'ka tema i problemy zhanru." In Marko Pavlyshyn, *Kanon ta ikonostas*, 185–93. Kyiv: Chas, 1997.

——— "Peredmova." In Izdryk, *Votstsek & votstsekurhia*, 7–30 (Lviv: Kal'variia, 2002).

Pokal'chuk, Iurii (Pokalchuk, Yuri). "Vershnyk letyt' nad svitom." In *Psy sviatoho Iura. Literaturnyi al'manakh*, 10. Lviv: Prosvita, 1997.

Polishchuk, Iaroslav. *Mitolohichnyi horyzont ukraïns'koho modernizmu. Literaturoznavchi studiï*. Ivano-Frankivsk: Lileia-NV, 1998.

Popil', Daryna (Popil, Daryna). "Ukraïns'kyi postmodernizm u dzerkali mediia." *Visnyk L'vivs'koho universytetu. Seriia zhurnalistyky* 34 (2011): 54–56.

Prokhas'ko, Taras (Prokhasko, Taras). "… botake," *Kurier Kryvbasu* 119–21 (1999): 338.

———. *NeprOsti*. Ivano-Frankivsk: Lileia-NV, 2002.

———. "Vid toho, shcho i iak my hovorymo, zalezhyt' nashe zhyttia i to, shcho zalyshyt'sia pislia n'oho …," interview by Oleh Kryshtopa. *Knyzhnyk review* 23, no. 56 (2002): 23.

———. *Inshyi format, Iurii Andrukhovych*. Ivano-Frankivsk: Lileia-NV, 2003.

———. *Inshyi format, Oksana Zabuzhko*. Ivano-Frankivsk: Lileia-NV, 2003.

———. "Vid chuttia pry sutnosti." In Taras Prokhas′ko, *Leksykon taiemnykh znan′*. Lviv: Kal′variia, 2003.

Roll, Serafima, ed. *Postmodernisty o postkul′ture. Interv′iu s sovremennymi pisateliami i kritikami*. Moscow: Elinin, 1998.

Rorty, Richard. *Contingency, Irony, and Solidarity*. Cambridge: Cambridge University Press, 1989.

Saint-Amour, Paul K. "Bombing and the Symptom: Traumatic Earliness and the Nuclear Uncanny." *Diacritics* 30, no. 4 (2000): 59–82.

Schmid, Ulrich. "Flowers of Evil: The Poetics of Monstrosity in Contemporary Russian Literature (Erofeev, Mamleev, Sokolov, Sorokin)." *Russian Literature* XLVIII (2000): 206.

Sedakova, Ol′ha (Olha). "Postmodernizm: zasvoiennia vidchuzhennia." *Dukh i Litera* 1–2 (1997): 375.

Sehed′-Masak, Mihai (Sehed-Masak, Mihai). "Postmodernizm i postkomunizm," *Krytyka* 5 (1998): 18.

Semkiv, Rostyslav. "Ironiia nepokirnoii struktury." *Krytyka* 5 (2001): 29.

Shcherbak, Iurii (Yuriy). "Chornobyl′. Documental′na povist′." *Vitchyzna* 4 (1988): 16–55, 5 (1988): 14–49, 9 (1988): 2–26, and 10 (1988): 78–118.

Shelazhenko, Iurii. "Solodka Darusia, heroi ukrsuchlitu." http://gazeta.univ.kiev.ua/actions.php?act=print&id=532. Accessed February 19, 2019.

Sherekh-Sheveliov, Iurii (George Shevelov). "Ho-Hay-Ho. Pro prozu Iuriia Andrukhovycha i z pryvodu." In Iurii Andrukhovych, *Rekreatsii. Romany*, 257–68. Kyiv: Chas, 1997.

Slyvyns′kyi, Orest (Slyvynsky, Orest). "Zapiznila myt′ prozrinnia." *Vitchyzna* 5–6 (2001): 81.

Smith, Anthony D. *Theories of Nationalism*, second edition. New York: Holmes-Meir Publishers, 1983.

Sontag, Susan. *Against Interpretation, And Other Essays*. New York: Octagon Books, 1982.

———. "One Culture and the New Sensibility." In Susan Sontag, *Against Interpretation and Other Essays*, 293–304. New York: Octagon Books, 1982.

Sorokin, Vladimir. "Literatura kak kladbishche stilisticheskikh nakhodok! (Interv′iu s Vladimirom Sorokinym)," In *Postmodernisty o postkul′ture. Interv′iu s sovremennymi pisateliami i kritikami*, edited by Serafima Roll. Moscow: Elinin, 1998.

Spanos, William. "The Detective and the Boundary: Some Notes on the Postmodern Literary Imagination." *Boundary* 2 (1972): 165–66.

Starikova, N., ed. *Postmodernizm v slavianskikh literaturakh*. Moscow: Institut slavianovedeniia, 2004.

Stasiuk, Andzhei (Andrzej), and Iurii Andrukhovych. *Moia Europa. Dva eseï pro naidyvnishu chastynu svitu*. Lviv: Klasyka, 2001.

Stefanyk, Vasyl′ (Vasyl). "U korchmi." in Vasyl′ Stefanyk, *Vybrane*. Uzhhorod: Karpaty, 1979.

Struve, Piotr. "Obshcherusskaia kul′tura i ukrainskii partikuliarizm." *Russkaia mysl′* 1 (1912): 65–86.

Syvachenko, Halyna. "Zrushennia kordoniv: postsotsializm chy postmodernizm?" *Slovo i chas* 12 (1991): 55–62.

Szeman, Imre. "Who's Afraid of National Allegory? Jameson, Literary Criticism, Globalism." *The South Atlantic Quarterly* 100, no. 3 (2001): 803–27.

Taran, Liudmyla, ed. *Zhinka iak tekst. Emma Andiievska, Solomiia Pavlychko, Oksana Zabuzhko: fragmenty tvorchosti i konteksty*. Kyiv: Fakt, 2002.
Tarnavs′kyi, Iurii (Tarnawsky, Yuriy). *6x0. Dramatychni tvory*. Kyiv: Rodovid, 1998.
Thompson, Ewa M. "Ways Out of the Postmodern Discourse." *Modern Age: A Quarterly Review* (Summer 2003): 196.
Toth, Josh, and Neil Brooks. *The Mourning After: Attending the Wake of Postmodernism*. Amsterdam: Rodopi, 2007.
Tsybul′ko, Volodymyr (Tsybulko, Volodymyr). *Main kaif. Knyha dla narodu*. Lviv: Kalvariia, 2000.
Uniłowski, Krzysztof. *Polska proza innowacyjna w perspektywie postmodernizmu: Od Gombrowicza po utwory najnowsze*. Katowice: Wydawnictwo Uniwersytetu Śląskiego, 1999.
Vimina, Alberto. "Reliatsiia pro pokhodzhennia ta zvychaï kozakiv." *Kyïvs′ka starovyna* 5 (1999): 64–69.
Virilio, Paul. *The Information Bomb*. Translated by Chris Turner. London and New York: Verso, 2005.
Virk, Tomo. "'Politicheskii' postmodernism v proze nekotorykh slavianskikh literatur." *Slavianovedenie* 6 (2006): 60.
Wachtel, Andrew. "The Legacy of Danilo Kiš in Post-Yugoslav Literature." *The Slavic and East European Journal* 50, no. 1 (2006): 135–49.
Weart, Spencer R. *Nuclear Fear: A History of Images*. Cambridge and London: Harvard University Press, 1988.
Zabuzhko, Oksana. "Postskryptum: monolog perekladacha pro podzvin pokynutykh khramiv." In Svitlana Aleksiievych, *Chornobyl′: Khronika maibutnioho*, translation and afterword by Oksana Zabuzhko. Kyiv: Fakt, 1998.
———. *Kazka pro kalynovu sopilku*. Kyiv: Fakt, 2000.
———. *Khroniky vid Fortinbrasa*. Kyiv: Fakt, 2001.
———. "Pol′s′ka 'kultura' i my, abo Malyi apokalipsys moskoviady." In Oksana Zabuzhko, *Khroniky vid Fortinbrasa*, 314–25. Kyiv: Fakt, 2001.
———. "'Meni poshchastylo na starti …,' Rozmova z Oksanoiu Zabuzhko." In *Zhinka iak tekst*, edited by Liudmyla Taran Kyiv: Fakt, 2002.
———. *Fieldwork in Ukrainian Sex*. Translated by Halyna Hryn. Las Vegas: Amazon Crossing, 2011.
Zhadan, Serhii (Serhiy). *Tsytatnyk (virshi dla kokhanok i kokhantsiv)*. Kyiv: Smoloskyp, 1995.
———. *Balady pro viinu i vidbudovu. Nova knyha virshiv*. Lviv: Kal′variia, 2001.
———. *Big Mak*. Kyiv: Krytyka, 2003.
———. *Depesh Mod*. Kharkiv: Folio, 2004.
———. *Depeche Mode*. Translated by Myroslav Shkandrij. London: Glagoslav Publications, 2013.
———. "Pavlik Morozov: mizh pobutovym heroïzmom i pobutovym tryperom." www.samvydav.net. Accessed February 19, 2019.
Zholdak, Bohdan. *Ialovychyna (Makabreska)*. Kyiv: Ros′, 1991.

Index

Abelard, Pierre, 67
Abrams, Meyer Howard, 10
Abuladze, Tengiz, 141
Adamovich, Ales, 13
Adorno, Theodor, 5, 92
Aeschylus, 264
Anderson, Benedict, 92
Anderson, Perry, 51
Andrukhovych, Yuri, 27, 68–70, 72–74, 76–77, 84–86, 91, 94, 96–97, 99–109, 112–154, 160–161, 166, 170, 199–206, 208, 210, 219, 224, 235–236, 251–252, 283, 291–296. See also Bu-Ba-Bu (Bubabists; Bubabism).
Andrusiak, Ivan, 99, 171, 297. See also "New Generation."
Antonych, Bohdan-Ihor, 70, 73, 129–134, 231
Ariosto, Ludovico, 67
Ascari, Maurizio, 283
Augustine, St., 153

Babak-Lymansky, Mykola, 248
Bakhtin, Mikhail, 54, 107, 115, 141–144, 148, 151, 153, 242
Barańczak, Stanisław, 251
Barnes, Julian, 292
Barth, John, xiii–xiv, 27, 64, 141, 278
Barthes, Roland, 34, 133, 222
Basara, Svetislav, 280

Baudrillard, Jean, 7, 12, 28, 33, 35–37, 42, 51, 141, 210–211, 215, 259, 297
Bazdulj, Muharem, 280
Beauvoir, Simone de, 190, 193
Beckett, Samuel, 263–264
Belbeoch, Bella and Roger, 44
Belpoliti, Marko, 284
Benjamin, Walter, 219
Benn, Gottfried, 67
Ber, Viktor, see Petrov, Viktor
Berg, Mikhail, 64
Berger, James, 11
Bertens, Hans, 94, 281, 285, 293
Biden, Joe, 12
Bilotserkivets, Natalka, 60, 297
Bilous, Dmytro, 57
Bismarck, Otto von, 211
Bitov, Andrei, 64, 293
Blanchot, Maurice, 67, 158
Bloom, Harold, 92
Bondar, Andriy, 65–67, 69, 94, 297
Borges, Jorge Luis, 65, 280, 296
Bortniansky, Dmytro, 247
Braidotti, Rosi, 273–274
Brezhnev, Leonid, 55, 142, 179, 249
Briukhovetska, Larysa, 298
Broch, Hermann, 215
Brooks, Neil, 283
Bu-Ba-Bu (Bubabists; Bubabism), 64–65, 69–73, 79, 94, 98–115, 119–120, 130, 135–154, 169, 175,

186, 200–204, 224, 242, 245–246, 292–294. *See also* Andrukhovych, Yuri; Irvanets, Oleksandr; Neborak, Viktor.
Burgos, Fernando, 277

Calvino, Italo, 296
Camus, Albert, 216
Celentano, Adriano, 31
Chabon, Michael, 288
Che Guevara, Ernesto, 231
Choi, Susan, 288
Chyzhevsky, Dmytro, 56, 68, 203, 212
Conrad, Joseph, 215
Crow, Thomas, 220

da Vinci, Leonardo, 16
Danielewski, Mark, 277
Dante Alighieri, 92
Danto, Arthur, 282
Danylenko, Volodymyr, 76–77, 99, 108
Debord, Guy, 149
Deleuze, Gilles, 158, 160
Derrida, Jacques, 8, 12, 22–25, 34–35, 51, 213, 215, 257, 260, 286
Diachenkos, Maryna and Serhii, 94, 196
Dibrova, Volodymyr, 65, 114, 175–185, 224, 293–295
Dobrenko, Evgeny, 59
Domontovych, V., *see* Petrov, Viktor
Doniy, Oles, 224
Dostoevsky, Fyodor, 179, 238–239
Drach, Ivan, 11, 16–18, 20–21, 221, 223
Drozd, Volodymyr, 63–64, 293
Dzerzhinsky, Felix, 146
Dziuba, Ivan, 93

Eagleton, Terry, 139, 154
Eco, Umberto, 269, 284, 292, 294
Eggers, Dave, 277
Eliot, T. S., 56, 202, 219, 254

Epstein, Mikhail, 87–88, 153, 239, 269, 272, 274–275, 295
Erofeev, Venedikt, 121, 140, 153
Escher, Morris Cornelis, xiii, 159
Etkind, Aleksander, 270–271, 272, 290
Euripides, 263–264

Fassbinder, Rainer Werner, 133
Faulkner, William, 81, 215
Ferguson, Frances, 14
Fiedler, Leslie, 281, 285
Fizer, John, 291
Flaubert, Gustave, 131–132
Foer, Jonathan Safran, 288
Fokkema, Douwe, 62, 284, 293
Foster, Hal, 278
Foucault, Michel, 222
Fowles, John, 279, 292
Franz, Kafka, 24, 79, 166
Freud, Sigmund, 52, 77, 139, 195, 262
Funes, Louis de, 55

Gachev, Georgy, 272
Gandlevsky, Sergei, 140
Giles, Paul, 288
Ginsberg, Allen, 219
Gogol, Mykola (Nikolai), 72, 102, 185, 249
Golomshtok, Igor, 59
Gombrowicz, Witold, 63
Gorbachev, Mikhail, 6, 115
Gorky, Maksim, 67
Grabowicz, George, 68, 72, 74
Greenberg, Clement, 282
Groys, Boris, 54–55, 59, 97

Habermas, Jürgen, 291
Habsburgs (*also* Habsburg Empire), 108, 116, 203
Hamilton, Richard, 220
Haraway, Donna, 277

Hassan, Ihab Habib, 60, 280–281, 285
Havel, Vaclav, 204, 233, 251
Hayles, Katherine, 277
Heidegger, Martin, 4, 193, 223
Heine, Heinrich, 102
Hemon, Aleksandar, 280
Herasymyuk, Vasyl, 82, 149, 226, 297
Hesse, Hermann, 65, 178
Heym, Georg, 67
Hnatiuk, Ola (Aleksandra), 115
Hnizdovsky, Jacques, xiii
Hoberek, Andrew, 288–289
Hoffmann, Ernst Theodor Amadeus, 52, 102
Holoborodko, Vasyl, 226
Homer, 92
Honchar, Nazar, 119, 201. *See also* LuHoSad.
Honchar, Oles, 59, 76, 141
Houellebecq, Michel, 220
Hrinchenko, Borys, 141
Hrycak, Alexandra, 140
Huizinga, Johan, 89, 242
Husserl, Edmund, 4, 158
Hutcheon, Linda, 281, 285

Iarovy, Oleksandr, 93
Ibanez, Blasco, 67
Iefremov, Serhii, 68
Ilnytzkyj, Oleh, 136, 276
Iovenko, Svitlana, 17–21, 186
Irvanets, Oleksandr, 102, 105, 109, 111–112, 114, 129, 139, 201, 238–239
Izdryk, Yurko, 27, 65, 84, 86, 99–100, 108, 130, 157, 161, 164–174, 190, 208, 210, 224, 235, 292–293

Jameson, Fredric, 40–41, 51, 60–61, 275–277, 281–282, 285, 288, 297, 299
Janaszek-Ivaničkova, Halina, 40
Jančar, Drago, 279, 293

Jencks, Charles, 281
Jenet, Jean, 67
John the Baptist, 127
John the Theologian, 29, 42
Joyce, James, 24, 81, 122, 202, 215, 292
Jung, Carl, 171, 178

Kapranovs, Vitaly and Dmytro, 74
Kautsky, Karl, 180
Kellner, Douglas, 14, 286
Kermode, Frank, 4, 25–26, 28
Keruac, Jack, 219, 224
Kharchuk, Roksana, 58, 63, 300
Kharkhun, Valentyna, 59
Khvylovy, Mykola, 201
Kincaid, Jamaica, 186
Kirov, Sergei, 146
Kiš, Danilo, 67, 198, 279–280, 293
Kisch, Egon Erwin, 67
Kiyanovska, Marianna, 296
Kobylianska, Olha, 191, 263, 290–291
Kochs, Yurko and Olha, 150
Kononenko, Yevheniya, 186
Konwicki, Tadeusz, 122
Korniichuk, Oleksandr, 215
Kosach, Larysa, *see* Ukrainka, Lesia
Kosach, Yurii, 293
Kosovych, Leonid, 167
Kostenko, Lina, 223
Kostetsky, Ihor, 139
Kostomarov, Mykola, 73
Kotliarevsky, Ivan, 62, 70–73, 75–76, 152, 227
Kotsiubynska, Mykhailyna, 220
Kotsiubynsky, Mykhailo, 45
Kozhelianko, Vasyl, 132, 161, 173
Kramer, Stanley, 29
Krauss, Rosalind, 281
Kravchenko, Ihor, 93
Kristeva, Julia, 89
Kruk, Halyna, 297

Kryshtopa, Oleh, 161
Kulish, Panteleimon, 70, 73, 79–80
Kundera, Milan, 198, 204, 215, 251–252, 284
Kunkel, Benjamin, 288
Kurylyk, Vasyl, 17
Kvit, Serhiy, 93
Kvitka-Osnovianenko, Hryhorii, 173

Lacan, Jacques, 81, 122, 207, 301
LaCapra, Dominick, 271
Lahiri, Jhumpa, 288
Lawrence, David Herbert, 222
Le, Ivan, 58
Lemarchand, Frederique, 38–39
Lenin, Vladimir, 51, 111, 146, 179
Lennon, John, 84, 208, 233–234
Lethem, Jonathan, 288
Levi-Strauss, Claude, 222
Levinas, Emmanuel, 3–4
Levytsky, Ivan, *see* Nechui-Levytsky, Ivan
Lipovetsky, Mark, 270–272, 290
Logvynenko, Olena, 92
"The Lost Letter," 108, 119. *See also* Lybon, Semen; Nedostup, Viktor; Pozayak, Yurko.
Lotman, Yuri, 54
Luchuk, Ivan, 119, 201. *See also* LuHoSad.
LuHoSad, 108, 119, 171, 201–202. *See also* Honchar, Nazar; Luchuk, Ivan; Sadlovsky, Roman.
Luperini, Romano, 283–284
Lybon, Semen, 119. *See also* "The Lost Letter."
Lyotard, Jean-François, 5, 7, 15, 51, 53, 60, 273, 281, 297
Lysheha, Oleh, 171, 226

Maidanska, Sofia, 186
Makanin, Vladimir, 13
Malenky, Ihor, 142
Malkovych, Ivan, 297
Mallarmé, Stéphane, 24
Mamleev, Yuri, 271
Mandelshtam, Osip, 180
Mao Zedong, 209, 227
Marcuse, Herbert, 222
Márquez, Gabriel García, 132
Matios, Maria, 75
McCarthy, Cormac, 34, 289
McLaughlin, Robert, 278–279
Medvid, Viacheslav, 79–82, 107–108, 149, 153, 190, 226, 292
Midianka, Petro, 296
Miłosz, Czesław, 63, 198, 204
Miniailo, Viktor, 64
Miroshnychenko, Mykola, 119
Moiseienko, Anatolii, 119
Molotov, Vyacheslav, 225–226
Morrison, Toni, 186, 288
Moskal, Mykhailo, 248
Mozart, Wolfgang Amadeus, 146
Mykhailenko, Anatolii, 17

Nahnybida, Mykola, 76, 120, 141
Narbikova, Valeria, 240
Neborak, Viktor, 67–68, 71–72, 74–75, 77, 79, 98, 100–106, 109, 112, 114, 138, 142, 145–146, 150, 150, 152, 224, 276
Nechui-Levytsky, Ivan, 68, 141, 182
Nedostup, Viktor, 119. *See also* "The Lost Letter."
Nestor-the-Chronicler, 76
"New Degeneration," 108. *See also* Andrusiak, Ivan; Protsiuk, Stepan; Tsyperdiuk, Ivan.
Nietzsche, Friedrich, vii, 87, 178, 196, 263, 291
Norris, Christopher, 32
Nostradamus, 29

Oliynyk, Borys, 17
Ong, Han, 288
Oord, Thomas Jay, 286–287

Pachlovska, Oxana, 92, 224, 298–300
Pahutiak, Halyna, 186, 190, 196–197
Pamuk, Orhan, 296
Panch, Petro, 58–59
Parpura, Maksym, 70
Pashkovsky, Yevhen, 77, 82, 96, 99, 108, 149, 153, 190, 210–217, 224, 236, 246, 250, 292
Paul the Apostle, 212
Pavić, Milorad, 292–295
Pavliuk, Ihor, 297
Pavlychko, Dmytro, 223
Pavlychko, Solomiia, 68
Pavlyshyn, Marko, 20–21, 45–46, 136, 170
Pelevin, Viktor, 137, 292
Perec, Georges, 165
Perloff, Marjorie, 281
Pervomaisky, Leonid, 76
Petrosaniak, Halyna, 108
Petrov, Viktor, xiii, 62
Petrovsky, Myron, 45
Picasso, Pablo, 256, 260
Pidmohylny, Valerian, 57, 68
Plato, 25, 257
Pluzhnyk, Yevhen, 58, 77
Poderviansky, Les, 65, 114, 175–185, 239, 293–295
Pokalchuk, Yuri, 137, 151, 161
Pollock, Jackson, 219
Poplavsky, Mykhailo, 297
Pozayak, Yurko, 66, 119. *See also* "The Lost Letter."
Prigov, Dmitri, 239
Prokhasko, Taras, 65, 84, 99, 108, 157–163, 168, 292–293
Proskurnia, Serhii, 136, 148, 150

Protsiuk, Stepan, 171. *See also* "New Generation."
Proust, Marcel, 212, 216
Pyrkalo, Svitlana, 132, 161

Rabelais, François, *see* Bakhtin, Mikhail
"Red Cart," 108. *See also* Zhadan, Serhiy.
Rich, Buddy, 234
Rilke, Rainer Maria, 127, 203
Rorty, Richard, 96, 136
Roth, Joseph, 203
Różewicz, Tadeusz, 63
Rublev, Andrei, 16
Rushdie, Salman, 186, 216, 283
Rylsky, Maksym, 231
Rymaruk, Ihor, 82, 149, 171, 297

Sacher-Masoch, Leopold von, 67
Sadlovsky, Roman, 119. *See also* LuHoSad.
Sambuk, Rostyslav, 76
Schmid, Ulrich, 271
Schulz, Bruno, 203
Sedakova, Olha, 137
Sehed-Masak, Mihai, 136, 150
Semenko, Mykhail, 62, 79, 109, 293
Semkiv, Rostyslav, 167
Shakespeare, William, 92, 110, 263
Shcherbak, Yuriy, 7, 17, 29
Shevchenko, Taras, 17, 71, 73–78, 93–94, 139, 141, 195, 222–223, 227, 231, 246
Shevchuk, Valery, 64, 68, 82, 107, 226
Shevelov, George, 102, 139
Shkurupii, Geo, 58
Shlemkevych, Mykola, 71
Shukshin, Vasily, 64
Skoryna, Liudmyla, 59
Skovoroda, Hryhory, 57, 68, 71, 201
Skriabin (music band), 3
Slyvynsky, Ostap, 297
Smith, Anthony, 204
Smith, Bernard, 281

Sniadanko, Natalka, 132
Socrates, 25
Sokolov, Sasha, 140
Solzhenitsyn, Aleksandr, 251
Sontag, Susan, 14, 31, 221, 284–285
Sorokin, Vladimir, 229, 239–240
Sosiura, Volodymyr, 111, 141, 231
Spanos, William, 285, 300
Stalin, Joseph, 93, 116, 144, 183, 272
Stasiuk, Andrzej, 55, 198–200
Stefanyk, Vasyl, 46–47, 261
Stelmakh, Mykhailo, 59, 76
Struve, Piotr, 61
Stus, Vasyl, 52, 223
Sverbilova, Tetiana, 59
Sverstiuk, Yevhen, 223
Svitlychny, Ivan, 222
Symonenko, Vasyl, 17, 221, 223

Tarnawsky, Yuriy, 186, 253–265, 292–293
Tarkovsky, Andrei, 30, 134
Thompson, Ewa, 274
Timmer, Nicoline, 277
Tolstoy, Leo, 92
Toth, Josh, 283
Trifonov, Yuri, 64
Tsybulko, Volodymyr, 65, 74, 98, 218, 237–252
Tsyperdiuk, Ivan, 171. *See also* "New Generation."
Tubaltseva, Nadiia, 186
Turgenev, Ivan, 215
Turner, Victor, 152
Tychyna, Pavlo, 17, 141, 227, 231

Ugrešić, Dubravka, 283
Ukrainka, Lesia, 139, 194
Ulianenko, Oles, 77, 108, 149, 153
Updike, John, 288–289

Vian, Boris, 67
Vianu, Tudor, 67
Vimina, Alberto, 153
Virgil, 70–71, 73
Virilio, Paul, 210
Virk, Tomo, 293
Viviani, Raphael, 67
Vul, Roman Mykhailovych, 67
Vynnychuk, Yuri, 65, 73, 161, 292–293
Vyshensky, Ivan, 57, 68, 212

Wallace, David Foster, 277, 279
Warhol, Andy, 218, 282
Wasson, Richard, 285
Winter, Jay, 279
Wittgenstein, Ludvig, 158
Wolfe, Tom, 67
Wolfe, Tomas, 67
Woolf, Virginia, 67

Yavorivsky, Volodymyr, 17
Yeshkilev, Volodymyr, 79, 82, 91, 108, 138, 161, 171, 242, 294
Yevshan, Mykola, 73

Zabuzhko, Oksana, 27, 46, 65, 68, 78, 82, 84, 96, 99–100, 122, 129, 186–197, 202, 224, 239, 293–294
Zahrebelny, Pavlo, 64
Zakharchuk, Iryna, 59
Zakhovai, Iurii, 67
Zborovska, Nila, 68
Zemliak, Vasyl, 63–64
Zerov, Mykola, 68, 71
Zhadan, Serhiy, 65, 74, 77, 84–86, 94, 99, 181, 190, 208–209, 218–236, 292–293. *See also* "Red Cart."
Zholdak, Bohdan, 45, 65, 96, 114, 161, 175–185, 239, 293–295
Žižek, Slavoj, 207